Computational Imaging

Computational Imaging

Ayush Bhandari
Achuta Kadambi
Ramesh Raskar

The MIT Press
Cambridge, Massachusetts
London, England

©2022 Massachusetts Institute of Technology

All rights reserved. No part of this book may be reproduced in any form by any electronic or mechanical means (including photocopying, recording, or information storage and retrieval) without permission in writing from the publisher.

This work is subject to a Creative Commons CC-BY-ND-NC license. Subject to such license, all rights are reserved.

The MIT Press would like to thank the anonymous peer reviewers who provided comments on drafts of this book. The generous work of academic experts is essential for establishing the authority and quality of our publications. We acknowledge with gratitude the contributions of these otherwise uncredited readers.

This book was set in LaTeX by the author(s). Printed and bound in the United States of America.

Library of Congress Cataloging-in-Publication Data

Names: Bhandari, Ayush, author. | Kadambi, Achuta, author. | Raskar, Ramesh, author.

Title: Computational imaging / Ayush Bhandari, Imperial College, London, Achuta Kadambi, University of California, Los Angeles, Ramesh Raskar, Massachusetts Institute of Technology.

Description: Cambridge, Massachusetts : The MIT Press, [2022] | Includes bibliographical references and index.

Identifiers: LCCN 2021037259 | ISBN 9780262046473 (hardcover)

Subjects: LCSH: Image processing—Digital techniques—Mathematics. | Imaging systems.

Classification: LCC TA1637.5 .B53 2022 | DDC 006.6—dc23
LC record available at https://lccn.loc.gov/2021037259

10 9 8 7 6 5 4 3 2 1

Contents

List of Figures	xi
List of Tables	xxix
Preface	xxxi

1 Introduction to Computational Imaging — 1

- 1.1 What Is Computational Imaging? — 1
- 1.2 Historical Roots of Computational Imaging — 3
- 1.3 Modern Uses of Computational Imaging — 4
- 1.4 Roadmap of the Book — 6

I PART I: TOOLKITS — 9

2 Imaging Toolkit — 11

- 2.1 Optics — 11
    - 2.1.1 Animal Eyes — 11
    - 2.1.2 Light, Waves, and Particles — 11
    - 2.1.3 Measuring Light with Rays — 13
    - 2.1.4 Pinhole Model — 16
    - 2.1.5 Ray Bending and Lenses — 19
    - 2.1.6 Lenses and Focus — 30
    - 2.1.7 Masks and Aperture Manipulation — 32
- 2.2 Image Sensors — 35
    - 2.2.1 Cameras, Rays, and Radiance — 35
    - 2.2.2 Digital Image Formation — 37
    - 2.2.3 Image Interpolation — 39
    - 2.2.4 Digital Imaging Pipeline — 44
- 2.3 Illumination — 48
    - 2.3.1 Duration and Intensity — 48
    - 2.3.2 Auxiliary Lighting — 49
    - 2.3.3 Modifying Color, Wavelength, and Polarization — 50
    - 2.3.4 Modifying Position and Orientation — 52
    - 2.3.5 Modifying Space and Time — 53

		Exercises	57
3	**Computational Toolkit**		**67**
	3.1	Modeling: Forward vs. Inverse Problems	67
	3.2	Mathematical Tools	68
		3.2.1 Signal Processing	68
		3.2.2 Linear Algebra	80
	3.3	Model-Based Inversion	85
		3.3.1 Examples of Ill-Posed Inverse Problems	86
		3.3.2 Tools and Techniques	90
		3.3.3 Examples of Model-Based Reconstruction	95
	3.4	Data-Driven Inversion Techniques	99
		3.4.1 Machine Learning	99
		3.4.2 Neural Networks and Deep Learning	102
		3.4.3 Convolutional Neural Networks and Computer Vision	110
	3.5	Hybrid Inversion Techniques (Data Driven and Model Based)	115
		3.5.1 Physics-Based Regularization	116
		3.5.2 Physics-Guided Network Initialization	116
		3.5.3 Physics-Based Network Architectures	116
		3.5.4 Hybrid Models	118
		3.5.5 Optical Neural Networks	118
		Exercises	121
II	**PART II: PLENOPTIC IMAGING**		**125**
4	**Spatially Coded Imaging**		**127**
	4.1	Coding the Aperture	128
		4.1.1 Physical Perspective	128
		4.1.2 Mathematical Perspective	130
		4.1.3 Noncoded Aperture	130
		4.1.4 Pinhole	130
		4.1.5 Coded Aperture	131
	4.2	Coding the Sensor	138
		4.2.1 Coded Sensors for Color Imaging	138
		4.2.2 Coded Sensors for High Dynamic Range Imaging	140
		4.2.3 Modulo Sensors for HDR Imaging	142
		4.2.4 Tone Mapping	145
		4.2.5 Exposure Metering	148
		4.2.6 Improving the Resolution	149
		4.2.7 Capturing Fast Phenomena	150
		4.2.8 Using Coded Sensors for Light Field Capture	151
	4.3	Coding the Illumination	151
		4.3.1 Coded Illumination Imaging with Flash	152
		4.3.2 Coded Illumination Imaging with Lasers	153
		4.3.3 Coded Illumination Imaging with LEDs	154

4.4	Further Research		155
	4.4.1	Compressive Imaging	156
	4.4.2	Ghost Imaging	158
	4.4.3	Spectrometry	160
	Exercises		162

5 Temporally Coded Imaging — 169

5.1	A Brief History of the Time-of-Flight Revolution		170
5.2	Optical Time-Resolved Imaging		173
5.3	Time-Resolved Image Formation Model		176
	5.3.1	Probing Function	176
	5.3.2	Scene Response Function	176
	5.3.3	Reflected Function	179
	5.3.4	Instrument Response Function	180
	5.3.5	Continuous-Time Measurements	181
	5.3.6	Discrete-Time Measurements	181
5.4	Lock-in Sensor–based 3D Imaging		181
	5.4.1	Continuous Wave Imaging	181
	5.4.2	Coded Time-of-Flight Imaging	182
5.5	Application Areas		185
	5.5.1	Diffuse Imaging	185
	5.5.2	Light-in-Flight Imaging	187
	5.5.3	Multidepth Imaging	188
	5.5.4	Fluorescence Lifetime Imaging	190
	5.5.5	Non-Line-of-Sight Imaging	191
5.6	Summary of Recent Advances and Further Applications		196
	5.6.1	Time-Resolved Imaging through Scattering Media	198
	5.6.2	Time-Resolved Imaging Systems	200
5.7	Related Optical Imaging Techniques		201
	5.7.1	Optical Coherence Tomography	201
	5.7.2	Digital Holography	202
	5.7.3	Time-Stretched Optics	202
	Exercises		206

6 Light Field Imaging and Display — 211

6.1	Historical Highlight: Lippmann Light Field Camera (1908)		212
6.2	Light Field Processing		212
	6.2.1	Light Field Formulation	213
	6.2.2	Refocusing	215
	6.2.3	Generating Novel Views	216
	6.2.4	Depth Estimation	218
	6.2.5	Further Research	221
6.3	Light Field Capture		225
	6.3.1	Camera Arrays	225
	6.3.2	Dappled Photography	229

		6.3.3	Microscopic Light Field Imaging	231

		6.3.3	Microscopic Light Field Imaging	231
		6.3.4	Further Research and Applications	234
	6.4	Light Field Displays	238
		6.4.1	Traditional 3D Displays	238
		6.4.2	Multilayer and Multiframe Displays	239
		6.4.3	Tensor Displays	242
		6.4.4	Open Problems with Light Field Displays	244
		Exercises	248
7	**Polarimetric Imaging**	**253**
	7.1	Principles of Polarization	253
		7.1.1	Formal Definition of Polarization	253
		7.1.2	Coding with Polarization	255
		7.1.3	Information in Polarization	258
	7.2	Full Stokes Imaging	260
		7.2.1	Parametrization of Polarization	260
		7.2.2	Measuring Stokes Parameters	261
	7.3	3D Shape Reconstruction	263
	7.4	Imaging through Scattering Media	267
		7.4.1	Underwater Imaging	268
		7.4.2	Imaging through Haze and Fog	270
		7.4.3	Polarization-ToF Fusion for Depth Maps	275
	7.5	Reflectance Decomposition Using Polarimetric Cues	276
		7.5.1	Specular vs. Diffuse Reflection	276
		7.5.2	Virtual vs. Real Image Decomposition	280
		Exercises	285
8	**Spectral Imaging**	**287**
	8.1	Spectral Effects on Light-Matter Interaction	287
		8.1.1	Formal Definition of Spectrum	287
		8.1.2	Absorption, Reflectance, and Transmittance	288
		8.1.3	Multispectral and Hyperspectral Imaging	290
		8.1.4	Applications of Nonvisible Light	292
	8.2	Color Theory	293
		8.2.1	Retinal Color	293
		8.2.2	Perceptual Color	295
		8.2.3	Information Loss in Human-Inspired Vision	295
	8.3	Optical Setups for Spectral Imaging	297
		8.3.1	Prisms, Gratings, and Scanners	297
		8.3.2	Multispectral Filter Arrays and Compound Imaging	299
		8.3.3	Spectrum-RGB Parallel Capture	300
		8.3.4	Coded Spectral Illumination	301
	8.4	Computational Methods for Analyzing Spectral Data	304
		8.4.1	Spatiospectral Matrix Representations	304
		8.4.2	Dimensionality Reduction	306

Contents ix

| | | 8.4.3 | Multispectral Demosaicking | 309 |
| | | Exercises | | 312 |

III PART III: SHADING AND TRANSPORT OF LIGHT   315

9 Programmable Illumination and Shading   317

- 9.1 Scene Reflectance and Photometry   317
  - 9.1.1 Albedo, Radiance, and Irradiance   317
  - 9.1.2 Lambert's Law   319
  - 9.1.3 Bidirectional Reflectance Distribution Function   319
- 9.2 Shape from Intensity   320
  - 9.2.1 Reflectance Maps and Gradient Space   322
  - 9.2.2 Calibrated Diffuse Photometric Stereo   323
  - 9.2.3 Uncalibrated Diffuse Photometric Stereo   326
  - 9.2.4 Dichromatic Reflection Model   328
  - 9.2.5 Shape from Interreflections   331
  - 9.2.6 Example-Based Photometric Stereo   334
- 9.3 Multiplexed Illumination   336
- 9.4 Applications in Graphics   338
  - 9.4.1 Light Stage   338
  - 9.4.2 Image Rendering and Relighting   338
  - 9.4.3 Local Shading Adaptation   341
- Exercises   343

10 Light Transport   357

- 10.1 Motivation   357
  - 10.1.1 Curse of Dimensionality   357
  - 10.1.2 Light Transport Addresses Curse of Dimensionality   358
  - 10.1.3 Forward vs. Inverse Light Transport   358
  - 10.1.4 Chapter Organization   359
- 10.2 Light Transport Matrix   359
  - 10.2.1 Light Transport Matrix: Forward Perspective   359
  - 10.2.2 Light Transport Matrix: Inverse Perspective   360
- 10.3 Relaxations of Inverse Light Transport   363
  - 10.3.1 Global and Direct Separation   364
  - 10.3.2 Optical Probing of the Light Transport Matrix   369
- 10.4 Non-Line-of-Sight Imaging   378
  - 10.4.1 Time-of-Flight Methods   379
  - 10.4.2 Intensity-Based Methods   397
- 10.5 Applications   401
  - 10.5.1 Applications in ToF Imaging   401
  - 10.5.2 Skin Imaging   404
  - 10.5.3 Imaging through Scattering Media   407
- Exercises   412

Glossary	415
References	421
Index	443

List of Figures

1.1 Motion deblurring: a comparison of traditional approach with computational imaging approach. Longer exposure period results in motion blurring because the object moves during the exposure. One can use a hardware-based solution that entails a short exposure period, with the disadvantage that the resulting image is noisy. Alternatively, one can use a computational approach that entails *deconvolution* and is inexact. The computational imaging alternative is based on coded exposure imaging. By coding the camera's shutter pattern, the resulting image can be recovered using an algorithm that results in superior performance.   2

1.2 Example of a room-sized pinhole camera.   5

1.3 Roadmap and organization of the book.   6

2.1 The ideal point source: a point source with radiant flux Φ and the irradiance/exitance for an imaginary sphere.   13

2.2 The steradian and the solid angle of a cone-shaped beam.   15

2.3 The area illuminated by a parallel beam as a function of the incident angle.   16

2.4 The pinhole camera. (a) The camera obscura, a darkened room with only a hole in a wall, is an example of the pinhole camera principle; (b) a diagram represents the pinhole camera principle.   17

2.5 The pinspeck camera: this imaging device is based on the opposite functioning principle of the pinhole camera, casting shadows that form a negative image.   18

2.6 The pinhole camera diffraction: when the pinhole size is comparable to the wavelength of the incoming light, a distant object is imaged as a circular disk with rings around it.   19

2.7 The refraction principle: a ray of light is bent at the boundary of two materials with an angle given by Snell's law.   20

2.8 The diagram of a thin lens: a point in the scene (black dot) is in focus if the light it reflects toward the lens converges on the sensor plane. When the camera refocuses on a different point, the sensor plane moves relative to the lens.   21

2.9 Lens-based camera obscura: the first cameras required a manual adjustment of the exposure time. Reprinted from (Raskar and Tumblin, 2011).   22

2.10 Ray bending at one surface of a thin lens: the ray reflected by the object (*red line*) intersects the lens surface at distance h from the optical axis. Snell's law governs the new direction of the ray relative to the line normal to the surface (*dotted line*).   24

2.11 The main categories of lenses, which are based on the shapes of the lenses' two surfaces.   26

2.12 The image curvature effect and the Kepler focal plane array: in using a single element spherical lens it is not possible to focus a whole object on the sensor plane, and thus the edges look out of focus. Unlike conventional digital sensors, the imaging sensor array used in the Kepler space observatory is curved so that the Petzval field curvature can be compensated. *Source*: NASA (NASA and Ball Aerospace, 2008).   27

2.13 Correcting chromatic aberration: the biconvex lens has a frequency-dependent focal length, which can be corrected by pairing it with a plano-concave lens.   28

2.14 Functioning principle of antireflective coating: the incoming light ray is reflected twice by the coating layer and the glass. The two reflections are in opposite phases and thus cancel each other. The sinusoidal curves do not represent the ray paths but rather their intensities.   29

2.15 Lens bending rays reflected by an object: cases in which the intersection between the rays from the upper and lower halves of the lens falls (a) onto the sensor plane, (c) in front of the sensor plane, and (e) behind the sensor plane; and the corresponding intensities along the sensor plane (b), (d), and (f). The green and red curves in (d) and (f) show how the two lens halves generate out-of-phase intensity functions when the object is out of focus. The intensity curves can then be measured independently and their phase difference used to compute the direction and distance of the lens motion. Reprinted from (Ramanath et al., 2005).   31

2.16 Example of lensless MURA photography: (a) mask used to capture the image, (b) detected image, (c) image reconstructed from the measurements, and (*bottom row*) MURA patterns with different matrix sizes.   33

2.17 Positions for placing a mask in a camera.   34

2.18 Imaging for different lens positions: (a)–(c) show an image captured with a gradually improved focus and (d) shows focus measure as a function of the lens position, which is maximized when the image is in focus. Reprinted from (Ramanath et al., 2005).   36

2.19 Nearest neighbor interpolation: the original continuous function $g(x) = \sin(x)$ is shown by the dashed line, and the interpolation $\widehat{g}(x)$ is shown by the solid line.   39

2.20 Linear interpolation: the original continuous function $g(x) = \sin(x)$ is shown by the dashed line, and the interpolation $\widehat{g}(x)$ is shown by the solid line.   40

2.21 Cubic interpolation: the original continuous function $g(x) = \sin(x)$ is shown by the dashed line, and the interpolation $\widehat{g}(x)$ is shown by the solid line.   41

2.22 Linear 2D interpolation: the original sample $\mathbf{I}_S(i,j)$ is shown by black dots, and the interpolation $\widehat{\mathbf{I}}(x,y)$ is shown by gray mesh. The samples were taken from the function $\mathbf{I}(x,y) = \sin\left(\sqrt{x^2+y^2}\right)/\sqrt{x^2+y^2}$.   42

2.23 Image downsampling without filtering: the original image (*left*) is downsampled by a factor of two (*right*). The aliasing effect is particularly visible for the high spatial frequencies at the top of the image.   43

List of Figures

2.24 Image filtering in the frequency domain: the Fourier transform of the original image (*left*), and of the image filtered with a boxcar function (*right*). 43

2.25 Image downsampling with filtering: the original image is filtered with a boxcar function (*left*) and subsequently downsampled by a factor of two (*right*). 44

2.26 The main steps in the digital imaging pipeline. Reprinted from (Ramanath et al., 2005). 45

2.27 The basic components of a CMOS camera sensor. 46

2.28 Spectral sensitivities in digital color cameras. Reprinted from (Ramanath et al., 2005). 47

2.29 The analog front end in the digital processing pipeline. 47

2.30 Removing artifacts from a flash image: the image gradients are used to locate the image artifact and remove it. Subsequently, the isolated artifact can be integrated to generate an image of the photographer. Reprinted from (Agrawal et al., 2005). 51

2.31 Generating synthetic colored lighting using conventional illumination: an image is captured using ambient light and subsequently with lighting from the left direction (*left*) and right (*center*). By subtracting the images with artificial lighting from the ambient-light image it is possible to generate synthetic colored lighting (*right*). Reprinted from (Haeberli, 1992). 51

2.32 3D object localization and the correspondence problem for a single camera and a projector. Several patterns are projected onto the scene object, which are detected by a single camera. The object 3D localization is computed via triangulation. 54

E2.1 Modeling the principle of refraction via Snell's law. 57

E2.2 The proposed thin lens setup. 57

E2.3 Simulating the sampling and quantization done by a sensor array. 59

E2.4 Increasing the resolution of a sampled and quantized image via interpolation. 60

E2.5 The adjustment of brightness and contrast in digital images. 61

E2.6 Images of the real world and of noise. 65

3.1 Block diagram for modeling inverse problems. 68

3.2 Complex exponentials are eigenfunctions of linear time-invariant systems. 72

3.3 Sampling theory addresses the problem of representing a continuous-time function with discrete samples. 76

3.4 (a) Bandlimited function, and (b) effect of Fourier spectrum periodization. 76

3.5 Two-dimensional bandlimited signal and periodization of its Fourier transform. 78

3.6 Discretization of information on one, two, and three dimensions, leading to mathematical objects vector, matrix, and tensor, respectively. In the case of images, a pixel maps to an element in the matrix. In the case of volume based data, a voxel maps to an element in the tensor. 80

3.7 Vectors and matrices: definitions and basic operations. 81

3.8 Classification of square, tall, and fat matrices and its link with rank deficiency. 83

3.9 Continuous-time linear system and its matrix representation. 86

3.10 Example of an ill-posed problem: deconvolution. (a) $f(t) = \varphi_{\sigma_f,0}(t)$ is filtered with a $h(t) = \varphi_{\sigma_h,0}(t)$; (b) when $\sigma_f \ll \sigma_h$, the measurement $g(t) = (f * h)(t) = \varphi_{\sigma_g,0}(t)$ is very similar to the filter $h(t)$; (c) Fourier transforms of the $f(t)$ and $h(t)$; and (d) reciprocal of the Fourier transform blows up, leading to instabilities. 87

3.11 Motion deblurring is an ill-posed problem that can be made well posed by using computational imaging methods. 89

3.12 Denoising by leveraging sparsity: (a) Time-domain samples of a sum of two sinusoids and its noisy measurements with 0 dB signal-to-noise-ratio (SNR). (b) Fourier domain representation. Because the data comprises a sum of two sinusoids, its Fourier domain representation is a 2-sparse signal comprising two spikes. Adding noise changes this, leading to a number of spurious spikes. (c) Reconstruction via sparsity. In the Fourier domain, we remove all but the two largest coefficients because we know that the data comprises a 2-sparse signal. Time domain reconstruction shows the effect of denoising and results in a nearly perfect reconstruction. 93

3.13 Soft-thresholding function as an inverse function. (a) Graph of the function $g_n = f_n + \lambda/2 \operatorname{sgn}(f_n)$. (b) To evaluate f_n given g_n, we invert the graph in (a), which yields the definition of the soft-thresholding function. The grid lines in gray represent $\pm \lambda/2$. 94

3.14 Computed axial tomography scanning: (a) scanner using parallel rays measured with sensor arrays, and (b) rays organized in a fan shape used by medical scanners. 97

3.15 Traditional programming vs. machine learning. 100

3.16 Examples of (a) clustering, (b) classification (KNN), and (c) linear regression. In (a) data is clustered into two groups; in (b) a new data point can be classified in either into class A or class B; and in (c) dots correspond to data points and the line corresponds to the linear fit. 101

3.17 Support vector machine model classification: (a) examples of nonoptimal hyperplane margins, and (b) optimal hyperplane separation. 102

3.18 A nonlinear model of a neuron. 103

3.19 Affine transformation produced by the presence of bias. 103

3.20 Various activation functions: (a) sigmoid activation function, (b) tanh activation function, and (c) ReLU activation function. 104

3.21 Fully connected feedforward network with one hidden layer and one output layer. 106

3.22 Recurrent network with no hidden neurons and no self-feedback loops. 107

3.23 A simple model of a perceptron. 108

3.24 A simple example to demonstrate the backpropagation algorithm: (a) the function represented using signal flow graph rules, and (b) the signal flow graph with backpropagation values updated. (The values in green are the input values, and the values in red are the values calculated using the backpropagation algorithm). 109

3.25 LeNet architecture. 111

3.26 VGG16 architecture. 112

3.27 Fast R-CNN architecture (Girshick, 2015). 112

3.28 Architecture of variational autoencoder: (a) the encoder, and (b) the decoder. 113

List of Figures

3.29 The reparameterization trick. 114

3.30 The generative adversarial network. 114

3.31 Phase retrieval under dominant effects of shot noise is shown to be more effective under a physics-based network, compared to standard deep learning, iterative, or model-based approaches (Goy et al., 2018). 117

3.32 Deep diffractive neural network. (a) Each point on the diffractive layers behaves as a point source, in accordance with Huygen-Fresnel's principle. (b) Light emanating from the input plane (the digit '5') is propagated through a classifier diffractive network. The classifier network focuses light into one of ten detectors, each corresponding to a digit from 0–9 (Lin et al., 2018). 119

4.1 Three main categories of spatially coded imaging modalities. The illumination is usually coded by obstructing partially or completely the light from a projector in a predefined pattern. A traditional lens integrates all light from a point in the scene. A coded aperture selects light arriving from a number of angles. The sensors can be coded by arranging pixels sensitive to certain wavelengths. A beam splitter can be used to project the incoming light beam onto several sensors with modified parameters. 128

4.2 Four aperture sizes and their corresponding f-stop values. 129

4.3 Three main types of apertures and their Fourier domain characteristics. 129

4.4 Pinhole camera with a large aperture. When the aperture size is increased, it can no longer be approximated with a Dirac delta function. As a consequence, different points in the scene are projected onto the same point in the projection plane, leading to a blurry image. 131

4.5 Coded exposure for objects in motion: (a) original blurred image, (b) rectification applied after estimating the vanishing point of motion lines, and (c) image deblurred using a camera with a fluttered shutter. 133

4.6 Fluorescence imaging for the nuclei of leukemia cells with different velocities. The motion blur of the cells, which is more pronounced with higher speeds, can be reversed with the fluttered shutter approach. Reprinted from (Gorthi et al., 2013). 134

4.7 Extending the depth of field: (a) focal stack measured with different focus points, and (b) extended depth of field by reducing aperture size. Reprinted from (Ng et al., 2005). 135

4.8 Reduced depth of field via blur estimation: (*left*) the original image; (*center*) the estimated blur; and (*right*) the image processed for depth reduction. Reprinted from (Bae and Durand, 2007). 136

4.9 Removing glare from the scene with a high frequency mask outside the camera. 136

4.10 Glare reduction using a high frequency mask near the sensor. The glare effect can be enhanced (*left*) or eliminated (*right*) by separating the light into global and direct components in the original image (*center*). 137

4.11 Prototype of a light field camera for glare reduction. 137

4.12 Three sensor architectures for color imaging. 139

4.13 A single-axis multiparameter (SAMP) camera; (a) diagram depicting the incoming light beam split sequentially into eight beams, each captured by different cameras with different settings; and (b) the SAMP camera setup. Reprinted from (McGuire et al., 2007). 140

4.14	Three sensor architectures for high dynamic range imaging.	141
4.15	HDR imaging from modulo samples. In (a), $\lambda = V_{max}$.	143
4.16	HDR tomography using modulo Radon transform.	144
4.17	Dynamic range compression with the bilateral filter method. The base and detail layers are computed on grayscale images. Color is treated separately by reducing the contrast on each of its components, and then recomposing into the new color.	146
4.18	Gradient-based local compression. Starting with five images taken with different exposure values (*top row*), a radiance map is computed with pixel gradient calculations. The gradient attenuations (*lower left*) indicate the attenuation at that pixel corresponding to the gradient value. The final result contains adequate detail in both the dark and bright parts of the scene (*lower right*). Reprinted from (Fattal et al., 2002).	146
4.19	Mapping the high dynamic range intensities onto pixel values.	147
4.20	Space-time superresolution technique. Using four video recordings at low frame rate and low resolution, a new higher resolution video is generated with frames at times where there is no physical measurement. Reprinted from (Shechtman et al., 2005).	150
4.21	The procedure of transferring details from the flash image to the ambient image: (a) flash image, (b) ambient image, and (c) ambient image processed with denoising and detail transfer. Reprinted from (Petschnigg et al., 2004).	153
4.22	A Fourier ptychography technique with multiplexed illumination. The LED pattern (*top*) illuminates the target object, leading to a different image (*center*). For each illumination, the resulting image has a spectrum computed in four subsets of the two-dimensional frequency domain, corresponding to the four LEDs (*bottom*). Reprinted from (Tian et al., 2014).	155
4.23	A single-pixel camera performing compressive sensing. Reprinted from (Baraniuk, 2007).	157
4.24	The ghost imaging paradigm.	159
E4.1	The aperture function (*top*) and the magnitude of its Fourier transform (*bottom*) in the case of aperture coded imaging.	162
E4.2	Simulating the out-of-focus blur effect by convolving with a kernel.	164
E4.3	A monochromatic sensor (*left*), the pixels arranged in the Bayer pattern (*center*), and breaking down the Bayer pattern in three components for each RGB color (*right*).	164
E4.4	Simulating an image captured by a sensor in the Bayer pattern.	165
E4.5	Reconstructing the original full resolution image via demosaicking with nearest neighbor and linear interpolation.	166
E4.6	Compressed image reconstruction using 30% fewer samples than the number of pixels in the original image.	167
E4.7	The reconstruction error as a function of the number of compressed samples M.	167
5.1	Conceptual diagram showing different mechanisms for creating diversity in measurements when capturing the high-dimensional plenoptic function.	170
5.2	Brief history of the time-resolved imaging revolution.	171

List of Figures

5.3 Example of a 3D image, showing the amplitude image (or the conventional digital image), the depth image, and 3D images seen from multiple viewpoints. 172

5.4 Time-resolved information at a single pixel in the following cases: (a) when a signal is backscattered from a single object; (b) when the signal is backscattered from two objects, such as in imaging through a window pane; and (c) similarly to (b) but in more challenging scenario, when reflections occur from closely spaced objects. Recovering individual light paths in (c) is known as superresolution. 174

5.5 Different time-resolved imaging sensors with their spatiotemporal parameters. Subnanosecond and picosecond range illumination resembles a spike; hence SPAD and streak tube-based sensors are known as impulse imaging devices. Lock-in sensors use a periodic waveform with frequencies in the range of a few megahertz to a few hundreds of megahertz; such sensors are known as continuous-wave imaging sensors. 175

5.6 Time-resolved imaging pipeline. 176

5.7 A scene with one light path. 177

5.8 Examples of scenes that lead to two light paths. 178

5.9 Fluorescence lifetime imaging. 179

5.10 Raw data samples based on a continuous wave, time-of-flight imaging sensor. 182

5.11 Bandlimited approximation of autocorrelated probing signal ($\phi = p * \overline{p}$) in time-domain ToF setup. The low-pass property is evident from its Fourier spectrum. This is a result of an experiment with $\Delta = 310\,\text{ns}$ and $M_0 = 30$. 184

5.12 Experimental setup for diffuse imaging. The goal is to be able to read the placard that is covered by a diffusive surface. Although conventional measurements seem corrupted by noise (due to specular reflection), in working with time-of-flight sensors, it is possible to recover the hidden information. 185

5.13 Coded time-of-flight measurements at a given pixel corresponding to the experimental setup in Figure 5.12. The probing signal used in this experiment is shown in Figure 5.11. 186

5.14 Recovering the unknown (sparse) scene parameters using the orthogonal matching pursuit algorithm. 187

5.15 Time slices of different scenes demonstrating light-in-flight imaging. (a) Adapted from (Heide et al., 2013). (b) Adapted from (Kadambi et al., 2013). 189

5.16 Continuous-wave imaging with two depths. (a) Scene response function in the time-domain. (b) Scene response function in the Fourier domain. (c) Multiple frequency measurements amount to a phasor addition, and the identification of the scene response function is equivalent to estimation of the phasor components. 190

5.17 Example of imaging with multiple depths. (a) The case when $K = 2$. Measurements are based on the Microsoft Kinect XBox One, adapted from (Bhandari et al., 2014b). (b) The case when $K = 3$. Measurements are based on the PMD sensor, and the experiment was adapted from (Bhandari et al., 2014). 191

5.18 The fundamental difference between depth imaging and lifetime imaging. (a) In depth imaging, the phase of the measurements is linearly proportional to the modulation frequency of the probing signal, whereas in lifetime imaging, the phase is nonlinearly dependent on the depth (d) and the lifetime (λ) parameters that arise due to the modified scene reflection function. (b) Time-of-flight measurements at different modulation frequencies. (c) Phase image at $40\,\mathrm{MHz}$. (d) Parametric curve fitting of observed phase for estimation of lifetime. 192

5.19 The NLOS imaging setup. 193

5.20 Reconstruction from NLOS imaging measurements. (a) Data collected for three different laser positions, where the object is a 2×2 cm white patch. Three of the pixels of the streak camera are denoted as p, q, and r. (b) The voxels that could have contributed to pixels p, q, and r are determined by the corresponding ellipses p', q', and r'. (c) The heatmap resulted from the backprojection algorithm, computed by super-imposing the elliptical curves corresponding to all pixels. (d) The heatmap resulted from 59 laser positions. (e) The final heatmap computed after filtering, representing the reconstruction of the patch. 195

5.21 Material classification setup using time-of-flight imaging. 197

5.22 The imaging prototype used in (Heide et al., 2014c). The cameras are imaging through a tank field with scattering medium placed frontally. Shown are an array of laser diodes and imaging sensor (*left*), arrangement diagram (*center*), and experiment setup (*right*). Reprinted from (Heide et al., 2014c). 199

5.23 Depth estimation in a scattering medium. A tank filled with water (*top*) and then with a gradual increase in milk volume up to 300ml (*bottom*). Imaging with a conventional camera (*left*) and with ToF correlation image sensors (*right*). Reprinted from (Heide et al., 2014c). 199

5.24 (a) Photonic time-stretch principle. (b) Imaging setup for time-stretch LiDAR. Reprinted from (Jiang et al., 2020). 203

E5.1 Producing depth values from a grayscale image. 206

E5.2 Example of depth recovery errors. 207

E5.3 Example of depth recovery errors for noisy measurements. 208

6.1 The 4D light field and two projections in Flatland. The 4D light field is quantified using the two-plane parameterization (*left*) and subsequently projected in Flatland to yield the light slab parametrization (*center*) and the spatio-angular parametrization (*right*). 212

6.2 Diagram of the composite eye (*left*) and the light field camera proposed by Lippmann (*right*). Reprinted from (Carpenter, 1856) and (Lippmann, 1908). 213

6.3 The plenoptic function in Flatland and 3D. The figure illustrates that the five-variable plenoptic function (*right*), or the equivalent three-variable function in Flatland (*left*) are both constant along light rays intersecting the origin. Hence the variables are not independent, and the plenoptic function can be expressed only as a function of four variables or in Flatland as a function of two variables. 214

6.4 Refocusing via light field processing. Image focused on the s line (*top*) and refocusing on a new line located at distance d (*bottom*). 216

6.5 The projection slice theorem applied for image refocusing. 217

List of Figures

6.6   Rendering new views from the light field. (a) The blue dot represents the new view, computed using rays captured by the cameras in the red dots; and (b) a ray that is not captured by any camera (red ray) can be estimated by interpolation using the sixteen closest rays. Reprinted from (Wu et al., 2017). 218

6.7   Generating novel views with a camera array. The view from one camera (*left*) and the synthetic aperture photograph generated with the views of the whole array (*right*). Reprinted from (Levoy, 2006). 219

6.8   Sampling an EPI function in the Fourier domain. (a) EPI function consisting of a single line. (b) The Fourier spectrum of the continuous EPI function. (c) The Fourier spectrum of the sampled EPI function, processed with a rectangular filter (blue). (d) Spectrum processed with a shear filter. Reprinted from (Wu et al., 2017). 220

6.9   Overcomplete dictionary of light field atoms. Light fields can be recovered in a very noise-robust way, mostly as the linear combination of very few light field atoms. 221

6.10 Multilayer perceptron. The connection weights are adjusted on a training data set such that it predicts the desired information from the scene on the basis of given 3D coordinates. 222

6.11 Optimization via gradient descent. The aim of the algorithm is to find the minimum value of the error function, plotted in yellow. At each iteration the parameters are changed in the direction in which the gradient descends fastest. Depending on the starting point, the algorithm might identify a local minimum instead of the global minimum. 223

6.12 Novel view synthesis using neural radiance fields: (a) neural network input consisting of 5D light field coordinates; (b) predicted RGB value; (c) rendered volume; and (d) rendering loss function between the predicted volume and ground truth. Reprinted from (Mildenhall et al., 2020). 224

6.13 Camera array recording system: (a) tightly packed configuration; and (b) widely spaced configuration. Reprinted from (Wilburn et al., 2005). 227

6.14 Images captured with a camera array, for which (a) the exposure time is equal for all cameras, and (b) the exposure time is adjusted individually for each camera. Reprinted from (Wilburn et al., 2005). 228

6.15 Collage computed from images generated with a flexible camera array: (a) the original images and the matching features, and (b) the collage generated through rotations, translations, and scalings. Reprinted from the presentation slides in (Nomura et al., 2007). 229

6.16 Dappled photography in the Fourier domain. The incoming light field is parametrized on the aperture plane and sensor plane (*top*). Illustration of the modulation theorem when the mask is on the aperture and is between the aperture and sensor (*bottom*). The chosen parametrization defines the sensor measurements as a horizontal slice from the modulated light field spectrum. 230

6.17 Dappled photography setup using two camera designs: the proposed cameras (*top*) and the associated masks (*bottom*). 232

6.18 Comparative diagram of the traditional and light field microscopes. 233

6.19 Image captured with a plenoptic camera. The image consists of small circular patches, each containing pixels with different perspectives of a point in the scene (Gkioulekas, 2018). 235

6.20 Computing 2D images from the plenoptic image. By picking only the pixels marked with red in each circular patch, one can simulate a desired viewing angle (*left*), aperture size (*center*), or focus depth (*right*) (Gkioulekas, 2018). 235

6.21 (a)–(d) Four of the typical light field cameras and examples of photographs sliced from each light field. Reprinted from (Levoy, 2006). 236

6.22 Light field sample interpolation using the commercially available Light L16 camera. The thirteen-megapixel images captured with its sixteen camera modules (*left*). Final fifty-two-megapixel image (*right*). Reprinted from (Sahin and Laroia, 2017). 237

6.23 Modern commercial light field camera: (a) front, and (b) back. *Source:* Lytro (https://en.wikipedia.org/wiki/Lytro). 237

6.24 Traditional and multilayer light field displays: (a) a traditional display based on a slitted barrier, (b) a traditional display using a front layer based on lenses, and (c) a multilayer display. Reprinted from (Wetzstein et al., 2012). 238

6.25 Content-adaptive light field displays. (a) A parallax barrier implemented with a dual-stacked LCD. The viewer sees only the light crossing the front LCD, and (b) a content-adaptive dual-stacked LCD, displaying several time-multiplexed frames corresponding to the viewing perspective. Reprinted from (Wetzstein et al., 2012). 240

6.26 Layered 3D display. The display design is based on five attenuation layers (*left*), the scene (*center*), and the corresponding light field and five optimal layers (*right*). Reprinted from (Wetzstein et al., 2011). 241

6.27 Tensor display with three layers. The light illuminating the rear LCD is attenuated cumulatively by each layer. 243

6.28 Viewing zone size for two-stacked displays. The display is a fixed distance ($d = 125$ cm) from the viewer, who has an interpupillary distance (IPD) set to 6.4 cm (*left*). The viewing zone can be computed as a function of the display resolution and interdisplay distance (*right*). A resolution beyond 600 dpi leads to significant blur. For a resolution below that, the interdisplay distance should be large enough that two views enter the same pupil (so that focus cues can be achieved). Reprinted from (Banks et al., 2016). 245

6.29 3D autostereoscopic light field display. Two perspectives of the display, each a pair for stereo vision (*left* and *right*). The object shown is photographed by a stereo camera system (*center*). Reprinted from (Jones et al., 2007). 246

E6.1 Turning a cell phone into a light field camera. (a) An all-in-focus image taken with a cell phone camera. (b) A light field stack is postprocessed to blur out the background. Notice how the helmet stands out from the background. 248

E6.2 Zigzag planar motion of the camera in front of the static scene to capture a video. 249

E6.3 Example of coordinate system and notation. The dashed plane is the virtual film plane, placed one focal length *above* the apertures located at $C^{(1)}, \ldots, C^{(k)}$. This is a common shorthand convention so that we do not have to flip the camera images. In reality, the actual film plane would be one focal length below the aperture location. This coordinate system is used as a guide, and you are welcome to modify it as needed. 252

List of Figures

7.1 Electromagnetic waves and polarization. Polarization describes the oscillation of the electric field of an EM wave over time as it propagates through space. 254

7.2 Vikings used what they called the *sunstone* for navigating the seas on cloudy days when the sun was out of sight. Historians believe that this navigation was enabled by the polarimetric properties of the stone, believed to be calcite. 256

7.3 Wire grid polarizer, by which light polarized perpendicular to the wires is transmitted. In other words, the transmission axis of the polarizer is perpendicular to the wires. 257

7.4 (a) Working principles of liquid crystal displays (LCD). (b) How 3D movies are projected. 258

7.5 Shape from polarization problem. We can determine the surface normal if we have information about the reflected polarization and the material's index of refraction. 259

7.6 Glare removal using cross-polarization. 260

7.7 Full Stokes imaging with a micropolarimeter array. Reprinted from (Zhao et al., 2010). 262

7.8 Poincaré representation of polarization. (a) Polarization measurements consistent with the definition of Stokes parameters vs. (b), a more robust measurement scheme for determining Stokes parameters with high SNR. 263

7.9 Azimuthal model mismatch in shape from polarization (Ba et al., 2020). 264

7.10 Shape reconstruction using polarization cues. Reprinted from (Kadambi et al., 2015). 265

7.11 Image formation model for passive polarization imaging. Reprinted from (Schechner and Karpel, 2005). 267

7.12 Snell's window (optical manhole). Total internal reflection past the critical angle creates only a small window visible from underwater. 269

7.13 Stereovision and polarization for underwater imaging. The use of stereo enables video rate capture of polarization images underwater. Reprinted from (Sarafraz et al., 2009). 269

7.14 Image formation model for dehazing. The airlight has certain polarimetric properties, which are leveraged to be removed from the image. 271

7.15 Contributions of airlight and direct transmission intensities. The polarization filter modulates the airlight and scattered light, but not the directly transmitted light. We leverage this fact to remove scatter and enhance the image. 273

7.16 Image dehazing using polarization and physics-based models. Reprinted from (Schechner et al., 2003). 274

7.17 Specular versus diffuse reflection. 276

7.18 Color constraints on specular/diffuse decomposition. 278

7.19 Image affected by semireflector. Reprinted from (Schechner et al., 1999). 280

7.20 Multiple reflections and refractions in a semireflector. Reprinted from (Schechner et al., 1999). 281

8.1 What is wavelength, and how do we use it in imaging? (a) Electromagnetic (EM) waves are characterized by a wavelength. (b) Electromagnetic spectrum. (c) A standard camera, similar to our eyes, captures visible light that is reflected by a scene, from which we extract photographs. However, images at different wavelengths capture different information about a scene. For example, a thermal image would be useful for heat seeking, while an infrared image would be useful for food quality inspection. (d) A spectral image samples scenes at a higher spectral frequency than normal RGB images. 288

8.2 Why is the sky blue? The interaction of the broadband beam coming from the sun with particles in the atmosphere is highly wavelength dependent. Blue light's shorter wavelength causes it to undergo Rayleigh scattering in the atmosphere, which enables our perception of a blue sky. 290

8.3 Interaction between light and matter. (a) When light interacts with an object, it will reflect off it, be absorbed by it, scatter through it, transmit through it, or do a combination of these. (b) Examining the interaction of light with an apple is a powerful, nondestructive method of analyzing the fruit's freshness. These interactions tend to be wavelength dependent, which is where spectral imaging is useful. 291

8.4 Multispectral versus hyperspectral imaging. 292

8.5 Seeing through walls with Wi-Fi. An interesting application of spectral imaging is in the use of non-traditional frequencies with Wi-Fi imaging (2.4 GHz) to image through walls. (a) Setup of Wi-Vi imaging module (Adib and Katabi, 2013) and (b) Wi-Vi image capturing different poses through a wall. Reprinted from online article by Adib. 292

8.6 (a) Retinal sensitivity to color. Our eyes have three types of cone cells: L-cones, M-cones, and S-cones (*left*). Each cone is optimized to sense light at different wavelengths. The spectral absorption of each cone is shown (*right*). (b) Illumination illusions. Our brain adapts to different illumination conditions to render a scene with spatial and color consistency. Reprinted from online artwork by Adelson. (c) Retinal vs. perceived color. Even with a blue overlay (*bottom*), our visual system is still able to correctly label each color. 294

8.7 Capturing a spectral image. A multispectral image can be captured by either (a) passive illumination, or (b) active illumination. With active illumination, external spectral light sources are used (either by placing several filters in front of one broadband source, or by using several narrow band sources). Passive illumination setups place several narrow band filters in front of the focal plane array. 296

8.8 Wavelength separation by (a) prisms, and (b) diffraction gratings. 297

8.9 Spatiospectral scanning. (a) An example of a pixelwise scan of an image. The scanner captures a spectrum for each pixel, then iterates to the next pixel, and repeats. (b) Satellite hyperspectral imaging using a push-broom camera. 298

8.10 Color filter arrays. Side-by-side comparison of (a) Bayer filter and (b) multispectral filter array, specifically a CMYG CFA. (c) Spectral sensitivity of C, M, Y, and G channels with a QBPF (solid line) and without a QBPF (dashed line). Reprinted from (Themelis et al., 2008). 299

List of Figures

8.11 (a) Multispectral compound imaging setup. A compound imaging setup consists of several units, each capturing an image at different wavelengths. The units capture spatially offset versions of the same scene. (b) Hybrid capture. An optically parallelized setup to capture an RGB and hyperspectral image simultaneously. 300

8.12 Multiplexed illumination. Methodically illuminating the scene with more than one spectral source at a time can enable efficient data capture and higher reconstruction accuracies. An example of the top three optimal illumination patterns are shown with two allowed measurements. 301

8.13 Dark flash photography. Capturing nonintrusive, high-quality images can be challenging in dimly lit environments. One way to get around this hurdle is by actively illuminating the scene with an infrared light source. We can then leverage the spectral proximity of red with infrared wavelengths to constrain the image reconstruction problem. Reprinted from (Krishnan and Fergus, 2009). 303

8.14 Principal component analysis (PCA). PCA seeks to represent data in a coordinate system as to maximize the variance of the data's projection onto each axis. Observe that by minimizing the least-squares error of the projection, the axis also maximizes the variance of the projections. 307

8.15 Statistical representation of spectral images. (*left*) PCA representation of patches in a hyperspectral image, and (*right*) log scale of variance of first 200 PCs. 308

8.16 Image demosaicking using (a) color difference interpolation, (b) residual interpolation, and (c) adaptive residual interpolation. 309

9.1 Lambert's law and foreshortening. When the incident light is at an angle with respect to the normal, the area of light incident on the surface is reduced, in what is known as foreshortening. This results in a reflected intensity proportional to the product of $\cos\theta_i$ and the incident intensity. 318

9.2 Specular or mirror-like reflection of light. 321

9.3 Phong BRDF model for specular highlights. 321

9.4 Geometry of image projection. (a) Perspective projection. (b) Orthographic projection. 321

9.5 Example of a reflectance map. 322

9.6 Mapping multiple intensities to surface orientations using a reflectance map. 324

9.7 Photometric stereo for Lambertian surfaces. (a) Light from the illumination source is incident on the object, with the source vector **s** known for each pixel. The light reflected to the sensor is approximately independent of the sensor location, due to the Lambertian approximation. (b) Multiple spatially offset light sources are used in photometric stereo, with a fixed camera position. 324

9.8 Important photometric angles: incident angle (i), view angle (e), and phase angle (g). 325

9.9	Scene interreflections. The most idealized model is the single-bounce model, in which light from the source bounces off the surface and directly reaches the sensor. However, the light can bounce off the surface n times, as shown for a two-bounce reflection and a three-bounce reflection. The total intensity measured at the sensor is the sum of the intensities for all possible number of bounces, from one to infinity. Reprinted from (Seitz et al., 2005).	331
9.10	Concave shape reconstruction using photometric stereo: (a) original shape, and (b) shape reconstructed with standard photometric stereo. Reprinted from (Nayar et al., 1991).	332
9.11	Direct and indirect illumination of surface points.	333
9.12	Iterative algorithm for extracting shape from objects with interreflections. Adapted from (Nayar et al., 1991).	334
9.13	(a) Standard photometric stereo. (b) Multiplexed illumination. Reprinted from (Schechner et al., 2007).	337
9.14	Light stage. (a) Light stage with a movable arm (Masselus et al., 2002). (b) Light stage based on several spatially offset light sources (Hawkins et al., 2001).	338
9.15	The coordinate system is defined such that a hemisphere completely contains the object of interest (Masselus et al., 2003).	339
9.16	Relighting based on discretized 4D light fields. Reprinted from (Masselus et al., 2003).	340
E9.1	By observing an object under different lighting conditions, we can extract the surface normals of the object, which are used as a proxy for local shape.	343
E9.2	An example optical setup for photometric stereo.	344
E9.3	Insert an object image and its corresponding specular sphere image.	351
E9.4	Insert your segmented specular sphere image.	352
E9.5	Label the point of specularity on your sphere image.	352
E9.6	Insert the normal map of the original object.	353
E9.7	Insert the reconstructed shading image and the error statistics.	354
E9.8	Insert the normal map and reconstructed shading image obtained with unknown lighting conditions.	355
10.1	Dual photography leverages the light transport matrix and Helmholtz reciprocity to swap camera and projector viewpoints. (a) The setup, with the projector viewing the card's face and the camera viewing its back. (b) Live photo of the setup. (c) The image produced using dual photography. Reprinted from (Sen et al., 2005).	360
10.2	Example of dual photography (a) The primal image. Lighting is from the perspective of the projector, and the photo has a resolution equal to that of the camera. (b) The dual image. Lighting is from the perspective of the camera, and the photo has a resolution equal to that of the projector. Reprinted from (Sen et al., 2005).	361

List of Figures

10.3 Primal and dual image matrices. The top diagram illustrates the primal setup in which light is emitted from the camera and captured by the projector. Helmholtz reciprocity, a consequence of conservation of energy, suggests that we can reverse this operation. For example, assume a ray from a projector pixel strikes the scene and is captured by a set of camera pixels. If those camera pixels were instead virtual projector pixels, the same amount of light would hit the scene and reach that single projector pixel (now a virtual camera). As illustrated in the bottom diagram, we can mathematically swap the location of the projector and the camera, in order to find out what the virtual camera would be capturing if it was in the projector's place. Reprinted from (Sen et al., 2005). 362

10.4 Separation of global and direct for a complex scene. (a) Original image of a scene with many optically complex objects. (b) Decomposed direct illumination image, scaled up by a factor of 1.25. (c) Global illumination image that includes diffuse and specular interreflections (wall wedge and nut), volumetric scattering (milky water), subsurface scattering (marble), translucency (frosted glass), and shadow (fruit on board). Reprinted from (Nayar et al., 2006). 364

10.5 Direct-global decomposition of concave and convex surfaces. Concave surfaces are curved inward, and convex structures are curved outward. 365

10.6 Failed direct-global decomposition. Failed separation due to the violation of the smooth global function assumption, when the checkerboard pattern is shifted. The highly specular reflections cause residual checkerboard patterns in each component. Reprinted from (Nayar et al., 2006). 368

10.7 Operations using probing matrix. (a) The light transport matrix can be rewritten as being multiplied elementwise by the probing matrix. This offers a greater degree of freedom in the light transport matrix. (b) This table outlines some potential probing matrix operations that we can do without knowing the full light transport matrix. Adapted from (O'Toole et al., 2012). 372

10.8 Optical probing pipeline. This diagram contains the full pipeline, with relation to the optical hardware, of the probing procedure. Reprinted from (O'Toole et al., 2012). 373

10.9 Optical probing algorithms. The two main algorithms used in the optical probing procedure: path isolation and optical matrix probing. Reprinted from (O'Toole et al., 2012). 374

10.10 Light transport matrix of a scene. (a) An image of the scene, containing various objects that have complex optical interactions. (b) A slice of the light transport matrix for the single highlighted row in (a). A point, (n, m) in the image, represents the light paths that were emitted by pixel m of the projected image and captured by pixel n of the camera (in the highlighted row). The diagonality of the slice implies that light was transported between projector and camera pixels that were close to each other. (c–f) Various notable aberrations in the light transport matrix slice and their causes. Reprinted from (O'Toole et al., 2012). 374

10.11 Stereo transport matrix using epipolar imaging. Diagram of the stereo light transport setup, where the matrix is subdivided into three groups of light: epipolar (green), non-epipolar (red), and direct (black). Reprinted from (O'Toole et al., 2014). 375

10.12 Michelson interferometer light transport probing. (a) An input beam is split by a beam splitter into two copies that reflect off the two mirrors at differing distances from the source; the two copies then recombine at the beamsplitter before being imaged by the camera. One of the mirrors is the target arm (scene), and the other is the reference arm. (b) Varying the source coherence properties, light of different lightpath decompositions can be captured. Reprinted from (Kotwal et al., 2020). 377

10.13 Types of NLOS detection methods. We discuss (a) time-of-flight-based, (b) coherence-based, and (c) intensity-based methods in this chapter. Reprinted from (Maeda et al., 2019). 379

10.14 Measuring the space time impulse response (STIR). (a) A single patch is illuminated at a time (p_1 in the upper image and p_2 in the lower image), and the times at which reflected light reaches the camera (p_0) are recorded for each patch. Adapted from (Kirmani et al., 2009). (b) Onset data collected from illuminating visible patches can be used to calculate the locations of hidden ones (assuming third bounces arrive before fourth bounces, no interreflections, and a known number of hidden patches). Adapted from (Kirmani et al., 2009). 381

10.15 Image capture procedure and geometry. The laser is aimed onto the wall via galvanometer and mirrors (a), and the camera takes a series of images in time (b). A confidence map (c) of the hidden object can be constructed from the results (Velten et al., 2012a). (d) The hyperbolic curves in the individual camera images result from the varying distances *(left)* and thus times *(right)* light travels to reach the sensor (Gupta et al., 2012). 383

10.16 Backprojection geometry. The set of possible hidden object locations corresponding to an image pixel form an ellipse, as each image corresponds to a set distance that light has traveled. Reprinted from (Gupta et al., 2012). 384

10.17 Examples of streak images. An occluded object (a) is probed indirectly with an ultrafast laser. (b) Many streak images of the hidden object are captured. The object can then be recovered via (c) backprojection and (d) filtering. Reprinted from (Velten et al., 2012a). 386

10.18 Data collection. Laser pulses bounce off a wall and hidden object to reach a single photon avalanche diode (SPAD) *(left)*, and a photon counter produces a graph of detector hits versus time *(right)*. 387

10.19 Image/camera setup. By now, this picture should seem familiar: (a) a relatively cheap laser and ToF camera (replace the faster lasers and expensive sensors of previous sections), with the goal of more accessibly capturing (b) the hidden scene. Reprinted from (Heide et al., 2014a). 387

10.20 Experimental results. The reconstructed depth *(left)*, albedo *(center)*, and hidden target *(right)* for both high *(bottom)* and low *(top)* ambient light. Reprinted from (Heide et al., 2014a). 389

10.21 Scene geometry. The familiar image capture diagram *(left)* remains the same, but here the hidden target is interpreted as a set of point sources *(left)* or reflectors *(right)*, and the wall itself is modeled as a sensor array *(right)*. Reprinted from (Kadambi et al., 2016). 389

10.22 Confocal NLOS setup. Confocal NLOS involves (a) simultaneously imaging and sensing the same point on a wall. (b) For each point, photon counts are measured versus time. (c) These measurements are then combined into streak images (O'Toole et al., 2018). 391

10.23   Object reconstruction. The steps of the reconstruction algorithm match the components of the convolution: (a)–(b) attenuation in time, (b)–(c) Wiener filtering, and (c)–(d) attenuation in space. Reprinted from (O'Toole et al., 2018).   392

10.24   Acoustic NLOS. With sound, walls act in a much more specular manner than they do with light, which results in a clearer virtual object *(left)*. This can be quantified by measuring the time delay of the return signal *(upper right)*, and then conducting a Fourier analysis *(lower right)*. Reprinted from (Lindell et al., 2019).   393

10.25   Comparing acoustic and visual NLOS imaging. The acoustic method *(right)* reproduces the L in the hidden scene *(left)*, whereas the visual method does not *(center)*. Reprinted from (Lindell et al., 2019).   394

10.26   Doppler radar NLOS. Using radar, as with using sound, means that various real-world surfaces become more specular. As in the previous section on acoustic NLOS, radar reflections are captured by an array of receivers positioned at the same location as the transmitter, and the outgoing and incoming signals are mixed. We can recover information about distance, velocity, and angle from the received signal. Reprinted from (Scheiner et al., 2020).   395

10.27   Fermat paths. (a) Experimental results for the reconstruction *(right)* of hidden topologies *(left)*. The objects on the left were 3D printed from ground truth meshes *(center)*, on top of which various reconstruction points (red) are overlaid. This is the reconstruction of a paraboloid object. (b) This is the reconstruction of the sigmoid object. (c) Transient light (as measured with a photon counter) exhibits discontinuities at Fermat pathlengths, which correspond to significant features on the hidden surface. Adapted from (Xin et al., 2019).   396

10.28   Hidden objects and shadows. In the presence of an occluding wall, objects hidden from the camera still influence the colors in the shadows cast by the wall *(left)*. Observations at a given angle from the wall *(upper right)* include light from only a portion of the background, resulting in a transfer matrix similar to that shown *(bottom right)*. Reprinted from (Bouman et al., 2017).   398

10.29   Motion from shadows. The (d) color-augmented version of (c) the shadow demonstrates the concept in Figure 10.28. This enables (e) a reconstruction of the motion of colored objects, or in this case, (b) people (Bouman et al., 2017).   399

10.30   Polarized NLOS. (c)–(d) The effective polarization axis of a polarizer changes based on viewing angle, as demonstrated by (a)–(b) the polarizer placed on top of a monitor. This occurs even when (e) two polarizers are placed at $90°$ angles. Light from a projector *(top right row)* is captured by a camera placed at the Brewster angle with respect to the screen *(top)*. Placing a polarizer in front of the camera leads to better results *(bottom row)* than without it *(center row)* (Tanaka et al., 2020).   400

10.31   Periscopy NLOS. (a) The classic NLOS setup, replete with occluder, hidden object, light source, and sensor. (b) The results of the reconstruction algorithm *(right)* on various scenes *(left)*, with the raw camera image in the center. Reprinted from (Murray-Bruce et al., 2019).   400

10.32 Multipath interference in ToF imaging. (a) An example of ToF in which a single light ray is emitted and captured after striking the scene surface at point p. (b) Here, a different light ray strikes the scene at point p after being reflected from q and also reaches the same sensor as the first light ray. This introduces interference in the ToF sensor computations. (c) Partial subsurface scattering of a light ray results in multiple light rays reaching the ToF sensor. (d) The first shape is the measured ground truth. The second shape is the error for the generated depth map using classical ToF imaging. Finally, we have the error of the depth map constructed using light transport optimization to lower noise. The corrected error is markedly lower than the original error. Reprinted from (Naik et al., 2015). ... 402

10.33 Epipolar ToF imaging. (a) In epipolar imaging, one row is imaged at a time, using a laser sheet. (b) Epipolar ToF imaging improves the depth measurements for even bright light bulbs. The errors caused by the surface reflectance of the light is suppressed in epipolar imaging. Reprinted from (Achar et al., 2017). ... 403

10.34 Epipolar scene sampling. (a) Capturing epipolar planes over time, in the manner of a rolling shutter camera. This reduces the effect of time varying motion blur, global illumination, and the ambient light on the image. (b) We can trade vertical resolution for higher temporal sampling by capturing every other epipolar plane. (c) Further optimization can be done for specific situations, by selectively increasing temporal resolution in different parts of the image. Reprinted from (Achar et al., 2017). ... 404

10.35 Optical behavior of skin. The epidermis, dermis, and subcutis skin layers all have their own unique optical behaviors based on their specific structures. Adapted from (Igarashi et al., 2007). ... 406

10.36 Separation of skin components. In (a), an RGB projector displays a green pattern on a hand, which is captured by a monochrome camera. In (b), we see the isolation of the veins and separation of global and direct components using the RGB and infrared spectra. Adapted from (Kadambi et al., 2013). ... 407

10.37 Confocal imaging and descattering. (a) Scene with objects in a 3D fish tank. (b) Original image of fish tank filled with diluted milk. (c) Partially descattered using confocal imaging. (d) Additional optimization removes more global scattering. (e) The recovered 3D structure is visualized for a different view. Reprinted from (Fuchs et al., 2008). ... 408

List of Tables

10.1 Overview of light transport: three interrelated views of light transport. The first view separates an image into $1, ..., n$ bounces of light (Seitz et al., 2005). The second view uses a smoothness relaxation to reduce the separation into 1-bounce and n-bounce transport (Nayar et al., 2006). The third view shows the ability to discriminate global light transport based on the distance from the diagonal of the transport matrix (O'Toole et al., 2014). 376

Preface

The motivation for writing this primer on *Computational Imaging* was to fill the need for an initial reference for the field. By laying down this foundation, the authors, who are also instructors, aim to provide students with a textbook that aligns with the titles of courses they teach on computational imaging. For established practitioners, we hope to provide a reference text that aligns with emerging journals and societies in engineering organizations as diverse as the IEEE, ACM, OSA, and SPIE. For instance, the IEEE Transactions of Computational Imaging was formally indexed in October 2018.

The timing of this book also coincides with an increase in the hiring of faculty who identify computational imaging as their core competency. Just a decade ago, only a few EE/CS departments had researchers who explicitly identified as computational imaging faculty (as in the case of author Raskar). Then, it may not have made sense to have a unified textbook for so few professors and classes. Today, nearly every top-tier university has at least one faculty member who identifies computational imaging as her or his core expertise. Many of these are junior faculty (as in the cases of authors Bhandari and Kadambi) who have been tasked by their departments to create new courses on computational imaging. We hope this textbook and associated set of homework assignments is handy not only for students and practitioners, but also for professors who are now teaching courses on computational imaging.

The push for increased hiring of computational imaging faculty is not a coincidence—it parallels advances in computer vision, machine learning, and signal processing tools. These tools, more powerful than ever, can enable imaging systems to see the invisible: cameras that operate at a trillion frames per second, microscopes that can see viruses long thought to be optically irresolvable, and, recently, telescopes that can image black holes. Closer to home, self-driving cars and smartphone cameras are powered in part by computational imaging techniques, impacting our everyday lives.

Scope of the Book

This book lays the foundations of computational imaging, which is a convergence of vision, graphics, signal processing, and optics. We may interest practitioners in any of these four fields; however, this is not a fundamental text for any of the four fields, nor is it intended to be, as there currently exist many excellent books. Recommendations for computer vision include (Forsyth and Ponce, 2012; Szeliski, 2011; Hartley and Zisserman, 2004); for graphics it could be (Gortler, 2012; Marschner and Shirley, 2018; Hughes et al., 2013); for signal processing, foundations, (Mallat, 2009; Vetterli et al., 2014)), sparse signal processing (Elad, 2010); numerical methods (Björck, 1996; Strang, 2016); convex optimization (Boyd and Vandenberghe, 2004). For optics and photonics, we recommend (Hecht, 2012; Goodman, 2005), and (Saleh and Teich, 1991), which all offer a vastly different treatise of the subjects. In contrast to these foundational books, our book discusses modern ideas that have captivated the field over the last decade such as the imaging of black holes, imaging at a trillion frames per second, light transport, and seeing around corners. These breakthroughs—seeming feats of physics—were led by computer scientists in key roles.

Acknowledgments

Writing a book on an exciting and emerging topic is a massive undertaking and this would not have been possible without the help and support of our friends, collaborators, colleagues, and the members of interdisciplinary communities. The authors gratefully thank these individuals for their feedback and comments on the earlier versions of the draft. In particular, we would like to thank Gordon Wetzstein for actionable comments on improving aspects of the book, particularly sections pertaining to light fields. Vishwanath Saragadam provided feedback on sections pertaining to multispectral and hyperspectral imaging. Kenichiro Tanaka and Teppei Kurita provided input on the polarization section. Bahram Jalali and Aydogan Ozcan have had numerous discussions with the authors at the seamline of AI and physics. Nick Antipa took the time to provide input on lensless imaging design. Discussions with Suren Jaysuriya date back several years and are interspersed in many aspects of the book.

On the typographical front, the authors also thank Kyle Icban (an editorial cartoonist at the DailyBruin) who contributed to some of the figures in the book and Amol Mahurkar for his help with LATEXtypesetting.

AB acknowledges Dorian Florescu for his help with review and preparation of materials, in particular, for the imaging toolkit, spatially coded imaging, as well as light field imaging. Logistic support from Dorian Florescu, Gal Shtendel and Humera Humeed was invaluable in maintaining tight timelines.

AK acknowledges the participants of the inaugural Computational Imaging class (ECE239) at UCLA. Questions, insights, and corrections raised by our students in lecture and office hours have enriched the contents of this book. Many students who were exposed to the class ended up contributing to the book. Siddharth Somasundaram made key contributions on book sections related to polarization, multispectral imaging, and programmable illumination. Alethea Sung-Miller, Chandra Suresh, Chinmay Talegaonkar, Madison Belk, Rajeshwari Jadhav, and Shreeram Athreya contributed to various aspects of light transport. Through many discussions, Siddharth and Chinmay contributed to the design and organization of the materials in the book.

RR thanks the members of the Camera Culture Group who, over the course of a decade, have not only contributed to the growing wealth of results in the area of computational imaging but also enabled this book by development of early coursework at MIT. Beyond MIT, RR acknowledges the collaboration with Jack Tumblin that resulted in the computational photography book (2007) which served as an early primer to the subject.

Many of the illustrations in this book are results developed by leading scientific groups around the world. We gratefully acknowledge our colleagues who, in the spirit of Open Science, have generously allowed us to use their original illustrations from their research— Amine Bermak, Aydogan Ozcan, Bahram Jalali, Diego Gutierrez, Dilip Krishnan, Ethan Schonbrun, Felix Heide, Fredo Durand, Georg Petschnigg, George Barbastathis, Gordon Wetzstein, Ioannis Gkioulekas, Jun Tanida, Kyros Kutulakos, Lei Tian, Marc Levoy, Masatoshi Okutomi, Matt O'Toole, Matthew Tancik, Michal Irani, Paul Debevec, Paul Haeberli, Philip Dutre, Qionghai Dai, Raanan Fattal, Rajeev Ramanath, Rajiv Laroia, Richard Baraniuk, Ross Girshick, Shree Nayar, Stanley Pau, Vasilis Ntziachristos, Vivek Goyal, and Yoav Schechner.

A.B., A.K., R.R.
September 2021

1 Introduction to Computational Imaging

1.1 What Is Computational Imaging?

Imagine that it was possible to photograph a black hole or create cameras that could image around corners. What if we invented a camera that could freeze light in motion or create new forms of light sensing that enable autonomous cars to "see" in fog?

These capabilities sound like fanciful superpowers—and they would be for the existing cameras that we are used to. However, the field of computational imaging seeks to transform the camera into something more, something that could achieve superpowered feats. The solution? To jointly design optics and computation to overcome longstanding limits of imaging.

> **Computational Imaging** Joint design of optical capture and computational algorithms to create novel systems.

In contrast to traditional imaging, computational imaging is distinguished by a heavy use of mathematical algorithms. For example, whereas an X-ray photograph is a conventional imaging system, the blending of multiple X-ray photographs to compute a 3D tomography model (CAT scan) is a computational imaging system. Another example consists of sharpening motion-blurred images. Traditionally, this is done purely through deconvolution software. The computational imaging alternative is *coded exposure imaging*, as shown in Figure 1.1.

A principled codesign of hardware and algorithms leads to overcoming some of the main limitations of traditional imaging. These include

Dynamic range All digital sensors, including imaging sensors, are limited in their dynamic range. Physical entities such as intensity or photon flux beyond a predefined threshold can cause sensor saturation, which results in a permanent loss of information.

Spatial resolution The resolution of a traditional camera is strictly determined by its sensor size. A classic example is the sensor resolution of consumer-grade three-dimensional

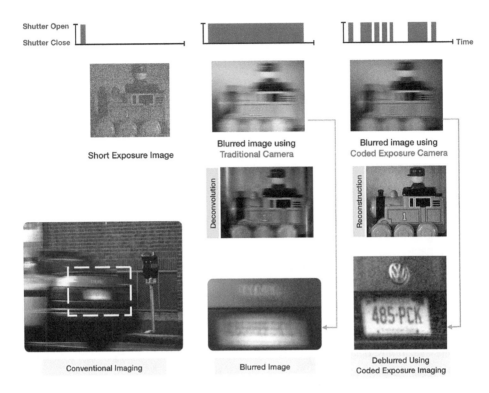

Figure 1.1 Motion deblurring: a comparison of traditional approach with computational imaging approach. Longer exposure period results in motion blurring because the object moves during the exposure. One can use a hardware-based solution that entails a short exposure period, with the disadvantage that the resulting image is noisy. Alternatively, one can use a computational approach that entails *deconvolution* and is inexact. The computational imaging alternative is based on coded exposure imaging. By coding the camera's shutter pattern, the resulting image can be recovered using an algorithm that results in superior performance.

> imaging sensors such as the Microsoft Kinect. Whereas consumer cameras offer spatial resolution on the order of tens of megapixels, this is not the case with three-dimensional imaging sensors.
>
> **Depth of field** Due to their inbuilt features, traditional imaging systems are subject to a number of trade-offs between the depth of field, field of view, and imaging parameters.

Computational imaging devices generate far more than an array of pixel values, amounting to additional scene information that may include data such as depth or spectral information. From a mathematical perspective, computational imaging systems capture a far larger class of signals or information that can be leveraged using advanced algorithms. For instance, apart from simply associating a red, green, and blue (RGB) component with every pixel, such systems can measure the 3D structure of the scene and the shape boundaries and can perform decompositions such as foreground and background objects, direct and indirect illumination, and layers of transmission and reflection.

We may also want to have more subtle contributions to the imaging process by highlighting small features that are not observable by the human eye. In order for a photograph to be photometrically accurate, that is, taking into account human perception, it is possible to use small pictorial elements that can be achieved with customized lighting or viewpoint adjustment.

Contributions to computational imaging have come from a diverse set of communities including signal processing, optics, computer vision, computer graphics, and applied mathematics. Computational imaging is not only a shared interest among these diverse communities but is a necessity to address current scientific applications.

1.2 Historical Roots of Computational Imaging

Ideas of combining imaging with computation date back to the beginning of the computing revolution. In the 1960s astronomers were interested in measuring X-ray radiation emitted from various astronomical objects such as the sun, neutron stars, and black holes. Traditionally, astronomical objects were viewed through a telescope, which used glass lenses to bend optical radiation. Unfortunately, X-ray astroimaging introduced a unique challenge: X-rays do not bend through glass.

To overcome this problem, scientists recognized the importance of using *straight-line* imaging techniques to photograph X-ray point sources. A straight-line imaging technique does not require the controlled bending of light rays to form an image. An early example is the *pinhole camera*. With origins in ancient Greek and Chinese civilizations, the pinhole camera enabled image formation using a light-blocking mask with a small hole. An example is a room with all the windows shuttered and light admitted only through a small hole, illustrated in Figure 1.2. This everyday example was so remarkable that the word *camera* is taken from the Latin word for *room*. For the X-ray imaging problem, pinhole camera techniques could indeed be used to form an image. A mask that blocks X-ray radiation can be fashioned out of an attenuating media, and a hole can be punched through the lead mask. The pinhole camera does not need to bend light to form an image. A pinhole camera seems like a terrific solution to the X-ray imaging problem!

Unfortunately, the pinhole camera does not solve the X-ray imaging problem. A pinhole is small and lets very little light through, causing very low signal-to-noise ratios. Attempts to enlarge the pinhole and let more light through are stymied by an increase in image blur. A fundamental trade-off between *signal-to-noise ratio* (SNR) and resolution is observed with the pinhole camera. For the X-ray astronomers this trade-off was insufficient, so they moved away from the pinhole in a quest to overcome the trade-off. (Mertz and Young, 1961) published a paper on the use of carefully coded mask patterns. As an abstraction, these masks behaved like an array of pinholes spread in a carefully chosen manner. Details of this approach are considered in chapter 4.

Coded apertures in X-ray imaging were just the beginning of the computational imaging revolution. Today we have access to a tool that lies outside the grasp of the ancient Assyrians (inventors of the first lens), ancient Greeks and Chinese (inventors of the pinhole camera), and Renaissance Europe (inventors of the telescope). This tool is the computer.

1.3 Modern Uses of Computational Imaging

The timing of this inaugural textbook on computational imaging aligns with the wide use of computational imaging systems in industrial and scientific practice. Here, we describe a few application areas of computational imaging and where they can be found in subsequent portions of the book.

- **Smartphone photography:** Many of the *computational imaging* techniques in this book can be directly prototyped or subsequently ported to smartphone imaging systems. For instance, the problem set complementing chapter 6 focuses on how to convert a smartphone to a lightfield camera. Industry practitioners may use the term *computational photography* to refer to cases in which a computational imaging method is used for the specific problem of photography. For consistency with this text, we stick to the term *computational imaging*. The dominant industrial use of computational imaging is on consumer smartphones. Large technology companies such as Apple and Google have dedicated computational photography teams. As readers are doubtless aware, smartphones have become omnipresent portable cameras. According to InfoTrends via Bitkom, since 2015 the number of photographs captured worldwide has been over 1 trillion, with a 10% increase every year. A staggering 85% of those were captured with mobile phones. Computational imaging techniques are particularly well-suited for smartphone applications, which are often hardware constrained. The thin body of a cellular phone does not enable specialized lenses. Bulk production costs limit exotic design. To generate aesthetically pleasing photographs, practitioners must innovate on the algorithmic side. Chapter 4 of this book discusses how a combination of imaging sensors, optical aperture, and illumination can be used in different ways to enhance imaging capabilities. An example is the high dynamic range imaging feature present in all modern smartphones, which uses the computational imaging approach.

- **Autonomous driving:** Computational imaging systems can be used to upgrade the visual acuity of autonomous cars to superhuman levels. An autonomous vehicle, also known as

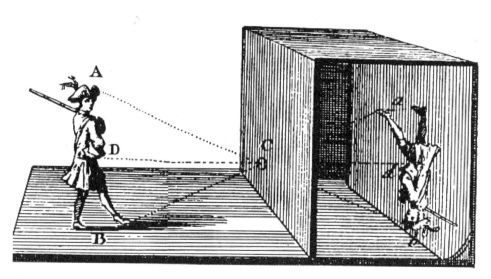

Figure 1.2 Example of a room-sized pinhole camera.

a self-driving car, is capable of actively sensing the surrounding environment and driving safely when human input is minimal or absent altogether. It is not a requirement that camera sensors on these cars mimic the human eye: they can surpass it.

The importance of vision sensors to autonomous driving is hard to overstate. An early joke about self-driving cars equipped with 3D LiDAR systems was along the lines of: "that LiDAR costs more than the car." Chapter 5 discusses time-resolved imaging, on which LiDAR is based. Although costs of LiDAR have come down tremendously, the exorbitant price in the early years illustrates the importance of LiDAR to safe navigation— engineers would not have used such expensive LiDARs if the data had not been critical for downstream performance.

Beyond the fundamental exposition in chapter 5, readers interested in autonomous driving may find chapter 10 of interest where this topic is further discussed from the perspective of light transport. The chapter discusses cases of multipath interference (e.g., driving through fog) as well as seeing around corners. Recent papers have integrated computational imaging systems that can see around corners with vehicular platforms.

- **Medical imaging:** In the beginning of this chapter we used the examples of X-rays and CAT scanners to introduce the notions of imaging and computational imaging, respectively. These examples only scratch the surface of the medical applications that use computational imaging.

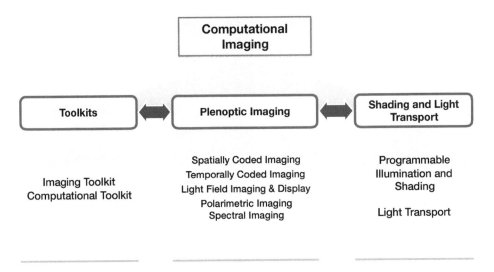

Figure 1.3 Roadmap and organization of the book.

In chapter 4 we discuss coded exposure imaging, which is used for imaging of flowing cells in flow cytometry. In chapter 5 we touch on how consumer-grade depth imaging sensors, such as the Microsoft Kinect, can be repurposed for bioimaging tasks such as fluorescence lifetime imaging. Finally, in chapter 10 we discuss how global and direct separation of light transport can be used as a technique to potentially see deeper inside the body without using X-rays.

1.4 Roadmap of the Book

With the twin goals of turning newcomers into practitioners and sharpening the skills of experts, the book has three parts, shown in Figure 1.3.

> **Part I.** The first part introduces basic knowledge required to study and innovate in computational imaging. Because the field is a codesign of optics and computation, we briefly review an optical toolkit (chapter 2) and a computational toolkit (chapter 3). For newcomers to the field, it is recommended to have some familiarity with part I before moving on to subsequent portions of the book.
>
> **Part II.** The discussion moves to the frontier of research, where the codesign of optics and computation is studied in the context of different modalities of light. Part II is unified by the plenoptic function, which describes the degrees of freedom of a light ray. The function describes how an image need not be a function

1.4 Roadmap of the Book

only of space (chapter 4) but also of angle (chapter 5), of time (chapter 6), of polarization (chapter 7), and of wavelength (chapter 8).

Part III. The book ends with a description of light transport techniques. By analyzing shadows and smartly coding illumination, it is possible to design imaging systems that obtain micron-scale 3D shape or optimize for noise-free imaging (chapter 9). We conclude by describing advanced techniques in computational light transport (chapter 10), including optical computing and non-line-of-sight imaging.

I TOOLKITS

2 Imaging Toolkit

In this chapter, our goal is to develop an understanding of the digital image formation model that is central to most current imaging devices. Building on the foundations of the image formation model and the various parameters associated with it allows us to understand the limitations of the conventional imaging pipeline. This becomes critical in subsequent chapters as we see how the computational imaging philosophy helps us go beyond what is possible with conventional imaging.

2.1 Optics

2.1.1 Animal Eyes

The human eye is a very sophisticated image-capturing device. It uses a lens to focalize the light reflected by an object onto the retina, which is made up of photosensitive cells called cones and rods. The eye's functioning principle is similar to that of a modern camera. However, during its evolutionary development, the animal eye was not always this complex. An earlier anatomy, still found in animals such as the marine mollusk called a nautilus, is the rudimentary pinhole eye, which is simply a sphere with a tiny hole in front and a layer of photoreceptors on the opposite side.

2.1.2 Light, Waves, and Particles

A light ray is modeled as a line describing the trace that a photon might leave behind. When capturing an image, each pixel captures the color of a ray of light. Therefore, the image allows us to detect the environment by mapping visible external points to points on the camera sensor.

However, light rays are used not only for measuring the environment; they can also be used to investigate optical systems (e.g., the lens surface or its coating). The light is generally attenuated from source to destination. The *reversibility* property of light rays means that the overall attenuation for a ray does not change when the source and destination are swapped.

Conceptually, light rays are infinitesimal in width and have an infinitesimal point of emergence. Therefore, measuring a single ray is challenging. To better understand light, we need to look at several models that describe it.

Firstly, light can be described as an electromagnetic wave. Therefore all light frequencies, from low-frequency radio signals to high-frequency cosmic rays, propagate through a vacuum at a constant rate
$$c = 299,792,458 \text{ m/s}.$$

The frequency v and wavelength λ are linked via the equation
$$v = \frac{c}{\lambda}$$

such that rays with high frequency, such as gamma rays, have a very small wavelength. It is important to point out that the light we typically measure propagates at speeds smaller than c because it is obstructed by the surrounding matter particles.

The wavelength has tremendous impact on the way the light wave propagates. For instance, when we drive underneath a bridge, we can always see our surroundings; however, the AM radio signal is likely to flicker. This is caused by the difference in wavelength between the two electromagnetic waves. The visible light has low wavelength compared to the bridge opening and therefore passes unobstructed. The wavelength of radio signals, however, is too large and therefore our antenna picks up only a noisy residue.

Whereas light was initially viewed as a waveform, Albert Einstein showed for the first time that light can be quantized as a stream of photon particles. Modeling light as a wave is convenient on a macroscopic scale, but more complex processing, such as analyzing light interference caused by diffraction through lenses, requires using the particle light model. This is useful if we are trying to understand, for instance, the maximum image resolution achieved with a given lens and why this is dependent on the lens size. In this course, light is primarily viewed using the wave or ray models.

In empty space, photons are well described by the *ray model*: single photon traces that do not interact with each other. Their wave-like behavior emerges in closed environments, such as when passing through a pinhole of a size comparable to the wavelength. As we will discuss subsequently, this causes diffraction, a wave-specific phenomenon.

The energy transported by a single photon is measured
$$E = h\frac{c}{\lambda} = hv$$

where $h = 6.62610 \times 10^{-34}$ J·Hz^{-1} denotes Planck's constant, v is the frequency, and λ is the wavelength.

This means that higher frequency photons carry significantly more energy. That is why higher frequency light (such as ultraviolet and x-ray) is more dangerous and can damage our bodies. One single photon carries an insignificant amount of energy (e.g., 4×10^{-18} J)

2.1 Optics

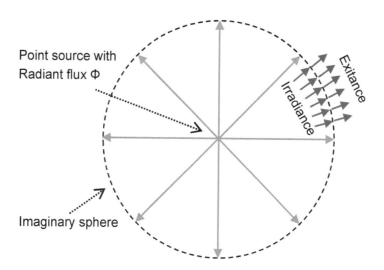

Figure 2.1 The ideal point source: a point source with radiant flux Φ and the irradiance/exitance for an imaginary sphere.

for visible light. Interestingly, when the human eyes are fully adapted to darkness, our rod cells can detect light bursts as small as eight to ten photons (Hecht et al., 1942).

2.1.3 Measuring Light with Rays

The light is affected by a range of factors such as

- the power transmitted,
- direction of radiation,
- area of real or imaginary surfaces,
- wavelength, and
- visibility.

The power transmitted is measured by the *radiant flux* Φ, which is defined as the energy emitted, reflected, transmitted, or received, per unit time, and is measured in watts, or J/s.

We introduce a simplified model of a light source called the ideal *point source* light, which is infinitesimal in size and radiates light outward uniformly in all directions. The point source is described by a radiant flux Φ. Let us consider an imaginary sphere centered in the point source, shown in Figure 2.1.

The point source has the following properties:

- All the rays in the point source arrive perpendicularly on the imaginary sphere at the same time.

- There is a one-to-one mapping between the sphere points and rays: every point on the sphere has a corresponding ray.
- The rays therefore form a continuum and their number is uncountably infinite (similar to the number of real values).
- The radiant flux is transmitted equally across the sphere's surface.

Thus each ray emitted by the point source carries 0 W, and one can measure only a 2D beam containing an uncountably infinite number of rays.

The **irradiance** is subsequently introduced to define the radiant flux incident to an area on the sphere for an ideal point source

$$R = \frac{\Phi}{4\pi r^2}$$

where Φ is the radiant flux, and r is the sphere radius. The irradiance is measured in W/m^2 and is inversely proportional to the square of the sphere radius. For example, increasing the radius ten times leads to an irradiance 100 times smaller for the new sphere. Given that the irradiance describes a particular spatial area on the sphere, we say that it measures the *spatial power density*.

The irradiance can be introduced in a more general context, where the surface is not necessarily a sphere. The value at a point on the surface is

$$R = \frac{d\Phi}{dA}.$$

For the radiance leaving the surface of interest we introduce the **exitance** M, measured in W/m^2.

In addition to the spatial power density, a thorough description of light requires introducing a way to measure the *angular power density*. In other words, we need to describe the radiant flux inside a beam of light. A 3D beam of light requires introducing a generalization of the 2D angle known as the *solid angle*.

Assume that we have a cone-shaped beam of light. The unit measure for the solid angle is the *steradian*, which is defined by a cone with the vertex in the center of a sphere of radius r whose base delimits a spherical *cap* of area r^2. The diagram of a steradian is depicted in Figure 2.2.

We now introduce the general solid angle Ω

$$\Omega = 2\pi \left(1 - \cos(\alpha)\right) \text{ sr}$$

where sr stands for steradians and α represents the half of the top planar angle of a cross-section of the solid angle, shown in Figure 2.2.

We point out that the solid angle corresponding to a whole sphere is $\Omega = 2\pi(1 - \cos(\pi)) = 4\pi$ sr. We can now introduce the **radiant intensity**, which measures the angular power density. For an ideal cone-shaped beam of light covering solid angle Ω with uniformly

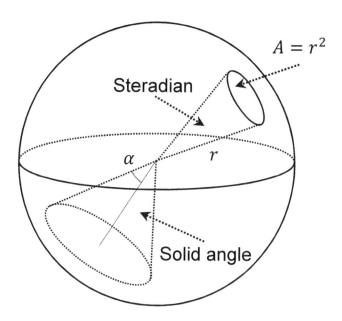

Figure 2.2 The steradian and the solid angle of a cone-shaped beam.

spread radiant flux Φ, the radiant intensity satisfies

$$I = \frac{\Phi}{\Omega}.$$

For example, a beam covering a whole sphere has a very low intensity $I = \frac{\Phi}{4\pi} = 0.08 \cdot \Phi$, so that one steradian contains a small part of the incoming light power. However, if we focus the same power in a beam covering the 1,000th part of a sphere, the intensity is significantly larger at $I = 80\Phi$.

The irradiance and radiant intensity allow us to model lighting phenomena such as the difference in heat between noon and dusk. At any time a beam of light from the sun, approximated here as a beam of parallel rays, illuminates an area on the ground that is proportional to $1/\cos(\alpha)$ where α is the beam incidence angle, as shown in Figure 2.3.

Therefore a given fixed area A on the ground is characterized by an irradiance that changes with the incident angle α

$$E(\alpha) = \frac{\Phi \cdot \cos(\alpha)}{A}$$

where Φ is the uniformly spread radiant flux of the beam of light covering the corresponding area. In this case we point out that the irradiance reaches a peak value when the incidence angle is 0, and the irradiance vanishes when the sun moves behind the horizon, corresponding to an incidence angle $\alpha = \pi/2$.

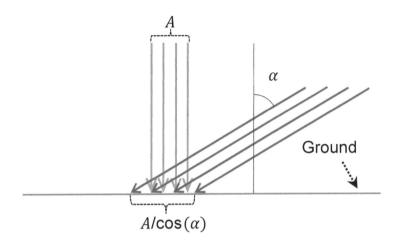

Figure 2.3 The area illuminated by a parallel beam as a function of the incident angle.

In a real-life application the objective is not to capture the irradiance or the radiance intensity but to capture the **radiance** L, which represents the ray strength, measuring the combined angular and spatial power densities.

The radiance is described by the equation

$$L = \frac{d^2\Phi}{dA\, d\Omega} \cdot \frac{1}{\cos(\alpha)}$$

where Ω is the solid angle, A is the area, and α is the incidence angle.

2.1.4 Pinhole Model

The principle underlying the biological eye is also the functioning principle of the first man-made cameras, called pinhole cameras (Young, 1989). The pinhole camera is based on a box with a half-millimeter hole and a photosensitive layer on the opposite side.

The functioning principle of the pinhole camera, which is identical to that of a camera obscura, is shown in Figure 2.4a. In a camera obscura, which is a dark room with only a tiny hole in one of its walls, the light is projected upside down on the opposite wall, called a projection plane. The axis passing through the hole perpendicular on the projection plane is called an optical axis. Because the light travels in straight lines and the hole is very small, each point on the projection plane is mapped uniquely to a point from the outside scene.

The diagram of the pinhole camera principle is shown in Figure 2.4b. Here the coordinate frame is placed with coordinate Z along the optical axis, coordinate Y perpendicular on the diagram plane and therefore not displayed, and the center in the pinhole. The distance d between the pinhole and the projection plane is called a focal distance, X_0, Y_0, Z_0 denote the coordinates of a point in the scene, and $-x$, $-y$, d denote the coordinates of

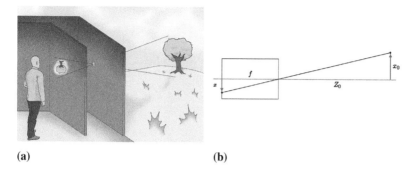

(a)   (b)

Figure 2.4 The pinhole camera. (a) The camera obscura, a darkened room with only a hole in a wall, is an example of the pinhole camera principle; (b) a diagram represents the pinhole camera principle.

the corresponding point in the projection plane. Then it follows that $x = -d \cdot X_0/Z_0$ and $y = -d \cdot Y_0/Z_0$. In a more compact form, the model of the *ideal pinhole camera* is

$$\begin{bmatrix} x \\ y \\ 1 \end{bmatrix} \sim \begin{bmatrix} d & 0 & 0 & 0 \\ 0 & d & 0 & 0 \\ 0 & 0 & 1 & 0 \end{bmatrix} \begin{bmatrix} X_0 \\ Y_0 \\ Z_0 \\ 1 \end{bmatrix}.$$

Here, $\sim$ means that the two quantities are proportional. If we want to produce a digital image, then the coordinates of a pixel on the projection plane x_p, y_p satisfy $x_p = s_x x$, $y_p = s_y y$ where s_x, s_y represent the scaling constants. In a real scenario, the coordinate system in the projection plane is not centered on the optical axis. Therefore we introduce constants u_0, v_0 to account for this

$$\begin{bmatrix} x \\ y \\ 1 \end{bmatrix} \sim \begin{bmatrix} s_x d & 0 & u_0 & 0 \\ 0 & s_y d & v_0 & 0 \\ 0 & 0 & 1 & 0 \end{bmatrix} \begin{bmatrix} X_0 \\ Y_0 \\ Z_0 \\ 1 \end{bmatrix}.$$

Moreover, to generate a more realistic model, we need to take into account the skew effect caused by the fact that the optical axis may not be perfectly perpendicular on the projection plane. This effect is modeled by the skew factor α, leading to the final *internal camera*

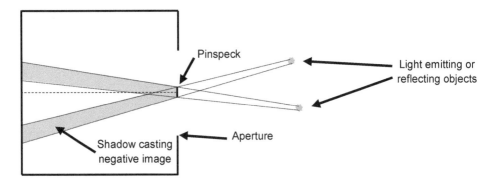

Figure 2.5 The pinspeck camera: this imaging device is based on the opposite functioning principle of the pinhole camera, casting shadows that form a negative image.

model:

$$\begin{bmatrix} x \\ y \\ 1 \end{bmatrix} \sim \begin{bmatrix} s_x d & \alpha & u_0 & 0 \\ 0 & s_y d & v_0 & 0 \\ 0 & 0 & 1 & 0 \end{bmatrix} \begin{bmatrix} X_0 \\ Y_0 \\ Z_0 \\ 1 \end{bmatrix}.$$

A pinhole camera creates an image by projecting light onto a plane. However, an image can be created using the opposite principle, by casting shadow. This is the functioning principle of the pinspeck camera (Cohen, 1982). Instead of a tiny hole, this camera is based on a wide aperture with a small speck in the middle. When objects are illuminating the camera, the speck is casting a shadow on the projection plane, effectively creating a negative image. A diagram of the pinspeck camera is shown in Figure 2.5.

Pinhole and pinspeck cameras are good mechanisms to study light properties. However, from a practical perspective they are subject to several problems, such as long exposure times, limited sharpness, and limited field of view. The exposure time is long because the pinhole permits only a small number of light rays to hit the sensor plane per time unit, which means that it takes longer for an image to be created. The image sharpness for a pinhole camera is inversely proportional to the hole size. However, holes that are too small cause diffraction, which is bending of light around the corners of the hole.

It is therefore important to find the right pinhole aperture size δ to capture good photographs. As we discussed previously, for larger pinholes each point on the projection plane is mapped to a point of the scene along a series of lines, as depicted in Figure 2.6. Therefore a distant object is imaged as a disk of radius δ. However, when δ is comparable with λ, the incoming light wavelength, the diffraction phenomenon causes the light rays passing close to the aperture boundaries to bend, leading to a circular disk with rings around it as shown in Figure 2.6. The diameter of the disk is given by $D = 2.44 \cdot \lambda \cdot d/\delta$.

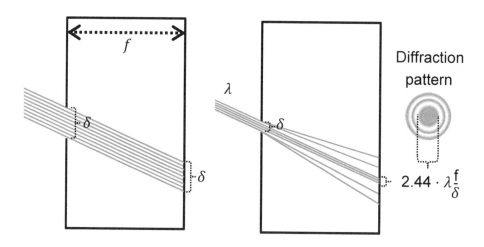

Figure 2.6 The pinhole camera diffraction: when the pinhole size is comparable to the wavelength of the incoming light, a distant object is imaged as a circular disk with rings around it.

The optimal aperture size is attained when the disk of the diffraction pattern has diameter δ. A smaller aperture would cause distortion due to diffraction, and a larger one would lead to blurry images due to loss in sharpness. Let us compute the optimal aperture size for the wavelength located in the center of the visible spectrum $\lambda = 500$ nm. Given that the focal distance d is measured in millimeters, after the appropriate conversions, the optimal aperture size is

$$\delta_{opt} = 2.44 \cdot \lambda \frac{d}{\delta_{opt}} \implies \delta_{opt} = 0.035\sqrt{d}.$$

However, this estimation assumes that all imaged objects are far from the aperture. Close objects would create disks larger than the aperture and thus distort the image. Moreover, the smaller the hole is, the more it limits the field of view—at the limit, a hole of infinitesimal width allows only rays perpendicular to the projection plane to enter the camera.

2.1.5 Ray Bending and Lenses

Given all the previously mentioned drawbacks of pinhole cameras, a better device is needed for successful photography. The lens, the component of choice in modern cameras, is based on the refraction principle. When a light ray passes through the smooth boundary of two different materials, it is bent by an angle depending on the indices of refraction of the two materials. The index of refraction is a constant characteristic of each material, and the

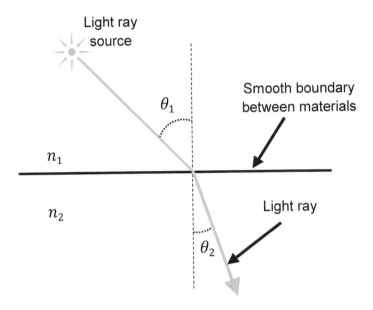

Figure 2.7 The refraction principle: a ray of light is bent at the boundary of two materials with an angle given by Snell's law.

refraction is governed by Snell's law

$$n_1 \sin(\theta_1) = n_2 \sin(\theta_2)$$

where n_1, n_2 are the indices of refraction, and θ_1, θ_2 are the angle of incidence and angle of refraction, respectively. The bending process, called refraction, is depicted in Figure 2.7.

A lens has a relatively generic definition. Any object that bends incoming rays into outgoing rays can be considered a lens. The number of lenses that can be generated on the basis of how they refract light according to Snell's law is very large. However, in optics it is common to use an idealized concept, called the *thin lens*. A thin lens, also called a paraxial, is a plane that bends light governed only by three parameters: the focal length, the aperture diameter, and the lens speed.

The *focal length* f is defined as the distance in millimeters between a thin lens and the point of convergence of a number of parallel rays passing through the lens. The inverse of the focal length 1/f is known as the focusing power and is measured in diopters. This parameter is an important characteristic of common eyeglasses.

2.1 Optics

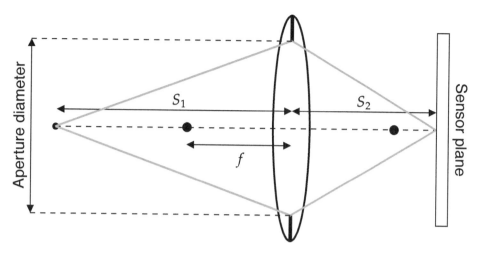

Figure 2.8 The diagram of a thin lens: a point in the scene (black dot) is in focus if the light it reflects toward the lens converges on the sensor plane. When the camera refocuses on a different point, the sensor plane moves relative to the lens.

The *aperture diameter D* is the diameter of the base of a conical shaped ray bundle passing through the lens. In other words, it is the diameter of the largest portion of the lens that is bending light.

The *lens speed*, or the f-number N, is the ratio between the focal length and the aperture diameter. It describes the ability of the lens to transmit light (i.e., the required exposure time for capturing an image).

The light bending principle with a thin lens is depicted in Figure 2.8. A scene point is in focus relative to a thin lens and a sensor plane if it obeys the thin lens equation:

$$\frac{1}{S_1} + \frac{1}{S_2} = \frac{1}{f}$$

where S_1 is the distance between the lens and the object in focus, S_2 is the distance between the lens and the sensor plane, and f is the focal length.

In the early days of photography, capturing an image required manually focusing the camera using a ground-glass viewer, inserting the light-sensitive plate inside the camera, and then manually controlling the exposure time by removing the cap from the lens for a predefined amount of time, as shown in Figure 2.9. Initially, this time was measured in hours, but it decreased to milliseconds thanks to the development of light sensors. Automatic shutters allowed for a fine control of the exposure time.

When we capture an image with a modern digital camera the device itself automatically tunes a large number of parameters to allow a crisp and detailed view of the scene. Those

Figure 2.9 Lens-based camera obscura: the first cameras required a manual adjustment of the exposure time. Reprinted from (Raskar and Tumblin, 2011).

parameters can relate to the camera as a whole, to the camera lens, to the shutter, or to the light sensor.

The camera itself has a certain *position* and *orientation* in space, allowing it to capture a portion of the scene. For a dynamic scene, the *time of the capture* is another important parameter. The scene *lighting*, either coming from the camera flash or from an external source, can affect the colors or visibility of the scene objects.

The light reflected by the scene first meets the opening of the camera shutter. This opening is called *aperture*, controlling how many light rays enter the camera at any given moment. This parameter is tightly connected with the *exposure time*, which measures how long the shutter is kept open. Therefore, a small aperture and short exposure time lead to darker images. However, apart from image brightness, the two parameters have different effects on the image, as subsequently explained.

After passing through the aperture, the light rays are bent by the camera lens. This is determined by the lens *focal length*, which describes how fast parallel light beams converge after passing through the lens. This parameter affects the width of the field of view. A lens with short focal length corresponds to a wide *field of view*, which allows it to capture a larger portion of the scene. At the same time, the short focal length lens in conjunction with a small aperture allows a much longer depth of field. Lenses with wide fields of view shrink scene features and exaggerate foreshortening (depth-dependent size). On the other hand, lenses with narrow fields of view, also known as telephoto lenses, enlarge scene features and reduce foreshortening.

Using lenses instead of pinholes vastly increases the final image brightness, as discussed previously. Therefore, fewer rays emitted by a distant object would reach the camera. This might suggest that this object might appear less bright in the image, but that is not true for the following reason.

If we double the distance between an object and the camera, the light beam detected by the camera is decreased by $1/2$ both horizontally and vertically, leading to a solid angle decreased by $1/4$, and therefore a radiant flux Φ smaller by $1/4$. However, the brightness of each point of the object projected on the sensor is given by the irradiance $R = d\Phi/dA$, where A denotes the area on the sensor that is illuminated by the point. Given that the solid angle of the light beam is $1/4$ smaller, it follows that the area A projected on the sensor is also decreased by $1/4$, and thus the average irradiance of the object in question $\overline{R} = \Phi/A$ stays the same.

By comparison with the human eye, we can define a *normal lens* that replicates approximately the eye's field of view. Given that the field of view is affected by the focal length, a normal lens has a focal length approximately equal to the diagonal dimension of the film or digital sensor that captures the photograph. However, in order to create a realistic perception, we need to take into account that we usually view images from a distance, which is why in practice the normal lens generates slightly larger fields of view than the biological eye. Wide angle lenses are used to capture larger areas of the scene that cannot be accommodated by normal lenses. This leads to distorted photographs, but the effect is generally addressed with larger prints.

When one attempts to photograph a tall object in a scene such as a building, the photographer notices the tilt effect, in which the object in the image seems to lean backward. This is due to the upward tilt in the camera required to include the whole building in the image frame. Professional photographers and architects are interested in capturing a tall object that appears straight in the final photograph. One option is the post-processing of the image. However the problem can be solved by capturing the image with a tilt-shift lens that compensates for this effect.

In order to understand the limits of the thin lens formula, let us consider one surface of a convex lens, as shown in Figure 2.10. When the ray intersects the lens surface, it is refracted according to Snell's law. However, because the lens surface is not flat, the incidence angle θ_1 is computed relative to the line normal to the surface. Therefore Snell's law is

$$n_1 \sin(\theta_1) = n_2 \sin(\theta_2).$$

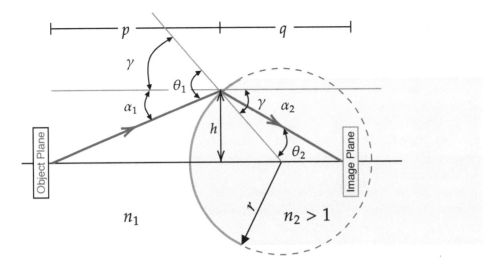

Figure 2.10 Ray bending at one surface of a thin lens: the ray reflected by the object (*red line*) intersects the lens surface at distance h from the optical axis. Snell's law governs the new direction of the ray relative to the line normal to the surface (*dotted line*).

Furthermore, we can derive the following:

$$\theta_1 = \gamma + \alpha_1, \quad \theta_2 = \gamma - \alpha_2,$$
$$\tan(\alpha_1) = \frac{h}{p}, \quad \tan(\alpha_2) = \frac{h}{q}, \quad \sin(\gamma) = \frac{h}{R}.$$

These trigonometric expressions lead to rather complex calculations. Therefore it is common to use the *paraxial approximation*, which assumes that the angle between the light ray and the optical axis is very small. From trigonometry, we know that when an angle β is very small, we can say that $\beta \simeq \tan(\beta) \simeq \sin(\beta)$. Using this approximation, the equations become

$$n_1 \theta_1 = n_2 \theta_2,$$
$$\theta_1 = \gamma + \alpha_1, \quad \theta_2 = \gamma - \alpha_2,$$
$$\alpha_1 = \frac{h}{p}, \quad \alpha_2 = \frac{h}{q}, \quad \gamma = \frac{h}{R}.$$

If we substitute the last two lines in the first equation, we have

$$n_1 \left(\frac{1}{p} + \frac{1}{R} \right) = n_2 \left(\frac{1}{R} - \frac{1}{q} \right) \iff \frac{n_2 - n_1}{R} = \frac{n_1}{p} + \frac{n_2}{q}.$$

2.1 Optics

We now combine the equations describing light bending on both sides of the lens to yield the lens equation

$$\left(\frac{n_2}{n_1} - 1\right)\left(\frac{1}{R_1} - \frac{1}{R_2}\right) = \frac{1}{p} + \frac{1}{q}$$

where R_1, R_2 are the radii of the two surfaces of the lens. Now let us assume that the object is located far away from the lens. In this scenario we have

$$p \to \infty, \; q \to f, \; \alpha_{1,2} \to 0, \; \Theta_{1,2} \to 0, \; \gamma \to 0$$

where f is the focal length of the lens, which is in line with the paraxial approximation. The lens equation then takes the following form, which is also known as the lens maker's equation:

$$\left(\frac{n_2}{n_1} - 1\right)\left(\frac{1}{R_1} - \frac{1}{R_2}\right) = \frac{1}{f}.$$

The two versions of the equation also lead to the previously introduced thin lens equation

$$\frac{1}{p} + \frac{1}{q} = \frac{1}{f}.$$

As we have previously shown, this equation describes how to change the distance between the sensor and the lens in order to keep an object in focus. However, with regard to its derivation, it is important to remember that it relies on the *paraxial approximation*. This means that the equation will no longer be precise for objects that are close to the lens and not on the optical axis. Similarly, the object may not be close, but if the lens is large, then the paraxial approximation will not hold for rays intersecting the lens at its outer boundaries.

In the preceding equations $1/f$ is the *focusing power* of the lens. The explanation of its name is intuitive. When $1/f$ increases, then q decreases, so that the rays converge faster, and thus we say that the lens has a higher focusing power.

Let us see what happens when we place two thin lenses in a sequence. The equations are

$$\frac{1}{p_i} + \frac{1}{q_i} = \frac{1}{f_i}, \; i = 1, 2.$$

In this case the rule is that the object plane of the second thin lens is located at $-q_2$, where q_2 is the focus plane of the first thin lens. Therefore $p_2 = -q_2$ and thus by adding these equations we obtain

$$\frac{1}{p_1} + \frac{1}{q_2} = \frac{1}{f_1} + \frac{1}{f_2} = \frac{1}{f_c}$$

where f_c is the focal length of the compound lens. We notice that the focusing power of the compound lens is the sum of the focusing powers of all the individual lenses. Conversely, we can work out the focal length of the compound lens f_c as $f_c = f_1 f_2 / f_1 + f_2$. Intuitively, the second lens makes the rays that pass through the first lens converge even faster, therefore leading to an accumulated focusing power.

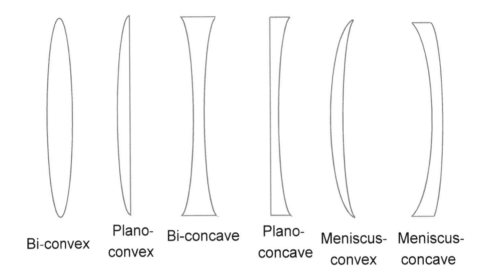

Figure 2.11 The main categories of lenses, which are based on the shapes of the lenses' two surfaces.

If the two lenses are separated by length d then the equation becomes (Ronchi and Rosen, 1991)
$$\frac{1}{p_1} + \frac{1}{q_2} = \frac{1}{f_1} + \frac{1}{f_2} - \frac{d}{f_1 f_2}.$$

A real lens does not need to be spherical. The main categories of lens shapes, defining how they bend the light, are shown in Figure 2.11.

One may wonder why it is necessary to use more than one lens. The reason is that in using a single element spherical lens, the image of the object is not created on a plane but instead on a sphere. This optical aberration is known as the *Petzval field curvature* or simply the *field curvature*. The result is that we cannot focus an entire object on a plane sensor, which causes the image to look blurred around the edges. Other than using complex lens designs, a hardware solution is to use a curved imaging sensor that compensates for this effect. An example of such an imaging sensor, NASA's Kepler focal plane array, is shown in Figure 2.12.

Another way to address this effect is to use the meniscus lens, which creates a much flatter image. However, this lens introduces chromatic aberration, that is, it focalizes different waveforms on different planes.

The paraxial approximation is a significant constraint on the lens size and object position relative to the lens. It assumes that $\beta \simeq \sin(\beta)$ for small values of β. This estimation

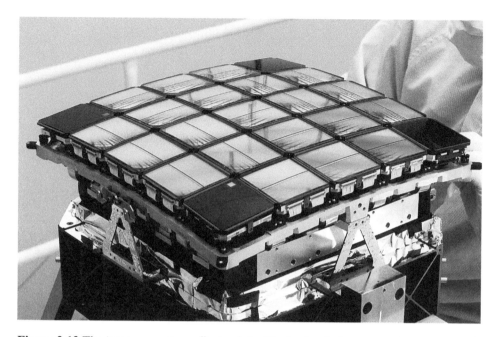

Figure 2.12 The image curvature effect and the Kepler focal plane array: in using a single element spherical lens it is not possible to focus a whole object on the sensor plane, and thus the edges look out of focus. Unlike conventional digital sensors, the imaging sensor array used in the Kepler space observatory is curved so that the Petzval field curvature can be compensated. *Source*: NASA (NASA and Ball Aerospace, 2008).

comes from the Taylor series expansion of the sine, which states that

$$\sin(\beta) = \beta + \frac{\beta^3}{3!} + \frac{\beta^5}{5!} + \cdots.$$

The estimation is more precise if we include more terms. For example, Ludwig von Seidel used the third-order approximation $\sin(\beta) \simeq \beta + \beta^3/3!$ to evaluate the imperfections of lenses, and concluded that there are five aberrations that make real lenses bend light differently from a perfect lens. Therefore one may wonder why we do not use very high-order optics to make better lenses? The reason is that making less regular lenses is very expensive, and it is much more affordable to use stacks of compound lenses of simple shapes.

A well known example of lens imperfection is the chromatic aberration, also called *dispersion*. The phenomenon, depicted in Figure 2.13, causes light rays of different wavelengths to focus at variable distances from the lens. In essence, it means that the lens has a wavelength-dependent focal length. One option to correct this is based on the observation that plano-concave lenses also bend different wavelengths differently but in the opposite

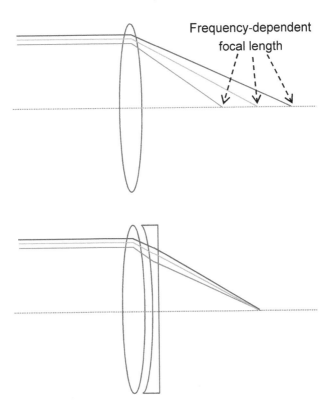

Figure 2.13 Correcting chromatic aberration: the biconvex lens has a frequency-dependent focal length, which can be corrected by pairing it with a plano-concave lens.

direction. Therefore the dispersion can be corrected by pairing the biconvex lens with a plano-concave lens as shown in Figure 2.13. Section 8.3.1 explains how the properties of dispersion are used in image capture setups to obtain spectral information.

As one might expect, using a stack of two lenses can correct the chromatic aberration for two frequencies, resulting in an *achromat* lens. The same principle can be applied by stacking up three (apochromat) and four (superchromat) lenses, which corrects three and four frequencies, respectively. However, this also leads to an increase in cost.

Another lens aberration occurs because a proportion of the light is reflected by each lens, bouncing back and causing flares and other undesired effects. The ratio of reflected light intensity between two materials is given by Fresnel's equation

$$r = \sqrt{\frac{|n_2 - n_1|}{n_2 + n_1}}$$

2.1 Optics

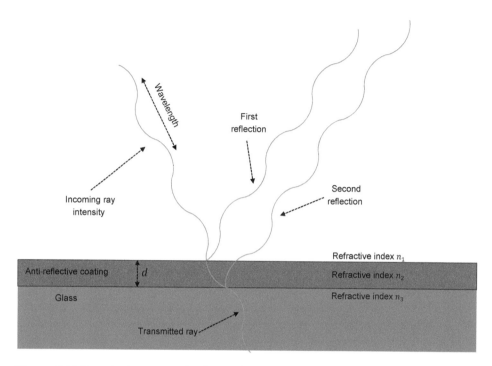

Figure 2.14 Functioning principle of antireflective coating: the incoming light ray is reflected twice by the coating layer and the glass. The two reflections are in opposite phases and thus cancel each other. The sinusoidal curves do not represent the ray paths but rather their intensities.

where n_1 and n_2 are the refractive indices of the two materials, and r is the amplitude ratio of the reflected light. For the case of the air/glass reflectivity, we have that $n_1 = 1$, $n_2 = 1.5$, and this leads to $r = 0.2$. This corresponds to a ratio of reflected light intensity of $r_2 = 0.04$ or 4%. It is important to point out that this ratio of light intensity is reflected at all boundaries between materials, and therefore it is specifically undesirable for large stacks of compound lenses.

The solution in this case is applying a layer of antireflective coating on each lens. This will not stop the reflection, but it will cancel out the reflected light to prevent the image from being distorted. The coating introduces a new boundary of reflection, and this results in two separate light reflections, one caused by the coating and the second by the glass, as depicted in Figure 2.14. The two reflected rays are subject to the following conditions:

1. The rays should have identical intensities.
2. The phases of the rays should be opposite.

These two conditions ensure that the reflections cancel each other, as shown in Figure 2.14. Let n_1, n_2, n_3 be the refractive indices of the three regions. Assuming that $n_1 < n_2 < n_3$, condition 1 can be written

$$\frac{n_2 - n_1}{n_2 + n_1} = \frac{n_3 - n_2}{n_3 + n_2} \iff n_3 n_2 - n_3 n_1 + n_2^2 - n_2 n_1 = n_3 n_2 + n_3 n_1 - n_2^2 - n_2 n_1.$$

It is easy to see that this condition is satisfied for $n_2 = \sqrt{n_1 n_3}$, which gives us the refractive index of the coating layer.

Condition 2 is satisfied by choosing the thickness of the coating layer $d = \lambda/4$, where λ is the wavelength of the incoming light. This ensures that by the time the second reflection travels through the coating layer twice, which amounts to half its wavelength, its phase is opposite to that of the first reflection and thus it cancels the first reflection.

As before, this correction works for one wavelength only. Typically lenses have two or three coating layers to cover a larger portion of the spectrum.

2.1.6 Lenses and Focus

In the previous subsection we examined how Snell's law determines when a point in the scene is in focus. However, Snell's law cannot be used on a regular basis to focus a camera in a real-life scenario, because the length S_1 from the lens to the object is usually unknown.

In the early days of photography, focusing was accomplished by moving the lens manually to maximize the image contrast.

One of the main methods to focus a camera is phase-based autofocus. This method was used in 1977 for the first autofocus camera, the Konica C35 AF. The autofocus system measures the light intensity on the sensor originating from the two halves of the lens. Each half generates an intensity curve, as shown in Figure 2.15. The lens moves relative to the sensor plane until the two curves are in phase, which ensures that the camera is in focus.

Current autofocus cameras embed the lens-translating motor in the lens itself. Let us look in more detail at the functioning principle of phase-based autofocus. As shown in Figure 2.15, the system requires measurements of the intensity curves from different parts of the lens. However, this is not possible using only one lens and a sensor.

The system uses a beam splitter to measure light phase coming from the opposite sides of the lens using one-dimensional (1D) sensor arrays. The intensity curves can then be measured independently and their phase difference used to compute the direction and distance of the lens motion.

Alternatively, a camera can use *contrast-based autofocus*. This mechanism involves a sensor that calculates the contrast as the difference in light intensity of nearby pixels. Unlike phase-based autofocus, the direction of movement cannot be derived immediately and requires a search routine. Therefore contrast-based autofocus is slower, as used in some smaller setups such as cell-phone cameras.

2.1 Optics

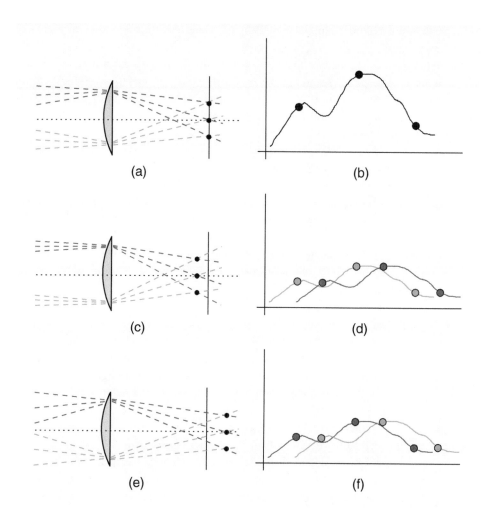

Figure 2.15 Lens bending rays reflected by an object: cases in which the intersection between the rays from the upper and lower halves of the lens falls (a) onto the sensor plane, (c) in front of the sensor plane, and (e) behind the sensor plane; and the corresponding intensities along the sensor plane (b), (d), and (f). The green and red curves in (d) and (f) show how the two lens halves generate out-of-phase intensity functions when the object is out of focus. The intensity curves can then be measured independently and their phase difference used to compute the direction and distance of the lens motion. Reprinted from (Ramanath et al., 2005).

A third category of focusing methods is *active autofocus*. This method involves measuring the distance to the object independently using ultrasound waves or infrared light. This principle does not require a minimal contrast in the scene to work, and it generally leads to lower performance.

In addition to choosing the best focusing mechanism, the photographer needs to choose the focus plane. A scene has several planes on which the camera can focus. Some cameras automatically focus on the objects that are closest, are brightest, or have the highest contrast. Modern cameras can perform live face detection to pick the focus plane, or allow the user to manually select the plane, typically by tapping on the desired scene point on a touchscreen.

2.1.7 Masks and Aperture Manipulation

In the previous subsection we saw that adding an obstacle in the light pathway, a mask with a rectangular window in its center, allowed separating the light coming from different parts of the lens and revealed information that otherwise would have been unknown, that is, the level of focus on an object in the scene. Generally speaking, masks represent planar elements that occlude or attenuate light rays in a spatially varying fashion.

Interestingly, it is possible to fully replace the lens with a mask. This allows extracting information from the image that is not available with lens imaging. Compared with a pinhole camera, which uses a mask with a single hole, this approach has a much higher light throughput, leading to brighter images. A drawback however is that the image requires postprocessing. This idea is elaborated for a variety of applications in chapter 4.

One example of such an application in photography is the modified uniformly redundant array (MURA) architecture, which gathers around 22,000 times more light than a pinhole camera by using a mask that is almost 50% empty (Gottesman and Fenimore, 1989). The mask is defined by a binary matrix $A_{i,j}$ such that a value 0 represents an occluder and a 1 denotes a gap allowing the light to pass through. The matrix $A_{i,j}$ is displayed in Figure 2.16a, in which white corresponds to 1 and black corresponds to 0.

The raw captured image, depicted in Figure 2.16b, does not reveal a lot about the scene. However, a decoding algorithm allows recovering a high quality image, as shown in Figure 2.16c. Further examples of MURA patterns with different matrix sizes are shown in Figure 2.16 (*bottom row*).

The concept was later extended by replacing the lens with a series of light-attenuating layers that can be controlled in both time and space (Zomet and Nayar, 2006). This setup allows extracting more information from the measurements, such as changing the viewing angle after the image has been captured.

Masks have also been used together with lenses to generate images. The mask can be positioned in one of three places relative to the imaging system, as shown in Figure 2.17:

1. on the camera aperture,
2. on the sensor, or
3. on the scene.

2.1 Optics

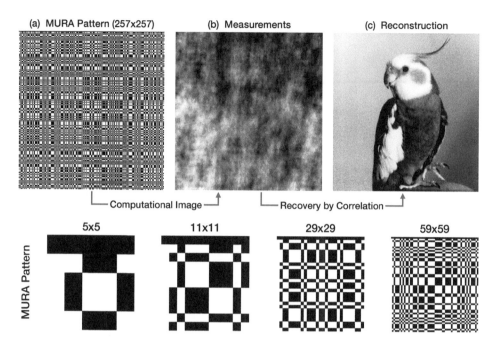

Figure 2.16 Example of lensless MURA photography: (a) mask used to capture the image, (b) detected image, (c) image reconstructed from the measurements, and (*bottom row*) MURA patterns with different matrix sizes.

Masks placed at the aperture level reveal many interesting properties of the scene. For example, (Farid and Simoncelli, 1998) computed the differentiation of the image intensity that is passed through a mask as the camera viewpoint changes. They demonstrated that two masks can be applied such that the derivatives calculated for each mask can be used to estimate the range of the scene. Later it was also shown that both conventional photographs and the corresponding depth maps can be recovered by placing a mask at the aperture of a consumer grade camera (Levin et al., 2007), therefore applying a technique known as *depth from defocus*. Single view depth estimation can also be achieved using an end-to-end design approach, jointly designing the mask and the reconstruction algorithm (Wu et al., 2019).

Rather than applying a mask, the camera aperture can be programmed to act like a mask. Taking multiple photos with various aperture sizes and shapes can be used to generate an image with much higher spatial resolution (Liang et al., 2008, 2007). (Sinha et al., 2017) used a lensless imaging system and model the inverse transform as a deep neural network to obtain the image phase. Further examples and several other state-of-the-art implementations are discussed in the context of spatially coded imaging in section 4.1.5.

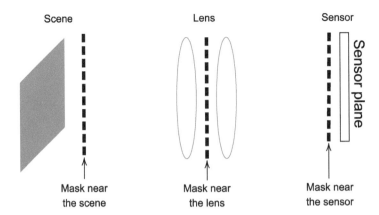

Figure 2.17 Positions for placing a mask in a camera.

The second option for positioning a mask is on the sensor. The most common example, existing on many consumer grade cameras, is the Bayer filter. A color image is made up of three images, each capturing the intensity of the color red, green, or blue. Using the Bayer filter all three images can be captured at once with one single sensor array, by using pixels sensitive to each of the RGB colors arranged in a pattern called the Bayer pattern.

It is possible to project light onto only selected regions of a sensor, the equivalent of placing a mask on a sensor, by using a digital micromirror device (DMD). A DMD is essentially an array of micromirrors that have two possible orientations, reflecting the light either toward the sensor array or away from it. A DMD is used for computing high dynamic range images and also for performing object recognition (Nayar et al., 2006). Using a DMD allows reducing the sensor array to a single pixel (Takhar et al., 2006). This is done by generating various patterns with the DMD and directing the accumulated light toward the pixel sensor. This system was used together with compressive sensing theory to generate images of reduced size (Takhar et al., 2006).

Placing a mask on the sensor or lens is easy to imagine, but placing a mask on the scene would be more difficult. In this case it is common to simply illuminate certain parts of a scene while leaving others in darkness, which has a similar effect. This has been done to extract the two sources of illumination in a scene: the direct illumination by the source and the global illumination from other points in the scene (Nayar et al., 2006). This separation of the sources has a practical significance because each one reveals different information about the scene. The direct component enhances the material properties of a given point, and the global component reveals the optical properties of the scene, indicating how a certain point is illuminated by other points in the scene.

2.2 Image Sensors

2.2.1 Cameras, Rays, and Radiance

A point in the scene is imaged by measuring the emitted/reflected light that converges on the sensor plane. However a real sensor detects the light irradiance, which is zero if the light is absorbed by the sensor only at a point. A single-pixel detector measures the irradiance in the vicinity of the point receiving light from the scene.

Therefore the detected brightness depends on the area on the sensor that a light beam covers. In the previous section we looked at how light falling at an angle leads to the *cosine falloff* effect, in which the irradiance decreases with the cosine of the light incident angle. One may then wonder why when one looks at a screen at an angle the image brightness does not appear to change. The explanation is that because light has an incident angle, the new perspective captures more light rays and therefore the increased radiant intensity compensates for the larger area covered by the light beam on the sensor.

After they are processed by the camera lenses, the light rays hit the sensor, which converts them into electrical signals. The sensor affects the appearance of the final image with a range of parameters. First, the *light sensitivity*, or ISO, can be used to brighten dark images. However, increasing the ISO also generates a noisier image. Moreover, the *dynamic range*, defined as the range of luminance the sensor is able to detect, has a strong impact on the clarity of the resulting image, especially if it contains both bright and dark areas. The *tonal range*, on the other hand, is given by the actual number of tones captured by the camera and is affected by other sensor parameters such as ISO. The sensitivity of the sensor to color is called *wavelength sensitivity* and can be adjusted on most cameras using the color balance or saturation settings. The clarity of an image is, of course, strongly influenced by the sensor *spatial resolution*, which is the number of pixels on the sensor.

To view an object clearly, it must be at *focusing distance* from the lens. The reason is that the light rays reflected by the points at focus distance converge on the sensor. When the camera focuses on a different object, it changes the distance between the sensor and the lens, which in turns changes the focusing distance. A focus measure can be used to determine the right lens position to keep an image in focus. Figure 2.18 shows the focus measure as a function of a lens's position and three captured images with different lens positions.

The depth of field is the distance between the closest and farthest points in the scene on which the camera can focus. The longest depth of field is achieved by a camera with short focal length and small aperture. Even if points in the distant part of the scene may be out of focus, a small aperture decreases how much the light rays diverge when they arrive at the sensor. The human eye is subject to the same effect. That is why squinting can help focus better on objects that are far away.

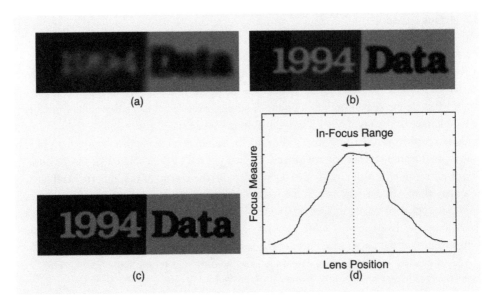

Figure 2.18 Imaging for different lens positions: (a)–(c) show an image captured with a gradually improved focus and (d) shows focus measure as a function of the lens position, which is maximized when the image is in focus. Reprinted from (Ramanath et al., 2005).

It can be observed that in a camera, achieving a long depth of field will typically lead to a darker image due to the small aperture. This can be fixed by increasing the exposure time. If the scene is dynamic, this would cause motion blur. Increasing the ISO instead addresses the brightness issue but adds noise to the measurements.

The sensor is affected by several types of noise, and each type is more prevalent in certain imaging environments. The *read noise* consists of the sensor pixel noise and the noise generated by the analog-to-digital converter. This noise determines the contrast of the captured image and affects the CMOS and the CCD sensors differently, because in the CCD the ADC is not part of the actual sensor. The *shot noise* results from the discrete nature of electrons captured by the potential well, and is more prevalent in brighter environments. A larger potential well is needed to address this type of noise.

Mathematically, the read noise is considered the Gaussian part of the noise, and is caused by stationary disturbances, and the shot noise is the Poissonian part, and is caused by the sensing of photons. Their names arise from the distributions that model their values. As the screens have higher and higher resolutions, the pixel sizes decrease and the sensitivity to photon noise increases for each pixel. Thus the shot noise is currently the main contributor to noise in imaging sensors.

The overall noise is thus signal dependent and very different from the usual additive white Gaussian noise that is common in image processing. The limited dynamic range of pixels

2.2 Image Sensors

leads in many cases to overexposure, or capturing light close to the maximum capacity per pixel. This effect further enhances signal-dependent noise.

The measurements $z(x, y)$ from a sensor at each pixel are given

$$z(x, y) = I(x, y) + \sigma(I(x, y))\zeta(x, y)$$

where x, y is the pixel position in two dimensions (2D), $I(x, y)$ is the light signal, $\zeta(x, y)$ is the Gaussian noise of 0 mean and standard deviation 1, and $\sigma(I(x, y))$ is the signal-dependent standard deviation. The aim of the preceding equation is to estimate $I(x, y)$ and $\sigma(I(x, y))$ from measurements $z(x, y)$.

To separate the influences of Gaussian and Poissonian noise, we write the measurements deviation (Foi et al., 2008)

$$z(x, y) = I(x, y) + \sigma(I(x, y))\zeta(x, y) = I(x, y) + \eta_p(I(x, y)) + \eta_g(x, y).$$

The noise distributions can be written as follows

$$I(x, y) + \eta_p(I(x, y)) \sim a \cdot P(a \cdot I(x, y)),$$
$$\eta_g(x, y) \sim \mathcal{N}(0, b),$$

where $P(r)$ is the Poisson distribution, and $\mathcal{N}(m, v)$ is the Gaussian distribution with mean m and variance v. The standard deviation has the expression (Foi et al., 2008)

$$\sigma(I(x, y)) = \sqrt{a \cdot I(x, y) + b}.$$

Denoising is a common process in signal processing. However, removing a signal-dependent noise is a more challenging task. In (Foi et al., 2008), the authors employ an algorithm in several steps to recover the image $I(x, y)$ and also estimate the varying standard deviation $\sigma(I(x, y))$. First, the image needs to be divided in smooth regions. To this end, they employ edge detection via segmentation. Second, they compute a *local estimation* of the standard deviation in the smooth regions. This estimation is based on the assumption that the changing standard deviation is relatively constant in a local small region. Last, a *global model* of the noise is fit using local measurements.

2.2.2 Digital Image Formation

The world as we see it using our eyes is a continuous three-dimensional function of the spatial coordinates. A photograph is a two-dimensional map of the *number of photons* that map from the three-dimensional scene. In film-based photography, this map is a continuous function. However, in referring to a digital image, the corresponding two-dimensional function is a discrete representation because the number of pixels used for imaging are discrete and finitely many. Hence, we can think of an image as a mathematical representation of a physical entity that describes a function over spatial coordinates. The individual pixels are the basic elements of the discrete representation of the continuous scene, and in analogy to the one-dimensional case, these are the samples of a function with reference to the Shannon-Nyquist sampling formula. Consequently, we must bear in mind that the image is merely a representation of the scene and not the continuous scene itself. To understand the basis of the image formation process, we must understand the physical laws that govern such a process.

From a mathematical standpoint, the image can be seen as a mapping from the spatial domain to the range of the imaging sensor. Let $\mathbf{r}$ refer to the spatial coordinates in the Cartesian plane; then the image $\mathbf{I} : S \rightarrow P$ is a mapping from the scene to the pixel domain such that each $\mathbf{r} \in S$ is mapped to $\mathbf{I}(\mathbf{r}) \in P$.

In the case of color imaging, for every point $\mathbf{r}$ in the two-dimensional space, we obtain three values per pixel, namely, the red, green, and blue values (in intensity). Hence, the resulting image can be represented as the following function that maps a vector to a vector,

$$\mathbf{r} \in \mathbb{R}^2 \rightarrow \mathbf{I}(\mathbf{r}) = \begin{bmatrix} \text{Red}(\mathbf{r}) \\ \text{Green}(\mathbf{r}) \\ \text{Blue}(\mathbf{r}) \end{bmatrix} \in \mathbb{R}^3.$$

In contrast, when working with monochromatic images, we have a simpler mapping of the form $\mathbf{r} \in \mathbb{R}^2 \rightarrow \mathbf{I}(\mathbf{r}) \in \mathbb{R}$.

For the data defined by $\mathbf{I}(\mathbf{r})$ to be stored on a computer, it needs to be processed in two stages: sampling and quantization. The luminance values, given by the values of $\mathbf{I}(\mathbf{r})$, are always positive and belong to a restricted interval.

In order to be stored or processed by digital devices, the luminance value is mapped to a set of finite values, typically $\{0, 1, \cdots, 255\}$, which is also known as quantization. In the case of color images, each color is mapped to one of the 256 possible values. In the case of monochromatic images, some sensors employ a higher resolution, with values encoded in 12 bits, that is, in the range $\{0, 1, \cdots, 4095\}$. The choice of resolution is dependent on two factors: the images captured and the processing to be performed. For example, computed tomography (CT) images use more than 10 bits, while a low-grade webcam around 6 bits per color. More complex processing, such as gradient calculations, also requires a higher resolution for good results.

Quantization guarantees that an image luminance can take only one of a finite number of possible values. However, the image has an infinite number of points. Therefore we need to sample the values of the images along each axis and define a new sampled image $\mathbf{I}_s(i, j) = \mathbf{I}(i\Delta x, j\Delta y)$, where $r_{i,j} = (i\Delta x, j\Delta y)$ denotes the sampled spatial coordinates, and Δx and Δy denote the sampling distances along x and y, respectively. Here, $\mathbf{I}_s(i, j)$ represents a pixel. It is important to point out that the pixel is a point sample taken from the image, and is not a small square of measurable dimensions as commonly misconstrued. When the image is 3D instead of 2D, the pixel is called a voxel, which is a point sample in a 3D space.

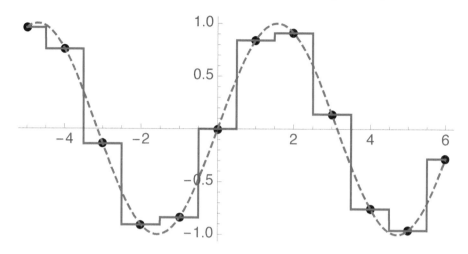

Figure 2.19 Nearest neighbor interpolation: the original continuous function $g(x) = \sin(x)$ is shown by the dashed line, and the interpolation $\widehat{g}(x)$ is shown by the solid line.

2.2.3 Image Interpolation

After the continuous image has been sampled and quantized, it is important to be able to compute the values of the original continuous image at any desired coordinates. The process of computing the value of $\mathbf{I}(x, y)$ at locations different from the sampling points is called interpolation. One may ask whether the interpolation can work for any selection of the sampling distances Δx and Δy? The answer is that the maximum sampling distances are a function of the image bandwidth, and the values represent an extension of the Shannon sampling theory, originally introduced for time samples. For now, assume that the sampling distances Δx and Δy are small enough that the samples are a good representation of the image. The luminance function $\mathbf{I}(x, y)$ is sampled along a two-dimensional space. Let us first look at some examples for interpolating one-dimensional functions $g(x)$ using the sampled function $g_S(k) = g(k)$. Notice that for simplicity we use a sampling interval of 1.

Nearest neighbor interpolation. As the name suggests, this method selects the sample value from the nearest sampling location without calculating a new value. This is a very computationally inexpensive interpolation method, but it does not always generate useful

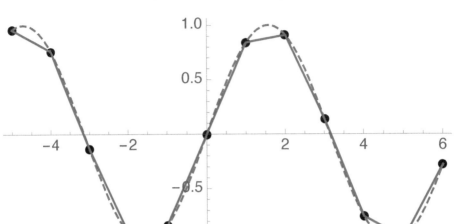

Figure 2.20 Linear interpolation: the original continuous function $g(x) = \sin(x)$ is shown by the dashed line, and the interpolation $\widehat{g}(x)$ is shown by the solid line.

results. The values of the new samples are computed $\widehat{g}(x) = g_S([x + 1/2])$, where $[\cdot]$ denotes the round function, and $[x + 1/2]$ denotes the integer closest to x. Figure 2.19 shows function $\widehat{g}(x)$ for $g(x) = \sin(x)$. Note that the function closely resembles $g(x)$ at the sample points and is very different from $g(x)$ between the sampling points.

Linear interpolation. This is a slightly more complex interpolation with improved results. Whereas nearest neighbor interpolation required one sample, here we use two samples to calculate the interpolation at a new location. This interpolated function is

$$\widehat{g}(x) = (1 - (x - k)) \cdot g_S(k) + (x - k) \cdot g_S(k + 1), \ x \in [k, k+1].$$

Here the interpolated function $\widehat{g}(x)$ is continuous in the mathematical sense, that is, it contains no jumps. However, it is not differentiable in the sampling points.

Higher order interpolation. The nearest neighbor and linear interpolations can be increased in complexity. Each of the two methods is essentially fitting a polynomial of degree 0 (nearest neighbor) and 1 (linear) to a number of samples, and the new sample value is computed as the fitted polynomial evaluated in the new sample location of interest. For

2.2 Image Sensors

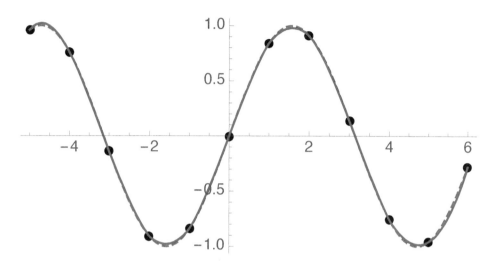

Figure 2.21 Cubic interpolation: the original continuous function $g(x) = \sin(x)$ is shown by the dashed line, and the interpolation $\widehat{g}(x)$ is shown by the solid line.

example, cubic interpolation refers to fitting polynomials of degree 3 of the form

$$\widehat{g}(x) = c_1 x^3 + c_2 x^2 + c_3 x + c_4, \ x \in [k, k+1].$$

The values of the coefficients c_1, c_2, c_3, c_4 can be computed analytically from the known samples

$$c_1 = \frac{1}{6}(-g_S(k-1) + 3g_S(k) - 3g_S(k+1) + g_S(k+2)),$$
$$c_2 = \frac{1}{2}(-g_S(k-1) - 2g_S(k) + g_S(k+1)),$$
$$c_3 = \frac{1}{6}(-2g_S(k-1) - 3g_S(k) + 6g_S(k+1) - g_S(k+2)),$$
$$c_4 = g_S(k).$$

Clearly, this is a more complex interpolation, but it leads to very good results, as shown by Figure 2.21.

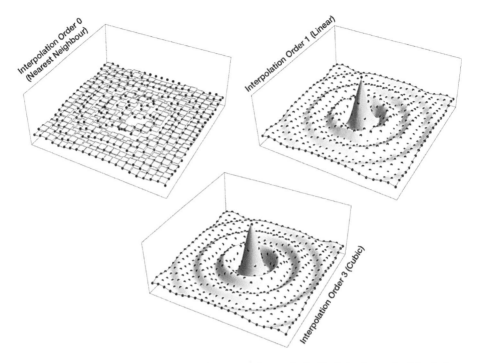

Figure 2.22 Linear 2D interpolation: the original sample $\mathbf{I}_S(i, j)$ is shown by black dots, and the interpolation $\widehat{\mathbf{I}}(x, y)$ is shown by gray mesh. The samples were taken from the function $\mathbf{I}(x, y) = \sin\left(\sqrt{x^2 + y^2}\right) / \sqrt{x^2 + y^2}$.

Image 2D interpolation. Once the 1D interpolation is understood, generalizing to 2D is straightforward. Image 2D interpolation is made up of two components: interpolating in the x direction and in the y direction. The linear interpolation of an image at points x, y, denoted as $\widehat{\mathbf{I}}(x, y)$, is computed

$$\begin{aligned}\widehat{\mathbf{I}}(x, y) &= (1 - (x - i))(1 - (y - j))\mathbf{I}_S(i, j) \\ &+ (1 - (x - i))(y - j)\mathbf{I}_S(i, j + 1) \\ &+ (x - i)(1 - (y - j))\mathbf{I}_S(i + 1, j) \\ &+ (x - i)(y - j)\mathbf{I}_S(i + 1, j + 1)\end{aligned}$$

where $x \in [i, i + 1]$ and $y \in [j, j + 1]$ are the coordinates of the interpolation point. An example of 2D linear interpolation is shown in Figure 2.22.

Let us see what happens to a real image during the process of sampling. To enhance the effect on high frequencies, we start with a checkerboard pattern with a tilted view, as shown in Figure 2.23.

2.2 Image Sensors

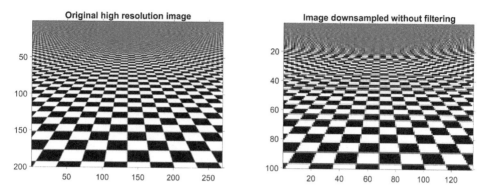

Figure 2.23 Image downsampling without filtering: the original image (*left*) is downsampled by a factor of two (*right*). The aliasing effect is particularly visible for the high spatial frequencies at the top of the image.

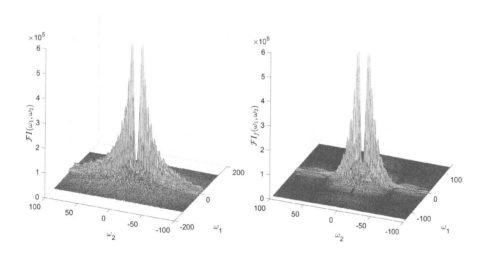

Figure 2.24 Image filtering in the frequency domain: the Fourier transform of the original image (*left*), and of the image filtered with a boxcar function (*right*).

Most of the image's high spatial frequencies are located at the top, where the squares are smaller and closer together. The figure shows that simply downsampling the image, by discarding one out of two pixels on both spatial dimensions, leads to distortions in the top part with high frequencies. These distortions, called aliasing, are explained theoretically by

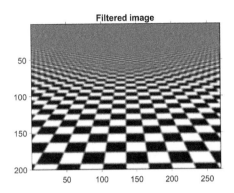 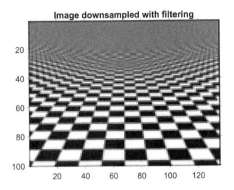

Figure 2.25 Image downsampling with filtering: the original image is filtered with a boxcar function (*left*) and subsequently downsampled by a factor of two (*right*).

Shannon's Nyquist rate formula, which states that the sampling frequency f_s should satisfy

$$f_s \geqslant 2 f_{MAX}$$

where f_{MAX} is the maximum frequency in the signal, and $2 f_{MAX}$ is known as the Nyquist rate (see section 3.2.1). Therefore, to obtain good results after sampling an image, we can filter the image. Figure 2.24 shows the Fourier transform of the original image $\mathcal{F}\left[\mathbf{I}(\omega_1, \omega_2)\right]$ and of the image filtered with a boxcar function $\mathcal{F}\left[\mathbf{I}_f(\omega_1, \omega_2)\right]$. The original image has frequency components near the edge of the frequency domain. A downsampled image has a reduced frequency domain, and therefore we need to remove the high frequencies up to half the maximum frequency in the signal via filtering to satisfy Shannon's Nyquist rate condition. The right-hand side of Figure 2.24 shows that the filtered spectrum is still surrounded by four small lobes. The reason is that the boxcar function is not an ideal low-pass filter, and its Fourier transform is a cardinal sine function.

If we perform downsampling on this new filtered image, we notice that the aliasing effect is barely visible, as depicted in Figure 2.25.

2.2.4 Digital Imaging Pipeline

The transformation stages from the light rays reflected by the scene to the final image files on our computers is called the digital imaging pipeline. It consists of a few major stages, depicted in Figure 2.26.

First, the light reflected by the scene is manipulated using the optical parameters such as aperture and exposure time, which bend the light and direct it toward the sensor. Sensors have evolved considerably over centuries, and today there are two main categories: charged coupled device (CCD) and complementary metal oxide semiconductor (CMOS). The CCD is based on a MOS capacitor and is used mainly in high-end cameras due to its high price

2.2 Image Sensors

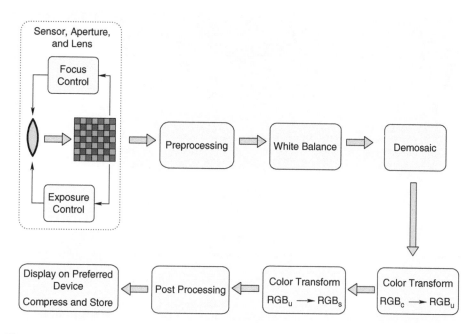

Figure 2.26 The main steps in the digital imaging pipeline. Reprinted from (Ramanath et al., 2005).

and power consumption. The CMOS is based on MOSFET transistors and is a lot more consumer friendly, with lower power consumption and a more affordable price. It is more prone to noise, but this can be reduced with digital denoising. Therefore we focus on CMOS sensors in this presentation.

The CMOS sensor is equipped with a microlens for each pixel, which has the effect of increasing the amount of light captured by that pixel. The light then passes through a color filter, which extracts the wavelengths relevant for each of the red, green, and blue colors. Then the filtered light hits the photodiode, which in response generates electrons that are then stored in the potential well. A diagram of a sensor is shown in Figure 2.27, which for simplicity shows only three pixels. The sensitivity of each color filter to each wavelength is depicted in Figure 2.28.

In a full commercial sensor, the color filters are not uniformly distributed. They follow a specific pattern, called the color filter array, which determines the final look of the image. One of the common color filter arrays is the Bayer filter, which contains 50% green filters, 25% blue, and 25% red. This proportion is inspired by the human retina, which during the daytime uses cone cells that are most sensitive to green light. The image generated by the sensor is called a *mosaiced* image due to its mix of pixels of different colors.

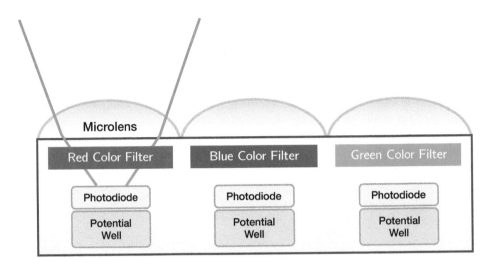

Figure 2.27 The basic components of a CMOS camera sensor.

The potential well of each pixel generates an analog voltage signal that enters the preprocessing pipeline, which consists of several stages. First, the voltage is processed with the analog front end, depicted in Figure 2.29. This converts the analog mosaiced image from the sensor outputs into the raw digital mosaiced image. First, the analog voltage is passed through an amplifier, whose gain is modulated by the ISO settings of the camera. The gain is larger for pixels farther from the image center, due to the vignetting effect, which darkens the extremities of an image. Second, the analog-to-digital (ADC) converter generates a digital signal, usually with sizes of 10–16 bits. Third, the sensor suffers from nonlinearities in the extremities of its range (very bright or very dark pixels), which is corrected using a lookup table. A lookup table, which simply maps an output value to any possible input value, is a very fast technique to process digital signals. (Gruev and Etienne-Cummings, 2002) describe implementing a pseudo-general image processor chip that enables steerable spatial and temporal filters at the focal plane.

The output image of the analog front end is called the raw image. Many consumer-grade cameras allow access to this format, because many applications, such as physics based computer vision, work much better on raw images than processed ones. However, they do not look very appealing due to a high level of noise and unsuitable color balancing.

The next processing stage adjusts the white balance of the raw digital images. This stage is necessary because what someone sees as white is significant in the viewer's perception of the scene. Therefore, the white balance is adjusted by imposing an assumption on the image coloring. One way to do that is to assume that the average color of an image is gray; this is called gray world assumption. A different method, called white world assumption,

2.2 Image Sensors

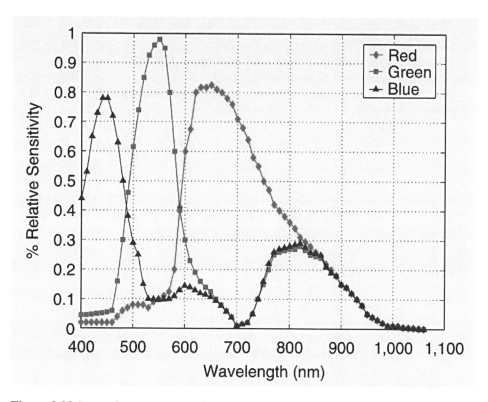

Figure 2.28 Spectral sensitivities in digital color cameras. Reprinted from (Ramanath et al., 2005).

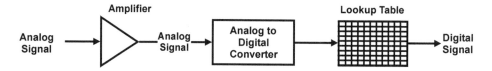

Figure 2.29 The analog front end in the digital processing pipeline.

assumes the brightest object in the scene to be white. However, modern cameras use histogram-based algorithms, assuming specific proportions of various colors.

Next, recall that at this stage the image is still a mosaic of colors, so that each pixel stores color-specific information. We need to turn this mosaicked image into three images, corresponding to the colors red, green, and blue. However, the red and green information is lost at the location of a blue pixel. So how can this information be recovered? This can be accomplished by interpolation, and even simple algorithms averaging the nearest neighbors can achieve good results. The three images at this stage are still strongly affected by noise.

Therefore, a denoising stage is often applied, such as averaging or computing the median of the neighboring pixels.

The human-perceived colors are mapped to light wavelengths using color spaces. Color spaces allow a reproducible representation of color. These mappings are denoted in the diagram as color transformations. Examples of color spaces are CIEXYZ and ISO-RGB (Ramanath et al., 2005).

After all these steps, the image still does not look natural. That is because light detection in the human retina is nonlinear as a function of the luminance, and is more sensitive to dark tones, whereas for a camera, this relationship is linear. To address this, a subsequent postprocessing step is employed. Because the nonlinear function in the case of the human eye resembles the mathematical function gamma, the process that compensates for this effect is called *gamma correction*. After this step, the image appearance to the human eye is significantly improved, but the image requires a lot of space. This motivates the final step, called compression, which decreases the image size by a third. The final result is an appealing image in a compressed format, such as jpeg or png.

2.3 Illumination

The use of lighting has not evolved a great deal since the beginnings of photography. It can be argued that lighting is the main thing that distinguishes an amateur photographer from a professional. A professional photographer measures the light intensity and then manually selects the optimal camera parameters such as ISO sensitivity, exposure time, and aperture. In automated cameras these parameters are selected automatically, but these choices do not always lead to the most pleasing picture.

Similarly to ISO or exposure time, several other parameters can be adjusted for the camera lighting:

- duration and intensity;
- presence or absence of auxiliary lighting;
- color, wavelength, and polarization;
- position and orientation; and
- modulation in space and time.

Each of the parameters above will be discussed separately below.

2.3.1 Duration and Intensity

Capturing fast-moving objects is possible using high shutter speeds. However, this approach is quite limited because it involves moving mechanical parts. Therefore, we cannot capture certain physical phenomena in this way.

2.3 Illumination

An alternative is to use fast bursting flashes of light with electronic devices called strobes. The image of a *bullet fired through an apple*, captured by MIT professor Harold Edgerton is an iconic example of a technique called *strobe photography*, developed in the 1930s, which uses light and sound to trigger the flash burst with precise timing. A natural continuation of strobe photography was high frame rate film. By combining the short light bursts of strobes with the high sensitivity of CCD and CMOS sensors developed in the 1980s, manufacturers developed cameras with ultrashort exposures. This technology has evolved to the point where today's affordable cameras reach up to 300 frames per second at 0.2 MB resolution, and high-end cameras reach 2,570 frames per second at full HD (2 MB) resolution.

Another way to exploit the capabilities of strobe photography is to generate several bursts in one camera exposure; this is called *sequential multiflash stroboscopy*. Typically this is done with a dark background, and the bursting frequency and duration is set so that the frames of the object in motion do not overlap.

2.3.2 Auxiliary Lighting

The flash illumination is built into most of today's cameras. By adjusting the illumination during image capture, it is possible to extract various features. (Dicarlo et al., 2001) have described recovering the object reflectivity using two snapshots with ambient lighting and flash, respectively. Taking photos with flash with various intensities makes it possible to simulate images captured with a continuous level of flash (Hoppe and Toyama, 2003).

The use of flash is clearly necessary when the ambient lighting is low. However, if there is adequate ambient light, is it recommended to use flash? Flash photographs are known to lead to images with clear high-frequency details and also with more noise robustness. However, ambient light is a part of the scene that we may want to capture. Moreover, flash photographs look rather unnatural due to the artificial lighting. It is possible to combine a flash photograph with one captured with ambient light and thus enjoy the advantages of both methodologies (Petschnigg et al., 2004; Raskar et al., 2004). The two methods generate a new image incorporating the details from the flash photo and the shadows from the ambient light photo. The separation is performed using an image processing technique called *joint bilateral filter*.

Similarly, a bilateral filter can be used to simply denoise the image captured without flash (Tomasi and Manduchi, 1998). Normally, when an image is filtered, the details are removed along with the noise. In using a bilateral filter, an intensity similarity measure cancels the filter effect in areas where there are image details, quantified as high frequencies in the flash image. This technique is prone to errors and artifacts when the flash image contains shadows that are interpreted as details by the bilateral filter, or when it is overexposed, so that the details are dimmed or removed altogether. This would cause the bilateral filter to remove details from the ambient image, or to leave unfiltered areas with no detail.

As mentioned previously, the bilateral filter method fails when the flash saturates portions of the image. Similarly, because sensors have predefined dynamic ranges, the flash might lead to colors too bright to be captured. In another scenario, objects could be located at different distances from the flash, and therefore a low intensity would not illuminate the distant objects, and a high intensity would saturate or *blow out* the nearby points in the scene. As in the case of the ambient-light image, the solution is to combine the beneficial characteristics of several images into one high-quality image. In a case of this kind, (Raskar et al., 2008) combined images captured with various flash intensities to generate a single high dynamic range (HDR) image.

Another way to address the artifacts in flash images is to compute the *gradient vector* for the flash and ambient light image (Raskar et al., 2008). The gradient vector in a pixel is the direction in which the intensity change is most abrupt. Therefore it is intuitive that the gradient at an edge is perpendicular on the edge for all pixels close to it. On the basis of this observation, an artifact is located at pixels where there is significant difference in gradient vector orientation between the ambient light and the flash image. This technique is called *gradient coherence* (Raskar et al., 2008).

An interesting research question is whether an image can be reconstructed from its generated gradient vectors. The gradient is typically implemented as a difference

$$\Delta \mathbf{I}(x, y) = [\mathbf{I}(x+1, y) - \mathbf{I}(x, y), \mathbf{I}(x, y+1) - \mathbf{I}(x, y)].$$

Therefore the problem of reconstructing $\mathbf{I}$ seems trivial, that is, recovering through the cumulative summation of $\Delta \mathbf{I}(x, y)$. However, complications arise when the gradient is not consistent and therefore the result is dependent on the path along which summation is done. There have been several methods addressing this issue (Agrawal et al., 2006). (Agrawal et al., 2005), used gradient vector projection to combine ambient and flash images into a high quality image with ambient-light features (see Figure 2.30). Interestingly, the residuals from the flash image gradients can be integrated to recover an image of the photographer that is not visible in the original flash image.

2.3.3 Modifying Color, Wavelength, and Polarization

So far we have looked at white illumination with varying intensity to achieve desired image characteristics. Choosing a flash containing specific colors allows performing programmable color manipulations on images. For example, two colors can look the same (see section 8.2.3) or different, depending on the type of lighting during capture, which could be countered by modulating the illumination wavelength (see section 8.3.4).

This approach is also used in fluorescence photography, which exploits the fact that fluorescent surfaces emit low frequency light in response to high frequency illumination. In this case, the light source emits ultraviolet light and the camera filters out nonvisible light, thus capturing only the reflection of fluorescent surfaces.

2.3 Illumination

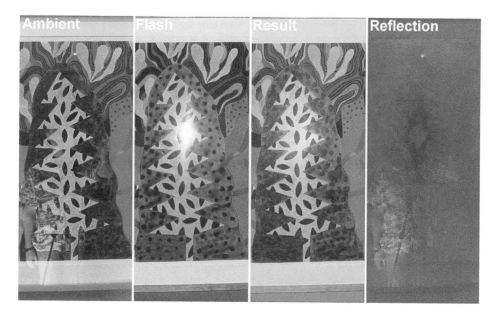

Figure 2.30 Removing artifacts from a flash image: the image gradients are used to locate the image artifact and remove it. Subsequently, the isolated artifact can be integrated to generate an image of the photographer. Reprinted from (Agrawal et al., 2005).

Figure 2.31 Generating synthetic colored lighting using conventional illumination: an image is captured using ambient light and subsequently with lighting from the left direction (*left*) and right (*center*). By subtracting the images with artificial lighting from the ambient-light image it is possible to generate synthetic colored lighting (*right*). Reprinted from (Haeberli, 1992).

Colored lighting can be simulated using photographs captured with conventional lighting. (Haeberli, 1992) used two lamps with white light positioned on either side of the subject, and captured three photographs: one with ambient lighting and two with lighting from either direction. By subtracting the ambient light image from the other two, it was possible to quantify the contribution of each lamp. Then, through software manipulation, the author simulated an image in which each light source has a different wavelength, as shown in Figure 2.31.

2.3.4 Modifying Position and Orientation

If we can alter the illumination of a scene, we can reveal different surface details otherwise hidden from view (see section 4.3). One example is locating shape discontinuities, which are depth differences between various patches of the scene. This connects closely with edge detection, because edges in an image are largely the cause of shape discontinuities.

In (Raskar et al., 2004) it was shown how to use multiple flashes to find silhouettes using depth discontinuities. Depth discontinuities, or edges, are identified via the shadow narrow strip, which is a sliver cast in the opposite direction of the lighting. This technique can also be used to generate shadow-free images.

The weak point of this method is that it does not accommodate small objects, or distant backgrounds. These lead to shadows that are detached from the subject. The method can be extended, however, to video footage by using a high-speed flash sequence (Raskar et al., 2004; Taguchi, 2014). This principle was also used to decode sign language input (Feris et al., 2004).

Generating synthetic lighting in images postcapture has also been proven useful for generating a painting interface for novices in photographic lighting design (Anrys and Dutré, 2004; Mohan et al., 2005). This allowed them to see the results after locally changing the lighting in images, which is much more convenient than retaking the photograph with different lighting each time.

If the image has only one lighting source, then the pixel brightness is linear with the intensity of that lighting. Assuming that the camera has a linear response, then the effect of more powerful lighting can be achieved simply by increasing the resulting pixel brightness (Nimeroff et al., 1995; Haeberli, 1992). If there are several light sources present, the final intensity is computed as a weighted sum of the corresponding intensities of each light source.

For maximum flexibility, one should ideally have access to photographs taken from any possible position. However, this is not possible when the lighting equipment is constrained inside a predefined area, such as inside a square. In a general framework, a scene is described by two four-dimensional (4D) functions known as light fields:

- the incident light field $L_i(u, v, \alpha, \beta)$, describing the irradiance of light incident on objects in space; and
- the radiant light field $L_r(u, v, \alpha, \beta)$ quantifying the irradiance created by an object.

This model was extended to define the eight-dimensional (8D) reflectance field, which measures irradiance at the sensor determined by incident light rays displayed by an arbitrary projector in space (Debevec et al., 2000). If we fix the viewpoint, the reflectance field can

2.3 Illumination

be reduced to be six-dimensional (6D). Even so, capturing and storing data of this high dimensionality creates problems in practical scenarios. The projector was also mounted on a robotic arm to acquire the reflectance field of a human face (Debevec et al., 2000). This can be viewed as a pixel translated over the surface of a sphere, leading to a dimensionality reduction of the incident light field, and therefore a reduced final 4D reflectance field.

By controlling the color and intensity of light from various positions around the subject, it is possible to seamlessly integrate the image of the subject into a new scene (Debevec, 2002). The reduction in size of the 4D reflectance field is very desirable in practice. To this end, (Malzbender et al., 2001) observed that in changing the lighting incident angle, the color of a pixel changes with a function that can be closely approximated with a biquadratic polynomial. This allowed them to store only the coefficients of the polynomial and subsequently use *compressive sensing* to greatly reduce the size of the reflectance field. However, as one might expect, specular reflections, which occur for only certain angles of the incident illumination, cause disturbances in the biquadratic polynomial approximation, and this remains an open problem in the field.

A 6D reflectance field is a better description of the scene. Even though the number of illumination setups is theoretically equal to the number of pixels in each projector multiplied by the number of projectors n, it was shown that it can be simplified significantly by illuminating the scene with a single projector moving in n positions (Masselus et al., 2003).

2.3.5 Modifying Space and Time

In order to control the radiance of each ray emitted, one can use projector-like light sources, which allow controlling each individual pixel and not just the overall brightness. It has been shown that using such a light source to assist capturing images with a camera allows extracting scene information that would be impossible to access using regular flash (Nayar et al., 2006). The projectorlike device, called a CamPro, is rather bulky to fully replace the traditional flash but could be promising if implemented with smart lasers.

Clearly an important task is recovering the 3D shape of a scene from 2D images. It turns out that the problem of recovering the 3D location of an object is closely connected to the *correspondence problem* of pixels in images with different views. This latter problem requires finding a correspondence between sets of points in each image, captured from a different angle, that matches the points in the 3D scene. The correspondence problem can be solved by using a projector with temporal multiplexing, which enables projecting a certain pattern at a time that can be identified by cameras recording different perspectives.

Once the correspondence problem is solved, the 3D location is recovered via triangulation between camera and projector. This can work with only a camera and a projector, as depicted in Figure 2.32.

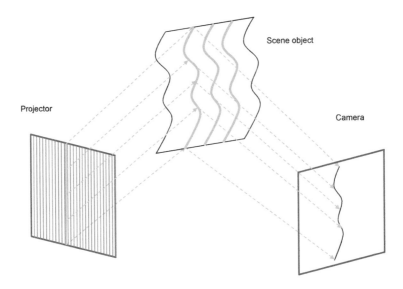

Figure 2.32 3D object localization and the correspondence problem for a single camera and a projector. Several patterns are projected onto the scene object, which are detected by a single camera. The object 3D localization is computed via triangulation.

This problem is similar to the problem of *stereo triangulation*, in which a set of 3D points in the scene are identified given the disparity map between the images captured from two or more viewing angles. In our case, instead of two passive cameras we have an active camera and a projector encoding the space via illumination, in a process called *active stereo triangulation*. For readers piqued by stereo imaging, we discuss some examples in which epipolar geometry is used in time-of-flight imaging for sequential acquisition of strips of the image scene in section 10.5.1.

The number of patterns generated by the projector can be reduced by coding the boundaries between the projected shapes (Rusinkiewicz et al., 2002). The projected light can also have a binary pattern, in which the pixels can either be turned off or have a fixed level of brightness (Posdamer and Altschuler, 1982).

The projector can be modulated in *space* such that at a given time it is illuminating differently the points in the scene, or in *time* such that the pattern changes in successive frames. The two modulations can also be used in conjunction.

As we briefly mentioned in section 2.1.7, a programmable flash can be used to separate the light scattered by the scene in two components:

1. The *direct illumination*, caused by the light source, which enhances the material properties at a given point; and

2.3 Illumination

2. The *global illumination*, determined by other points in the scene, which reveals the optical properties of the objects such as how a certain patch in the scene is illuminated by the scene itself.

One way to separate the two components was proposed in (Nayar et al., 2006), in which the projector was spatially encoded with a checkerboard binary pattern. The scene was divided into square patches that were lit and unlit intermittently by the projector. The technique is based on the observation that if one uses a high-frequency checkerboard pattern, the patches left unlit contain only global illumination components (light reflected from the lit patches). On the other hand, the lit patches contain both global and direct illumination components.

This means that, theoretically, it is enough to capture two frames: one illuminated with the checkerboard pattern and one with its complement illumination pattern. This ensures that it covers the whole scene and is enough to recover each illumination component. However, due to the common leakage effect in off-the-shelf projectors, it is necessary to capture five times more images to compensate for this imperfection (Nayar et al., 2006). The global and direct illumination merely separates one to several bounces of the lighting emitted by a projector. It is possible to go one step further, and model the individual bounces of an optical ray (Seitz et al., 2005).

Apart from modulation in space, the projector can be modulated in *time* by using high-frequency strobes to acquire snapshots of the scene periodically in a predefined pattern. An interesting effect is that the illumination frequency is different from the frequency of a periodic movement in the scene. In this case the captured images are characterized by a *perceived frequency*, which is the difference between the two frequencies.

Consequently, if the two frequencies are the same, the captured footage will show the scene object stagnating. This is very useful in applications such as distortion detection in vocal cords. By illuminating the cords with predefined frequencies, a physician can tell whether there is a physiological distortion in the cord movement.

Chapter Appendix: Notation

Notation	Description
c	Speed of light through vacuum
ν	Frequency
λ	Wavelength
E	Energy
h	Planck's constant
Φ	Radiant flux
R	Irradiance

M	Exitance
Ω	Solid angle
I	Radiant intensity
L	Radiance
d	Distance between the pinhole and the projection plane
α	Skew factor
δ	Aperture size
n_1, n_2	Indices of refraction
θ_1, θ_2	Angle of incidence and angle of refraction
f	Focal length
D	Aperture diameter
N	Lens speed, f-number
$\overline{R}$	Average irradiance
$1/f$	Focusing power of the lens
f_c	Focal length of the compound lens
r	Ratio of reflected light intensity
$I(x, y)$	Light signal
$\zeta(x, y)$	Gaussian noise of zero mean and standard deviation 1
$\sigma(I(x, y))$	Signal dependent standard deviation
$z(x, y)$	Sensor measurements at pixel (x, y)
$P(r)$	Poisson distribution
$\mathbf{I}$	Image
$\mathbf{I}_s$	Sampled image
f_{MAX}	Maximum frequency in the signal
$\mathcal{F}$	Fourier transform operator
$\mathbf{I}_f$	Filtered image

Exercises

1. Light ray bending

    a) Snell's law

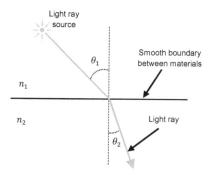

Figure E2.1 Modeling the principle of refraction via Snell's law.

Assume that a light ray passes the smooth boundary between air and water with an incident angle $\theta_1 = \pi/6$. Calculate θ_2 to two decimal places knowing that the refraction index for air is $n_1 \simeq 1$ and for water $n_2 \simeq 1.33$.

   b) Thin lens

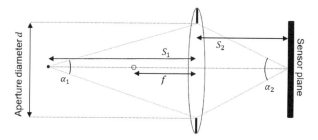

Figure E2.2 The proposed thin lens setup.

Consider the thin lens setup in Figure E2.2. An object is located on the lens optical axis such that its reflected light rays that pass through the lens aperture cover an angle α_1. The object is in focus and is projected onto the sensor plane at a point where the field of view is given by angle α_2. Assuming that the aperture diameter d is known, derive analytically the expressions of lens focal length f and the distances from the lens to the object S_1 and to the sensor plane S_2.

2. Capturing images via sampling and quantization

   a) Image sampling
   When an image is captured by a digital camera, it is sampled spatially by the sensor array, and then each pixel value is coded with a number of bits in a process called quantization.

   Let us simulate this process by starting with a high-resolution grayscale image, with size around 2000×2000 as shown in Figure E2.3a.

   For demonstration purposes, let us assume we have a sensor array with size 80×80 that captures the image.

   To simulate the spatial sampling process, we need to divide the original image pixels into 80×80 blocks and then average the pixels in each block, just as a sensor pixel would average all incident light intensities. This should lead to an image similar to the one shown in Figure E2.3b. Plot a result using your own image.

   What is a quick procedure to sample the image in the described way? Hint: it may involve convolving with a kernel.

   b) Image quantization
   Next, the captured pixel values need to be quantized. Of course, in your image they already are quantized (most probably in the range 0–255), but here we implement a course quantization that would enable a good visualization of the process. For example, a 4-bit quantization would involve mapping each pixel in an image to a value in $\{0, 1, \ldots, 15\}$. Is there a quick way to implement quantization without loops?

   Generate your own images after 4-bit and 3-bit quantization. The results should look similar to those shown in Figure E2.3c,d.

   c) Image interpolation
   Subsequent processing tasks could require a higher resolution image. How could the new pixel values be computed? The most straightforward way is to apply interpolation, as shown in Figure E2.4a. The distortion of the image due to quantization is still present 0–255, but it can be addressed via filtering, as shown in Figure E2.4b.

Exercises

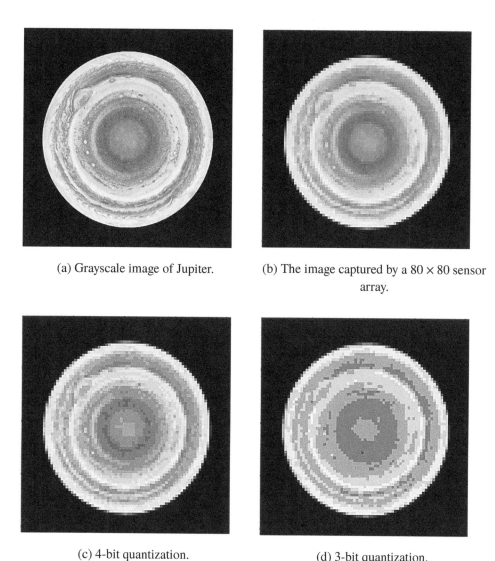

(a) Grayscale image of Jupiter.

(b) The image captured by a 80 × 80 sensor array.

(c) 4-bit quantization.

(d) 3-bit quantization.

Figure E2.3 Simulating the sampling and quantization done by a sensor array.

Now apply these steps to your quantized image, increasing the resolution to match the original image. Then filter the high-resolution quantized image with a 2 × 2 matrix of ones. The results should be similar to those shown in Figure E2.4.

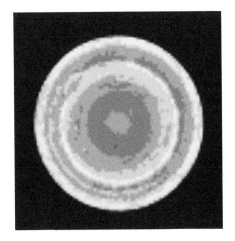 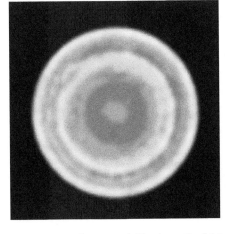

(a) Interpolation of a 3-bit quantized image.   (b) Interpolation and filtering of a 3-bit quantized image.

Figure E2.4 Increasing the resolution of a sampled and quantized image via interpolation.

What differences do you notice between the aspect of the low-resolution and the high-resolution quantized image? What drawback do you see in filtering? What do you imagine would happen if the filtering were repeated a large number of times?

Let $I(x, y)$ denote a continuous image, and let $I_S(k, l)$ denote the samples taken with period 1. Demonstrate that the linear interpolation of $I(x, y)$ at samples $I_S(k, l)$ is

$$\widehat{I}(x, y) = (1 - (x - k))(1 - (y - l)) I_S(k, l) + (1 - (x - k))(y - l) I_S(k, l + 1)$$
$$+ (x - k)(1 - (y - l)) I_S(k + 1, l) + (x - k)(y - l) I_S(k + 1, l + 1).$$

d) Brightness and contrast adjustment

In this example we examine the concepts of brightness and contrast by manually adjusting them for a chosen image. The image pixels with adjusted contrast and brightness are defined

$$\widetilde{I}(k, l) = B \cdot I(k, l)^C$$

where $B, C \in \mathbb{R}$ denote the brightness and contrast adjustments, respectively. Now choose a color image and change its brightness and contrast to enhance its features, as shown in Figure E2.5.

What is the drawback that prevents capturing an image with arbitrary high sharpness with a pinhole camera? How is the general quality of the brightness of pinhole camera

Exercises

(a) Original image.

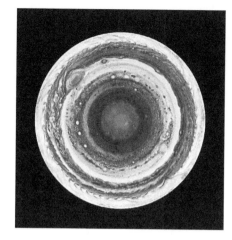
(b) Image with adjusted brightness and contrast.

Figure E2.5 The adjustment of brightness and contrast in digital images.

photos, and what trade-off is involved in attempting to adjust it? Can we say that the pinspeck camera solved the brightness problem of the captured images?

3. Image deconvolution

   a) Problem setup
      Under LSI conditions, deblurring can be posed as follows:

      $$J(k,l) = h(k,l) * I(k,l) + e(k,l) \qquad (2.1)$$

      where $J(k,l)$ is the observed image; $I(k,l)$ is the original image; $h(k,l)$ is the kernel; and $e(k,l)$ is the noise.

      To recover the original image, we usually apply deconvolution algorithms to the observations. In this exercise, we implement several deconvolution algorithms and compare their performances. To evaluate the performance quantitatively, we adopt two image similarity metrics: (1) peak signal-to-noise ratio (PSNR), and (2) structural similarity (SSIM) index.

      The PSNR between two images, $I_1(k,l)$ and $I_2(k,l)$, can be calculated as follows:

      $$\text{PSNR}(I_1, I_2) = 10 \log_{10} \left(\frac{R^2}{\text{MSE}(I_1, I_2)} \right)$$
      $$\text{MSE}(I_1, I_2) = \frac{1}{KL} \sum_{k=0}^{K-1} \sum_{l=0}^{L-1} [I_1(k,l) - I_2(k,l)]^2 \qquad (2.2)$$

      where R is the data range of the images, and (K, L) is the image size.

      The SSIM between $I_1(k,l)$ and $I_2(k,l)$ is calculated on the basis of three measurements, comprising luminance ($\mathscr{L}$), contrast ($\mathscr{C}$), and structure ($\mathscr{S}$):

      $$\begin{aligned}
      \text{SSIM}(I_1, I_2) &= [l(I_1, I_2)]^\alpha \cdot [c(I_1, I_2)]^\beta \cdot [s(I_1, I_2)]^\gamma \\
      \mathscr{L}(I_1, I_2) &= \frac{2\mu_{I_1}\mu_{I_2} + C_1}{\mu_{I_1}^2 + \mu_{I_2}^2 + C_1} \\
      \mathscr{C}(I_1, I_2) &= \frac{2\sigma_{I_1}\sigma_{I_2} + C_2}{\sigma_{I_1}^2 + \sigma_{I_2}^2 + C_2} \\
      \mathscr{S}(I_1, I_2) &= \frac{\sigma_{I_1 I_2} + C_3}{\sigma_{I_1}\sigma_{I_2} + C_3}
      \end{aligned} \qquad (2.3)$$

      where α, β, and γ are the weights for the three measurements; μ_{I_1}, μ_{I_2}, σ_{I_1}, σ_{I_2}, and $\sigma_{I_1 I_2}$ are the local means, standard deviations, and correlation coefficient for images $I_1(k,l)$ and $I_2(k,l)$; and C_1, C_2, and C_3 are three variables to stabilize the division.

Refer to PSNR and SSIM functions in the scikit-image package for more details.[1] Make sure you have changed the corresponding parameters of the SSIM function in scikit-image to match the implementation of (Wang et al., 2004). Remember to normalize the images before reporting the PSNR and SSIM scores.

b) Image preparation

To test the performance of different deconvolution algorithms, we need to generate a blurry image using a known blur kernel. The procedures are the following:

1. Pick an image from the BSDS500 dataset, and convert it to gray scale.[2] Crop a 256×256 region from the selected image. Denote this image $I(k,l)$.
2. Perform 2D convolution (with zero padding) on the cropped image using an identity matrix of size 21 as the kernel, $h(k,l)$. Remember to normalize the kernel to make sure that it sums up to 1 before convolution. Denote the obtained image $I_{\text{noiseless}}(k,l)$.
3. Calculate the standard deviation of $I(k,l)$ and add Guassian noise with zero mean and standard deviation of $0.01 \cdot \text{STD}(I(k,l))$ to $I_{\text{noiseless}}(k,l)$. Denote the noisy image as $I_{\text{noisy}}(k,l)$.

  i. Blurry image

  Plot $I(k,l)$, $I_{\text{noiseless}}(k,l)$ and $I_{\text{noisy}}(k,l)$. What are the sizes of $I_{\text{noiseless}}(k,l)$ and $I_{\text{noisy}}(k,l)$?

  ii. Metric baseline

  Briefly describe the differences between PSNR metric and SSIM metric. Report PSNR(I, $I_{\text{noiseless}}$); SSIM(I, $I_{\text{noiseless}}$); PSNR(I, I_{noisy}); and SSIM(I, I_{noisy}). You can crop the center 256×256 regions from $I_{\text{noiseless}}(k,l)$ and $I_{\text{noisy}}(k,l)$ when calculating PSNR and SSIM.

c) Naive deconvolution

Implement a function to conduct naive deconvolution. The function should take the blurry observation, $J(k,l)$ and the blur kernel, $h(k,l)$ as the input parameters and return the recovered image, $\widehat{I}(k,l)$. The discrete Fourier transform functions in NumPy might be useful.[3]

  i. Naive deconvolution algorithm

  Apply your naive deconvolution algorithm to both $I_{\text{noiseless}}(k,l)$ and $I_{\text{noisy}}(k,l)$.

---

[1] https://scikit-image.org/docs/dev/api/skimage.metrics.html.

[2] https://www2.eecs.berkeley.edu/Research/Projects/CS/vision/grouping/resources.html.

[3] https://docs.scipy.org/doc/numpy/reference/routines.fft.html.

Plot the recovered images, and report their PSNR and SSIM scores with $I(k,l)$. Remember to crop the boundaries of the recovered images.

ii. Naive deconvolution results

Why are the outputs of the preceding two cases different? You need to derive the Fourier transform of the recovered images to answer this question.

iii. Naive deconvolution analysis

Express the recovered image from Wiener deconvolution in frequency domain, and implement your own Wiener filter function based on it.

d) Wiener filter

i. Wiener filter algorithm

Express the recovered image from Wiener deconvolution in frequency domain, and implement your own Wiener filter function based on it.

ii. Ideal Wiener filter results

Apply your Wiener filter to $I_{\text{noisy}}(k,l)$. Plot your recovered image, and report its PSNR and SSIM scores. You can use the actual frequency-dependent SNR (ω_1, ω_2) in this question.

iii. Power spectral density

Normally, we do not have access to the frequency-dependent SNR (ω_1, ω_2) in real applications. Therefore, people usually approximate the SNR (ω_1, ω_2) from a predefined function. To explore how to estimate SNR (ω_1, ω_2), we first analyze the power spectrum of noise and real images. Plot the power spectral density of $I(k,l)$ and your added noise $e(k,l)$ in log scale.[4] Pick two other images with different scenes in the BSDS500 dataset, and plot these two images together with their log-scale spectral density.

iv. SNR approximation

On the basis of Figure E2.6, describe the features of real images and noise. Which function would you use to approximate SNR in this case?

v. Approximation results

Plot the deconvolution result using the preceding SNR approximation, and report your PSNR and SSIM scores.

---

[4] Remember to shift the zero-frequency component to the center of the spectrum by using fftshift in NumPy.

Exercises

Figure E2.6 Images of the real world and of noise.

3 Computational Toolkit

Distinct from the conventional notion of imaging, computational imaging heavily relies on mathematical tools and techniques that facilitate the required *computation* for recovery of images from measurements. In this way, computational imaging draws on the wealth of available knowledge from the areas of signal processing, optimization theory, and inverse problems. The purpose of this chapter is to combine known and recent tools from these areas in a holistic fashion, using the following organization. We start with a brief introduction to inverse problems. Then we recall basic tools and ideas from signal processing and linear algebra that help us develop mathematical models for imaging problems. For solving inverse problems, we rely on (1) model-based computational and numerical methods and (2) data-driven methods inspired by advances in machine learning literature. In the chapters that follow, these tools are used as mathematical algorithms in stand-alone configuration or as a combination of tools for image recovery.

3.1 Modeling: Forward vs. Inverse Problems

An end-to-end computational imaging system operates in two stages. The first stage describes an image formation model. This stage is concerned with the interaction of a physical entity with hardware. When this interaction is written in mathematical terms, it is known as the forward model. The forward model is a mathematical description of the hardware, that is, the measurements arising from a computational imaging system can be abstractly related to the output of the forward model. Then the task of recovering the underlying physical phenomenon from the measurements is accomplished by solving the inverse problem. This solution is typically composed of a series of mathematical steps, and the resulting procedure is the recovery algorithm. The block diagram shown in Figure 3.1 explains the key idea.

To place things in context, consider the problem of estimating a singer's age from audio samples of that singer's performances taken over a course of years. For instance, given a collection of all Michael Jackson's songs, can we estimate his age at recording of a given

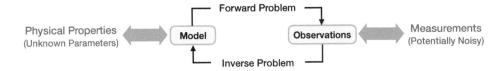

Figure 3.1 Block diagram for modeling inverse problems.

song? To solve this problem, we first need to derive a mathematical model that mimics the physical process of generating voice in human beings. This relies on the forward model for this case study. The more accurate the forward model is, the better it relates to the audio samples of the song. Clearly, one would expect the forward model to be parameterized by the age; the vocal cords change as one ages. With audio samples from a given era, the goal of estimating the unknown parameter, which in this case is the age, amounts to solving an inverse problem using a mathematical algorithm. The same analogy applies to a diverse set of problems spread across various applications in science and engineering. Some commonplace examples of inverse problems include the following:

- **Denoising**   Removing noise from measurements (Chatterjee and Milanfar, 2010).
- **Superresolution**   Recovering high-resolution features from low-resolution measurements (Park et al., 2003).
- **Deconvolution**   Removing the influence of a measurement device from the measurements (Kundur and Hatzinakos, 1996).

We revisit this aspect in detail in section 3.3.1. In this chapter we use foundational theory from topics such as signal processing and linear algebra to study forward models. Then we revisit some classical solutions to the inverse problems that will lay the groundwork for the rest of this book.

3.2 Mathematical Tools

3.2.1 Signal Processing

Signal processing consists of decomposing complicated phenomena into simpler constituents with the goal of revealing hidden information. Signals can be functions or sequences mapping a physical phenomenon into continuous and discrete-valued observations. Processing refers to the appropriate mathematical operations that reveal information of interest. One of the most interesting observations in this regard dates to the work of Joseph Fourier, who in 1807 claimed that any periodic phenomenon can be described as a combination of harmonic waves. Over the course of centuries, this seemingly simple observation has had significant implications across science and engineering.

3.2 Mathematical Tools

Here, we briefly visit the basics of signal processing that allow us to represent captured data (or measurements) as signals so that known tools can be applied for recovering information.

Linear systems A system is a conceptual object (physical, mathematical, or even computational) that maps an input to an output. As shown in the following, a system accepts both continuous-time functions and discrete sequences as an input. The output, however, can be either a function or a sequence. For instance, a CD player maps a sequence of binary *bits* to continuous-time sound.

$f(t) \longrightarrow \boxed{\text{System}} \longrightarrow g(t)$ $\qquad$ $f[m] \longrightarrow \boxed{\text{System}} \longrightarrow g[m]$

$\qquad$ Continuous-time system $\qquad\qquad\qquad\qquad$ Discrete-time system

Continuous-time functions are denoted by $f(t)$ where the time variable t takes real values, that is, $t \in \mathbb{R}$. Its discrete counterparts are represented as a sequence,

$$\cdots \quad f[-1] \quad f[0] \quad f[1] \quad \cdots$$

that accepts discrete-valued argument $m \in \mathbb{Z}$. For example, a continuous-time function can be represented by discrete *samples* $f[m] = f(mT)$ where $T > 0$ is the sampling rate.

In this section we restrict our discussion to *linear systems*. Beyond the practical utility of this subclass of systems, linear systems are appealing because their mathematical analysis is an understood topic. Linear systems defined by operator $\mathcal{L}$ satisfy two properties: (1) principle of additivity or sum at the output is the sum of the inputs, and (2) scaling or scaled input produces a scaled output. This leads to the definition of a linear system.

Definition 1 (linear system) *Let u and v be the input the linear system $\mathcal{L}$. Then, $\mathcal{L}$ is a linear system when*

$$\mathcal{L}[au + bv] = a\mathcal{L}[u] + b\mathcal{L}[v], \qquad a, b \in \mathbb{C}.$$

For stability purposes, operator $\mathcal{L}$ is required to be continuous, implying that small perturbations in the input produce small perturbations at the output.

Many of the well known systems including computational imaging systems are modeled as linear time-invariant systems. Stated simply, time invariance is a property in which delayed input produces delayed output. Let us define $f_\tau(t) = f(t - \tau)$, that is, f delayed by τ. Linear time-invariant or LTI systems satisfy the following property:

$$g(t) = \mathcal{L}[f](t) \Rightarrow g_\tau(t) = \mathcal{L}[f_\tau](t). \qquad (3.1)$$

In the context of discrete-time signals, this property is known as the linear shift-invariance property where *shift* is the discrete counterpart of continuous *delay*.

Impulse response of a linear system An important property of an LTI system is its impulse response. This is the response of an LTI system when the Dirac impulse is fed to the system. For any function f continuous around zero, the Dirac impulse is defined

$$f(0) = \int f(t)\,\delta(t)\,dt. \tag{3.2}$$

Expression (3.2) can also be written $f(t) = \int f(\tau)\,\delta_\tau(t)\,d\tau$. When $\mathcal{L}$ is continuous and linear, we have

$$\mathcal{L}[f](t) = \int f(\tau)\,\mathcal{L}[\delta_\tau(t)]\,d\tau. \tag{3.3}$$

Letting g be the impulse response of $\mathcal{L}$ or $g(t) = \mathcal{L}[\delta](t)$, time invariance property (3.1) implies that $\mathcal{L}[\delta_\tau](t) = g(t-\tau)$. Therefore, by combining time invariance property and (3.3), we obtain the definition of convolution or filtering,

$$\begin{aligned}\mathcal{L}[f](t) &= \int f(\tau)\,\mathcal{L}[\delta_\tau(t)]\,d\tau \\ &= \int f(\tau)\,g(t-\tau)\,d\tau = \int g(\tau)\,f(t-\tau)\,d\tau \\ &= (f*g)(t). \end{aligned} \tag{3.4}$$

Some useful properties of the convolution operator include

- Commutativity, which means that the convolution operator commutes, $(f*g)(t) = (g*f)(t)$.
- Associativity, which means that the convolution order does not matter, $(f*g*h)(t) = (h*f*g)(t) = (g*h*f)(t)$.
- Differentiation property that asserts that the derivative of a convolution is equivalent to convolution with the derivative of one of the functions

$$\frac{d}{dt}(f*g)(t) = \left(\frac{df}{dt}*g\right)(t) = \left(f*\frac{dg}{dt}\right)(t).$$

- Dirac convolution property, which amounts to shifting of functions, $f*\delta(t-\tau) = f(t-\tau)$.

These principles extend to the case of sequences. Whether working with continuous functions or sequences, one should take care in dealing with the properties of the convolution operator in that the corresponding integrals and summations should be absolutely convergent.

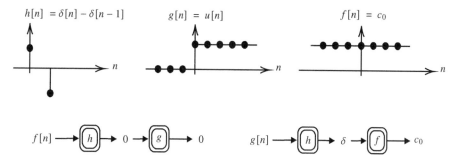

In the context of the associative property, we highlight a pathological example in the preceding, where $f[n] = c_0$ is a constant sequence; $g[n] = u[n]$ is the Heaviside sequence; and $h[n] = \delta[n] - \delta[n-1]$ is the finite difference filter. Clearly $(f * h * g)[n] \neq (g * h * f)[n]$ because on one hand, $f * h = 0$ (difference of constant), and on the other hand, $g * h = \delta$, and this leads to output c_0.

Causality and stability of a linear system. A linear system is known to be causal if the output depends only on the past and current inputs (and not the future inputs). Mathematically, this boils down to saying $\mathcal{L}[f](t)$ does not depend on $f(t'), t' > t$. Another important property of linear systems is that of stability. Practically speaking, this property asserts that bounded input produces bounded output. Suppose that $\mathcal{L}$ is characterized by impulse response $g(t)$ or equivalently $\mathcal{L}[f](t) = (f * g)(t)$. When the input is bounded or $\max_\tau |f(\tau)| < \infty$, the output is bounded

$$|\mathcal{L}[f](t)| \leqslant \int |f(\tau)| |g(t-\tau)| d\tau \leqslant \max_\tau |f(\tau)| \int |g(t)| dt \quad (3.5)$$

whenever $\int |g(t)| dt < \infty$ or the impulse response is absolutely integrable. This is known as bounded input and bounded output (BIBO) stability. There are further mathematical considerations (Unser, 2020) that make the BIBO stability statements precise. The stability property has profound implications in assessing the resilience of an algorithm to noise. Ideally, there is no noise or uncertainty at the input that would blow away the output, and this is where stability analysis is helpful.

Eigenfunctions of linear time-invariant systems and the Fourier transform. One of the key features of linear time-invariant systems is that the complex exponentials defined by $e^{j\omega t}$ are eigenvectors of the convolution operator. Recall that for any linear system $\mathcal{L}$, when the following is true

$$\mathcal{L}[e](t) = \lambda e(t)$$

we say that λ is the eigenvalue and $e(t)$ is the eigenfunction. As in the previous example, let $\mathcal{L}$ be characterized by impulse response $g(t)$. In view of (3.4), we have that $\mathcal{L}[f](t) = (f * g)(t)$. Consequently, when the input to this linear time-invariant system is $f_\omega(t) =$

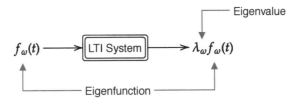

Figure 3.2 Complex exponentials are eigenfunctions of linear time-invariant systems.

$e^{J\omega t}$ the output is

$$\mathcal{L}[f_\omega](t) = \int g(\tau) f_\omega(t-\tau) d\tau = \int g(\tau) e^{J\omega(t-\tau)} d\tau$$

$$= \underbrace{e^{J\omega t}}_{f_\omega(t)} \underbrace{\int g(\tau) e^{-J\omega\tau} d\tau}_{\lambda_\omega} = \lambda_\omega f_\omega(t) \qquad (3.6)$$

implying that $f_\omega(t) = e^{J\omega t}$ is indeed the eigenfunction of linear time-invariant system because

$$\mathcal{L} f_\omega(t) = \lambda_\omega f_\omega(t).$$

This is explained schematically by Figure 3.2. Regardless of angular frequency ω, $|f_\omega(t)| = 1$. In view of the stability condition, provided that the impulse response is absolutely integrable, that is,

$$\int |g(\tau)| d\tau = \|g\|_{L_1} < \infty, \qquad \text{in other words, } g \in L_1,$$

the output $\mathcal{L}[f_\omega](t)$ is always bounded and hence the output is well defined.

Note that the eigenvector λ_ω defines the quantity

$$\lambda_\omega = \int g(t) e^{-J\omega t} dt$$

which is widely known as the Fourier integral or the *Fourier transform*. Its formal definition is the following:

Definition 2 (Fourier transform) *The Fourier transform of a function $f \in L_1$ is defined by*

$$\forall \omega \in \mathbb{R}, \quad \widehat{f}(\omega) = \int f(t) e^{-J\omega t} dt. \qquad (3.7)$$

Whenever $\widehat{f}$ is absolutely integrable or $\widehat{f} \in L_1$, the inverse Fourier transform is defined by

$$\forall \omega \in \mathbb{R}, \quad f(t) = \int \widehat{f}(\omega) e^{J\omega t} d\omega. \qquad (3.8)$$

The Fourier transform is a mathematical tool that measures the number of oscillations ω (in radians per second) present in $f(t)$. For example, in the case of sinusoidal waveforms, the Fourier transform is a Dirac impulse (3.2) present at the oscillation frequency. This is because Euler's formula allows us to write

$$\cos(\omega_0 t) = \frac{e^{J\omega_0 t} + e^{-J\omega_0 t}}{2}$$

which shows the presence of two frequencies at $\omega = \pm\omega_0$. More formally, this translates to

$$\int \cos(\omega_0 t) e^{-J\omega t} dt = \frac{1}{2}\int e^{-J(\omega-\omega_0)t} dt + \frac{1}{2}\int e^{-J(\omega+\omega_0)t} dt$$
$$= \underbrace{\frac{1}{2}\left(\delta(\omega - \omega_0) + \delta(\omega + \omega_0)\right)}_{\text{Fourier transform of }\cos(\omega_0 t)}. \quad (3.9)$$

On the other hand, it takes all frequencies from $-\infty$ to ∞ to constitute a Dirac impulse defined in (3.2). This is because

$$\int \delta(t) e^{-J\omega t} dt = e^{-J\omega t}\big|_{t=0} = 1.$$

Note that both sinusoids and Dirac impulses violate the assumption of absolute integrability or boundedness. However, with certain technical safeguards, it is possible to extend Fourier analysis to wider classes of signals. This aspect is beyond the scope of this book but we refer the interested readers to the book by (Mallat, 2009).

The Fourier transform is a unitary transform in that it preserves lengths. Let us denote the inner product by

$$\langle f(t), g(t)\rangle = \int f(t) g^*(t) dt$$

where * is the complex conjugate. When working with Fourier transforms, we have that,

$$\langle f(t), g(t)\rangle = \int \widehat{f}(\omega) \widehat{g}^*(\omega) d\omega = \left\langle \widehat{f}(\omega), \widehat{g}^*(\omega)\right\rangle.$$

This is known as Parseval's theorem. Substituting $g = f$, we obtain

$$\langle f(t), f(t)\rangle = \|f(t)\|_{L_2}^2 = \left\|\widehat{f}(\omega)\right\|_{L_2}^2$$

which is known as the Plancherel theorem, and the quantity $\|f(t)\|_{L_2}^2$ is a measure of the energy in $f(t)$.

Fourier transforms and convolutions. In the context of linear systems an interesting property relates to the interaction between convolution or filtering and Fourier transforms. The convolution theorem states that convolution or filtering in the canonical domain amounts to multiplication of Fourier transforms in the transform domain.

Theorem 3.1 (convolution theorem) *Let f and g to be two given functions and let $h(t) = (f*g)(t)$. Then, the Fourier transform of $h(t)$ is given by $\widehat{h}(\omega) = \widehat{f}(\omega)\widehat{g}(\omega)$. On the other* hand, let $p(t)$ be the product of functions, that is, $p(t) = f(t) g(t)$. The Fourier transform of $p(t)$ is the convolution of Fourier transforms of $f(t)$ and $g(t)$, respectively, that is, $\widehat{p}(\omega) = \left(\widehat{f} * \widehat{g}\right)(\omega)$.

The implication of theorem 3.1 is that

$$\mathcal{L}[f](t) = (f * g)(t) = \int \widehat{f}(\omega) \widehat{g}(\omega) e^{+j\omega t} d\omega.$$

For instance, when working with sinusoids (which is the case in many applications, including time-of-flight imaging, which is discussed in chapter 5), we directly obtain

$$\cos(\omega_0 t) \longrightarrow \boxed{g} \longrightarrow |\widehat{g}(\omega_0)| \cos(\omega_0 t + \angle \widehat{g}(\omega_0))$$

Specifically, in the case of discrete sequences, it allows for fast convolution or filtering operations via Fourier transforms to be implemented very efficiently using the fast Fourier transform (FFT) algorithms.

Frequency response. In signal processing jargon, *spectrum* refers to the Fourier transform of any given function, and *frequency response* defines the Fourier transform of the impulse response of a linear time-invariant system. Again, given a linear time-invariant system with $\mathcal{L}\delta(t) = g(t)$, its frequency response is defined by $\widehat{g}(\omega) = |\widehat{g}(\omega)| e^{j\angle \widehat{g}(\omega)}$. Here, $|\widehat{g}(\omega)|$ is known as the *magnitude response*, which is always non-negative and real-valued. The real-valued quantity $-\pi \leqslant \angle \widehat{g}(\omega) \leqslant +\pi$ is known as the *phase response*.

Fourier series (representing periodic functions). We say a function f is periodic if it repeats itself periodically; $f(t) = f(t + T_p)$ where $T_p > 0$ is the period. For example, $\sin(\omega_0 t) = \sin\left(\omega_0 \left(t + \frac{2\pi}{\omega_0}\right)\right)$ and hence $T_p = 2\pi/\omega_0$. Aperiodic functions can be periodized using the periodization operation,

$$f_{T_p}(t) = \sum_{k \in \mathbb{Z}} f(t + kT_p). \tag{3.10}$$

For example, the Dirac impulse (3.2) can be converted into a picket fence or a Dirac comb using

$$\text{III}_{T_p}(t) = \sum_{k \in \mathbb{Z}} \delta(t + kT_p).$$

The family of functions $\{e^{jk\omega_0 t}\}_{k \in \mathbb{Z}}$ with $\omega_0 = 2\pi/T_p$ forms an orthonormal basis of T_p-periodic functions

$$f_{T_p}(t) = \frac{a_0}{2} + \sum_{k=1}^{\infty} a_k \cos(k\omega_0 t) + b_k \sin(k\omega_0 t) = \sum_{k \in \mathbb{Z}} \widehat{f}_k e^{jk\omega_0 t} \tag{3.11}$$

where

$$\widehat{f}_k = \frac{1}{T_p} \int_{T_p} f_{T_p}(t) e^{-jk\omega_0 t} dt = \frac{1}{T_p} \widehat{f}(k\omega_0) \tag{3.12}$$

are the Fourier series coefficients, which amounts to observing the Fourier transform at frequencies $\omega = k\omega_0$. Because the Fourier transform of Dirac impulse is unity or $\widehat{\delta}(\omega) = 1$, we have $\widehat{\delta}(n\omega_0) = 1$ and hence

$$\sum_{n \in \mathbb{Z}} \delta(t + nT_p) = \sum_{k \in \mathbb{Z}} e^{jk\omega_0 t}.$$

Many of the properties of the Fourier series are similar to those of Fourier transform as the former is a specific case of the latter. For example, by using the convolution theorem in theorem 3.1, we see that

$$f_{T_p}(t) = \left(f * \mathrm{III}_{T_p}\right)(t) = \sum_{k \in \mathbb{Z}} \widehat{f_k} e^{jk\omega_0 t}$$
$$\Downarrow \qquad\qquad\qquad\qquad (3.13)$$
$$\sum_{n \in \mathbb{Z}} f(t + nT_p) = \sum_{k \in \mathbb{Z}} \widehat{f_k} e^{jk\omega_0 t}$$

which is the well known Poisson sum formula that is at the heart of studying analog-to-digital conversion.

Analog-to-digital conversion and sampling theory. At the heart of digital data acquisition systems is Shannon's sampling theorem, which bridges the continuous and the discrete realms. The sampling theorem has profound implications and has led to the *digital revolution*, also known as the *Third Industrial Revolution*. At the core of this fundamental topic is the question: when can a continuous function be represented by a sequence of discrete numbers? As shown in Figure 3.3, a continuous-time function $f(t)$ can be represented by its discrete counterpart, samples, defined by $f[m] = f(mT)$, $m \in \mathbb{Z}$. The key question is how big can T be? If T is too small, the amount of data needed to be stored will be huge. If T is too big, we may never be able to reconstruct or recover the continuous-time function. For instance, let $f(t) = \sin(\omega_0 t)$. Setting $f[m] = f(mT)$ with $T = \pi/\omega_0$ yields $f[m] = 0$, and there is no information in the discrete samples. From this thought experiment it is clear that the faster a signal fluctuates, the more quickly one should record its samples. To measure the *slowness* or the *rapidity* of a function, Shannon used the idea of bandlimitedness; the maximum frequency contained by a function. The tool to measure this object is the Fourier transform.

Definition 3 (bandlimitedness) *We say a function is Ω-bandlimited if the largest frequency contained by its Fourier transform is Ω, or*

$$f \in B_\Omega \Leftrightarrow \widehat{f}(\omega) = 0, |\omega| > \Omega.$$

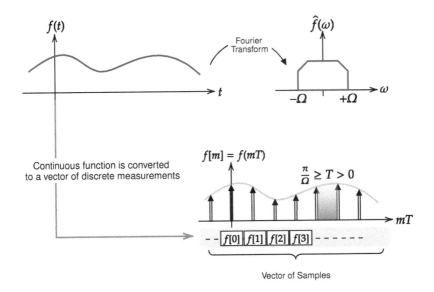

Figure 3.3 Sampling theory addresses the problem of representing a continuous-time function with discrete samples.

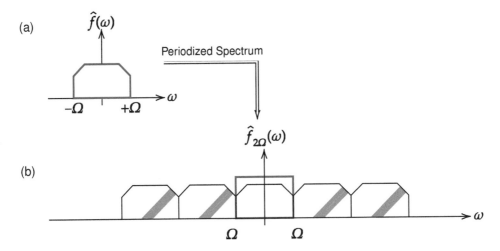

Figure 3.4 (a) Bandlimited function, and (b) effect of Fourier spectrum periodization.

In Figure 3.4a, we show the Fourier transform of a Ω–bandlimited function. Given Ω, we can periodize the Fourier transform of f using (3.10) to obtain

$$\widehat{f}_{\Omega_0}(\omega) = \sum_{n \in \mathbb{Z}} \widehat{f}(\omega + \Omega_0 n). \tag{3.14}$$

3.2 Mathematical Tools

which is shown in Figure 3.4b. Note that we must enforce $\Omega_0 \geqslant 2\Omega$ or else the successive bands $\{\widehat{f}(\omega + n\Omega_0)\}_n$ will overlap, leading to loss of information. This is known as *aliasing*. Due to the periodic nature of $\widehat{f}_{\Omega_0}(\omega)$, we can now represent this function as a Fourier series with harmonic frequency $T_0 = 2\pi/\Omega_0$ or

$$\widehat{f}_{\Omega_0}(\omega) = \sum_{k \in \mathbb{Z}} z_k e^{-jkT_0\omega}, \qquad T_0 = \frac{2\pi}{\Omega_0} \qquad (3.15)$$

where the Fourier series coefficients are given by

$$z_k = \frac{1}{\Omega_0} \int_{\Omega_0} \widehat{f}_{\Omega_0}(\omega) e^{jkT_0\omega} d\omega = \frac{1}{\Omega_0} f(kT_0).$$

Note that the (3.15) is equivalent to the result of (3.13). In order to reconstruct $f(t)$ from $\widehat{f}_{\Omega_0}$ we perform frequency domain filtering

$$f(t) = \int \widehat{f}_{\Omega_0}(\omega) \mathbb{1}_{[-\Omega,\Omega]}(\omega) e^{j\omega t} d\omega = \int_{-\Omega}^{+\Omega} \widehat{f}_{\Omega_0}(\omega) e^{j\omega t} d\omega. \qquad (3.16)$$

Using $\Omega_0 = 2\Omega$ and substituting (3.15) in (3.16), we see that

$$f(t) = \int_{-\Omega}^{\Omega} \widehat{f}_{2\Omega}(\omega) e^{j\omega t} d\omega = \frac{1}{2\Omega} \sum_{k \in \mathbb{Z}} f(kT_0) \int_{-\Omega}^{\Omega} e^{j\omega(t-kT_0)} d\omega$$

$$= \frac{1}{2\Omega} \sum_{k \in \mathbb{Z}} f(kT_0) \frac{2\sin(\Omega(t-kT_0))}{(t-kT_0)} = \sum_{k \in \mathbb{Z}} f(kT_0) \text{sinc}\left(\frac{t}{T_0} - k\right)$$

where $\text{sinc}(t) = \sin(\pi t)/(\pi t)$ is the *sinus cardinalis* function. In summary, we can write

$$f \in B_\Omega, \quad f(t) = \sum_{k \in \mathbb{Z}} f[k] \text{sinc}\left(\frac{t}{T_0} - k\right)$$

which shows that continuous-time, Ω–bandlimited signal $f(t)$ can be represented by discrete samples $f[k] = f(kT_0)$. This also defines the sampling distance T_0

$$\Omega_0 \geqslant 2\Omega \Rightarrow \frac{\pi}{T_0} \geqslant \Omega.$$

The upper limit π/T_0 on the maximum frequency Ω contained in f is known as the *Nyquist frequency*, commonly used in the context of digital communications. We are now ready to state the sampling theorem formally.

Theorem 3.2 (Shannon's sampling theorem) *If a function $f(t)$ is bandlimited to Ω (in radians per second), it is completely characterized by equidistant samples $f[k] = f(kT_0)$ spaced $T_0 = \pi/\Omega$ seconds apart.*

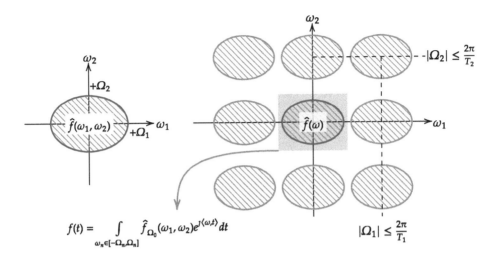

Figure 3.5 Two-dimensional bandlimited signal and periodization of its Fourier transform.

In almost all practical applications, the analog-to-digital conversion or sampling takes place at the level of the sensor and the discrete samples are obtained. However, at this stage of the data capture pipeline, the signals are discrete-time, so that a function of continuous time variable is converted to a sequence of samples that is defined for discrete-time instances. The amplitudes $f[k]$ are still real-valued and require infinite precision and storage capacity. Depending on the budget of the analog-to-digital converter (number of bits), the amplitude is assigned a finite memory by rounding its values. For instance $f[k] = \pi$ is converted to $f[k] = 3.142$. This process is known as *quantization*. Once the quantization has been performed, the samples have a discrete representation on both time and amplitude axes. Such a representation is known as *digital samples*.

The sampling theorem can be extended to the case of multidimensional signals such as images and volumes. To do this, we briefly revisit the N-dimensional Fourier transform. To represent a point in $\mathbb{R}^N$ as an N-tuple, let us denote $t = \begin{bmatrix} t_1 & \cdots & t_N \end{bmatrix}$. The N-dimensional Fourier transform

$$f(t) \text{ or } f(t_1, t_2, \ldots, t_N)$$

is then defined

$$\widehat{f}(\omega) = \int_{\mathbb{R}^N} f(t) e^{-J\langle \omega, t \rangle} dt = \int_{\mathbb{R}^N} f(t) e^{-J \sum_{n=1}^N \omega_n t_n} dt \quad (3.17)$$

where $\langle \omega, t \rangle$ is the conventional inner product or the dot product. A useful property that allows us to extend one-dimensional Fourier transform to N-dimensional Fourier transform

3.2 Mathematical Tools

is *separability*. A separable function f of N variables can be written as N-functions of one variable

$$f(t_1, t_2, \ldots, t_N) = f_1(t_1) f_2(t_2) \cdots f_N(t_N) \text{ or } f(\mathbf{t}) = \prod_{n=1}^{N} f_n(t_n).$$

This extends to the N-dimensional Fourier transform, in that

$$\widehat{f}(\omega) = \prod_{n=1}^{N} \widehat{f_n}(\omega_n).$$

For instance, Dirac impulses and the sinc function used in sampling theory are separable functions, and hence their Fourier transforms are separable, too. For simplicity, consider a two-dimensional function, bandlimited function $f(t_1, t_2)$, whose Fourier transform satisfies the bandlimitedness condition (cf. Figure 3.5)

$$\widehat{f}(\omega_1, \omega_2) = 0, \quad |\omega_1| > \Omega_1 \text{ and } |\omega_2| > \Omega_2.$$

In analogy to (3.14), its two-dimensional periodic version takes the form

$$\widehat{f}_{\Omega_0}(\omega_1, \omega_2) = \sum_{\mathbf{k} \in \mathbb{Z}^2} \widehat{f}(\omega_1 + \Omega_{0,1} k_1, \omega_2 + \Omega_{0,2} k_2)$$

and hence, the Fourier series representation of this function is

$$\widehat{f}_{\Omega_0}(\omega_1, \omega_2) = \sum_{\mathbf{k} \in \mathbb{Z}^2} z_{\mathbf{k}} e^{-J(k_1 T_1 \omega_1 + k_2 T_2 \omega_2)}, \quad z_{\mathbf{k}} = \frac{1}{\Omega_{0,1} \Omega_{0,2}} f(k_1 T_1, k_2 T_2).$$

This is shown in Figure 3.5. As previously, performing frequency domain filtering (3.16) with $\Omega_{0,1} = 2\Omega_1$ and $\Omega_{0,2} = 2\Omega_2$ and using the separability of the box function, we obtain the sampling formula

$$f(t_1, t_2) = \sum_{\mathbf{n} \in \mathbb{Z}^2} f(n_1 T_1, n_2 T_2) \operatorname{sinc}\left(\frac{t_1}{T_1} - n_1\right) \operatorname{sinc}\left(\frac{t_2}{T_2} - n_2\right)$$

which is exactly satisfied as long as

$$\frac{\pi}{T_n} \leqslant \Omega_n, \quad n = 1, 2,$$

which is the Nyquist criterion for two-dimensional signals.

The discussion of multi-dimensional sampling theory sets the foundation for multidimensional discrete representation of continuous functions. In this context, the dimensionality of the data representation leads to the definition of vectors, matrices, and tensors, which are mathematical objects in linear algebra that play key roles in designing mathematical algorithms for solving inverse problems.

As shown in Figure 3.6, one-dimensional data is stored as vectors. Two-dimensional data is stored as matrices, which is the case of images. Each element of a matrix corresponds

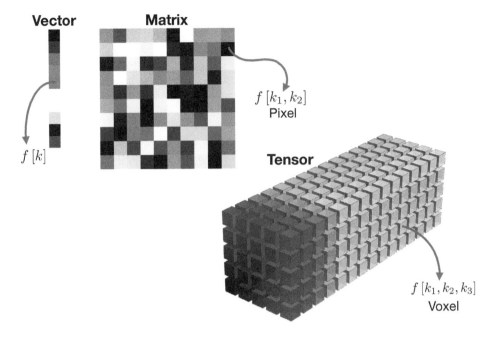

Figure 3.6 Discretization of information on one, two, and three dimensions, leading to mathematical objects vector, matrix, and tensor, respectively. In the case of images, a pixel maps to an element in the matrix. In the case of volume based data, a voxel maps to an element in the tensor.

to a pixel in the image sensor. Two-dimensional data is stored as tensors (or three-dimensional matrices). Such data structures typically arise in the context of time-resolved and hyperspectral imaging applications.

3.2.2 Linear Algebra

Stated simply, when a linear system is discretized, it can be interpreted as a linear system of equations. For example, when the Fourier transform defined in (3.7) is discretized, the integral takes the form of a sum. Similarly, discretization of the convolution/filtering operation in (3.4) can also be represented as operations with discrete sequences. In all such cases, the representations take the form

$$g[m] = \sum_{n=0}^{N-1} a[m,n] f[n], \quad m = 0, \ldots, M-1.$$

3.2 Mathematical Tools

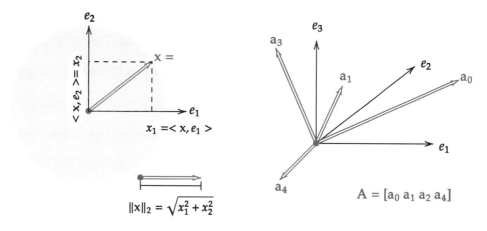

Figure 3.7 Vectors and matrices: definitions and basic operations.

which is a linear system of M equations,

$$g[0] = a[0,0]f[0] + a[0,1]f[1] + \cdots + a[0, N-1]f[N-1]$$
$$g[1] = a[1,0]f[0] + a[1,1]f[1] + \cdots + a[1, N-1]f[N-1]$$
$$\vdots$$
$$g[M-1] = a[M-1,0]f[0] + a[M-1,1]f[1] + \cdots + a[M-1, N-1]f[N-1].$$

Given measurements $\{g[m]\}_{m=0}^{M-1}$ and $\{a[m,n]\}_{m \in [0,M-1], n \in [0,N-1]}$, in many problems of interest, the goal is to estimate or recover $\{f[n]\}_{n=0}^{N-1}$.

With some of the first results dating to the work of Gauss, linear algebra seeks to handle the preceding cumbersome equations in a compact, systematic, and efficient fashion.[5] This is a mature topic, and many of the results are well understood. For a first course on this topic, we refer readers to (Meyer, 2000) and (Strang, 2016). Here we revisit basic tools and techniques that are used in subsequent chapters.

Notation and basic definitions. Four mathematical objects appear throughout this book: scalars, vectors, matrices, and tensors.

1. Scalars are single numbers that may be from a set of natural, integer, real, or complex valued numbers. They are named by lowercase, italic letters, for instance, $m \in \mathbb{Z}$, $x_1 \in \mathbb{R}$, and $z \in \mathbb{C}$.

---

[5] One of the classical approaches for solving a linear system of equations is credited to German mathematician, Johann Carl Friedrich Gauss. This approach is known as Gaussian elimination.

2. Vectors are arrays of ordered and indexed numbers denoted by boldface letters such as **x**. The number of elements in a vector defines the dimension of the vector. For example, $\mathbf{x} \in \mathbb{C}^N$ defines a vector with N ordered elements $x_1, x_2, \ldots, x_N$ that are complex-valued. In computations involving vectors, we explicitly list the entries in column form,

$$\mathbf{x} = \begin{bmatrix} x_1 \\ x_2 \\ \vdots \\ x_N \end{bmatrix} \in \mathbb{C}^N.$$

The transpose of a vector converts it into row format,

$$\mathbf{x} \in \mathbb{R}^N, \qquad \mathbf{x}^T = \begin{bmatrix} x_1 & x_2 & \cdots & x_N \end{bmatrix}$$

whereas the conjugate-transpose or *Hermitian* transpose converts a vector into row format with complex-conjugate operation on each element

$$\mathbf{x} \in \mathbb{C}^N, \qquad \mathbf{x}^H = \begin{bmatrix} x_1^* & x_2^* & \cdots & x_N^* \end{bmatrix}.$$

From a geometrical perspective, a vector is a point in space with each element representing its Cartesian coordinate. This is shown in Figure 3.7.

The length of a vector is measured using the Euclidian norm

$$\|\mathbf{x}\|_2 = \sqrt{\sum_{n=1}^{N} x_n^2} \equiv \sqrt{\mathbf{x}^T \mathbf{x}}.$$

This quantity is also known as the ℓ_2-norm. Square of this quantity defines the energy of a signal and generalizes the notation of the Pythagorean theorem to N dimensions. Normalizing a vector with its ℓ_2-norm results in a *unit vector*, also called *unit-norm vector*.

The contribution of a vector along each coordinate is known as *orthogonal projection*. This is measured using the inner product or dot product denoted by $\langle \mathbf{x}, \mathbf{y} \rangle = \mathbf{x}^H \mathbf{y}$, which always results in a real-valued scalar. This operation is defined

$$\left\langle \begin{bmatrix} x_1 \\ \vdots \\ x_N \end{bmatrix}, \begin{bmatrix} y_1 \\ \vdots \\ y_N \end{bmatrix} \right\rangle = \mathbf{x}^H \mathbf{y} = \sum_{n=1}^{N} x_n y_n^*.$$

In view of Figure 3.7, note that

$$\langle \mathbf{x}, \mathbf{e}_1 \rangle = \left\langle \begin{bmatrix} x_1 \\ x_2 \end{bmatrix}, \begin{bmatrix} 1 \\ 0 \end{bmatrix} \right\rangle = x_1 \quad \text{and} \quad \langle \mathbf{x}, \mathbf{e}_2 \rangle = \left\langle \begin{bmatrix} x_1 \\ x_2 \end{bmatrix}, \begin{bmatrix} 0 \\ 1 \end{bmatrix} \right\rangle = x_2.$$

3.2 Mathematical Tools

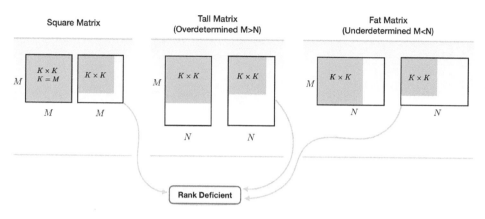

Figure 3.8 Classification of square, tall, and fat matrices and its link with rank deficiency.

Two vectors are orthogonal if their inner product is zero. In view of Figure 3.7, we have $\langle \mathbf{e}_1, \mathbf{e}_2 \rangle = 0$.

3. Matrices are formed by two-dimensional arrays of scalars, obtained by stacking vectors. Let $\{\mathbf{a}_n\}_{n=1}^{N} \in \mathbb{C}^M$, that is, N vectors each comprising M elements. Then we have a matrix

$$\mathbf{A} = \begin{bmatrix} \mathbf{a}_0 & \mathbf{a}_1 & \cdots & \mathbf{a}_N \end{bmatrix} \in \mathbb{C}^{M \times N}$$

In working with measurements in practice, there may be scenarios in which the number of unknowns is equal to, more than, or less than the number of measurements. These cases can be classified in terms of the dimension and the rank of the matrix used for modeling the physical phenomenon. In the following section, we first discuss the idea of full-rank and rank-deficient matrices. Depending on the classification, we develop solutions for inversion of the matrix.

Matrix dimension, rank, and inversion. The size of a matrix (or matrix dimension) and its rank play a key role in solving different classes of inverse problems. Whenever a number K is smaller than the smallest dimension of the matrix, the matrix is known to be *rank deficient*. The cases shown in Figure 3.8 are as follows:

- Square matrix ($M \times M$), in which the number of unknowns is the same as the number of equations. When $K = M$, the matrix is said to be a full-rank matrix and the matrix inverse is well defined, that is, $\mathbf{H}^{-1}\mathbf{H} = \mathbf{I}$ (identity operation). For a simple 2×2 matrix $\mathbf{H}$

$$\mathbf{H} = \begin{bmatrix} a & b \\ c & d \end{bmatrix}$$

its inverse is defined

$$\mathbf{H}^{-1} = \frac{1}{ad - bc} \begin{bmatrix} d & -b \\ -c & a \end{bmatrix}$$

and it is easy to see that such a matrix is always invertible provided that $ad - bc \neq 1$. A matrix is rank deficient when $K < M$ and the number of independent equations is smaller than the number of unknowns.

- Tall matrix ($M > N$), in which the matrix contains more rows than columns and the system of equations is known to be *overdetermined*. In this case, there are more equations than unknowns. When $K < N$, the matrix is rank deficient.

Inversion. Suppose that we are given a system of equations $\mathbf{g} = \mathbf{Hf}$ such that $M > N$ with linearly independent columns. It is natural to seek the best strategy for estimating $\mathbf{f}$ given measurements $\mathbf{g}$. In this setting, a desirable feature of the estimated solution $\mathbf{f}^\star$ is that when plugged in, the resulting $\mathbf{Hf}^\star$ should be close to the observed vector $\mathbf{g}$. Hence it makes sense to minimize the quantity $\|\mathbf{g} - \mathbf{Hf}\|_2^2$ that measures the distance between $\mathbf{Hf}$ and $\mathbf{g}$. In the literature, this is typically known as the cost function

$$C(\mathbf{f}) = \|\mathbf{g} - \mathbf{Hf}\|_2^2 = (\mathbf{g} - \mathbf{Hf})^\top (\mathbf{g} - \mathbf{Hf}) = \|\mathbf{g}\|_2^2 - 2\langle \mathbf{g}, \mathbf{Hf} \rangle + \langle \mathbf{Hf}, \mathbf{Hf} \rangle \quad (3.18)$$

and

$$\mathbf{f}^\star = \min_{\mathbf{f}} C(\mathbf{f}).$$

Here we assume that the vectors and matrices involved are real valued but that they apply to complex-valued system of equations. In either case, all terms that appear in $C(\mathbf{f})$ are scalars, and the first term $\|\mathbf{g}\|_2^2$ is independent of $\mathbf{f}$. To find the minimizer of $C(\mathbf{x})$, we set its derivative to zero. To this end, we have

$$\frac{\partial}{\partial \mathbf{f}} C(\mathbf{f}) = -2\mathbf{H}^\top \mathbf{g} + 2\mathbf{H}^\top \mathbf{Hf} \text{ and hence } \frac{\partial}{\partial \mathbf{f}} C(\mathbf{f}) = 0 \Rightarrow \mathbf{H}^\top \mathbf{Hf} = \mathbf{H}^\top \mathbf{g}.$$

Provided that $\mathbf{H}^\top \mathbf{H}$ is invertible, we have

$$\min_{\mathbf{f}} C(\mathbf{f}) = \min_{\mathbf{f}} \|\mathbf{g} - \mathbf{Hf}\|_2^2 \Rightarrow \mathbf{f}^\star = \left(\mathbf{H}^\top \mathbf{H}\right)^{-1} \mathbf{H}^\top \mathbf{g} \quad (3.19)$$

which is the least-squares solution for an overdetermined system.

- Fat matrix ($M < N$), in which the matrix contains more columns than rows and the system of equations is known to be *underdetermined*. In this case, there are more unknowns than unknowns. When $K < M$, the matrix is rank deficient.

Inversion. Due to the underdetermined nature of $\mathbf{g} = \mathbf{Hf}$, there are many solutions that satisfy this equation. In this case, a common practice is to seek a solution with the smallest energy (minimum norm). At the same time, we want to ensure that the solution $\mathbf{Hf}^\star$ fits the observed vector $\mathbf{g}$ or $\mathbf{g} = \mathbf{Hf}^\star$. This can be posed as the following constrained problem:

$$\min_{\mathbf{f}} \|\mathbf{f}\|_2^2 \text{ such that } \mathbf{g} = \mathbf{Hf}.$$

The least-squares constrained minimization problem can be solved using the method of Lagrange multipliers. To this end, we define the Lagrangian cost function,

$$C_\lambda(\mathbf{f}) = \|\mathbf{f}\|_2^2 + \lambda^\top(\mathbf{g} - \mathbf{H}\mathbf{f}) \equiv \langle \mathbf{f}, \mathbf{f} \rangle + \langle \lambda, \mathbf{g} - \mathbf{H}\mathbf{f} \rangle \qquad (3.20)$$

where λ is a vector of weights. In comparison to the cost function in the overdetermined case in (3.18), our next step is to minimize $C_\lambda(\mathbf{f})$ with respect to both $\mathbf{f}$ and λ. To this end, we have

$$\frac{\partial}{\partial \mathbf{f}} C_\lambda(\mathbf{f}) = 2\mathbf{f} - \mathbf{H}^\top \lambda \quad \text{and} \quad \frac{\partial}{\partial \lambda} C_\lambda(\mathbf{f}) = \mathbf{g} - \mathbf{H}\mathbf{f}.$$

By setting these derivatives to zero (for minimization), we obtain a simultaneous system of equations yielding λ

$$\left. \begin{array}{l} \dfrac{\partial}{\partial \mathbf{f}} C_\lambda(\mathbf{f}) = 0 \Rightarrow \mathbf{f} = \dfrac{1}{2}\mathbf{H}^\top \lambda \\ \dfrac{\partial}{\partial \lambda} C_\lambda(\mathbf{f}) = 0 \Rightarrow \mathbf{g} = \mathbf{H}\mathbf{f} \end{array} \right\} \Rightarrow \lambda = 2(\mathbf{H}\mathbf{H}^\top)^{-1}\mathbf{g}.$$

Provided that $\mathbf{H}\mathbf{H}^\top$ is invertible, we have

$$\mathbf{f}^\star = \mathbf{H}^\top(\mathbf{H}\mathbf{H}^\top)^{-1}\mathbf{g}. \qquad (3.21)$$

3.3 Model-Based Inversion

One of the key aspects of computational imaging is *computation*, which finds its way into the process involving recovery of information from captured data. In almost all cases, the sensor discretizes the measurements using the sampling process leading to the discrete representation of the mathematical model, which can be analyzed using the vector-matrix representation, as shown in Figure 3.9. Given a linear operator $\mathcal{L}$, in many problems of interest the goal is to recover or estimate the input $f(t)$ from the output

$$g(t) = \mathcal{L}[f](t) = \int f(\tau) h(t, \tau) \, d\tau.$$

Whenever the system is characterized by translation invariance, that is, $h(t, \tau) = h_\tau(t)$ (c.f. (3.1)), the preceding representation leads to a convolution/filtering or $g(t) = (h * f)(t)$. In either case, the discrete representation (via sampling or analog-to-digital conversion) of the problem can be written in vector-matrix form

$$\mathbf{g} = \mathbf{H}\mathbf{f}.$$

When solving for the input signal, we say the problem is *well posed* if

- a solution exists;
- the solution is unique; and

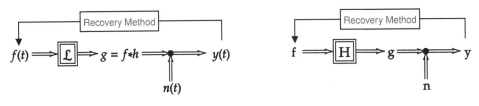

Figure 3.9 Continuous-time linear system and its matrix representation.

- the inversion is stable (or the inverse of the operator $\mathcal{L}$ in the context of $h(t, \tau)$ or $\mathbf{H}$ is continuous).

In the literature, these basic requirements are known as the *Hadamard criteria*. In working with finite-dimension data (the case with vector-matrix notion), although existence and uniqueness can imposed, the inversion of the discrete system may turn out to be a highly *ill-posed problem*.

3.3.1 Examples of Ill-Posed Inverse Problems

In this section we discuss some common place applications that lead to ill-posed problems.

1. Example of an ill-posed problem: deconvolution A classic example of a highly ill-posed problem is that of *deconvolution*, which arises in the context of filtering and is revisited a number of times in this book. Stated simply, deconvolution is the problem of recovering sharp features from smooth measurements. More concretely, consider the normalized Gaussian function

$$\varphi_{\sigma,\mu}(t) = \frac{1}{\sqrt{2\pi\sigma^2}} e^{-\frac{(t-\mu)^2}{2\sigma^2}} \xrightarrow{\text{Fourier}} \widehat{\varphi}_{\sigma,\mu}(\omega) = e^{-\frac{\omega^2\sigma^2}{2} - j\omega\mu}.$$

Suppose that the input signal $f(t) = \varphi_{\sigma_f,0}(t)$ is filtered with $h(t) = \varphi_{\sigma_h,0}(t)$ such that $\sigma_f \ll \sigma_h$ (see Figure 3.10a). In this case, the output is

$$g(t) = (f * h)(t) = \varphi_{\sigma_g,0}(t), \quad \sigma_g = \sqrt{\sigma_f^2 + \sigma_h^2}.$$

This is shown in Figure 3.10b. The goal of the deconvolution problem is to recover $f(t)$ given measurements $g(t)$ and filter $h(t)$. Clearly, when $\sigma_f \ll \sigma_h$, then $\sigma_g \approx \sigma_h$ and the problem is ill posed because the measurement and the filter h are relatively similar. This aspect becomes clear when we look at the Fourier domain representation of the problem. From the convolution-multiplication theorem (cf. theorem 3.1), we have

$$g(t) = (f * h)(t) \xrightarrow{\text{Fourier}} \widehat{g}(\omega) = \widehat{f}(\omega)\widehat{h}(\omega) \Rightarrow \widehat{f}(\omega) = \underbrace{\frac{\widehat{g}(\omega)}{\widehat{h}(\omega)}}_{\text{Deconvolution}}.$$

3.3 Model-Based Inversion

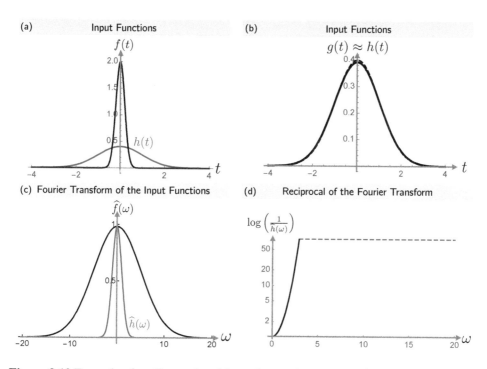

Figure 3.10 Example of an ill-posed problem: deconvolution. (a) $f(t) = \varphi_{\sigma_f, 0}(t)$ is filtered with a $h(t) = \varphi_{\sigma_h, 0}(t)$; (b) when $\sigma_f \ll \sigma_h$, the measurement $g(t) = (f * h)(t) = \varphi_{\sigma_g, 0}(t)$ is very similar to the filter $h(t)$; (c) Fourier transforms of the $f(t)$ and $h(t)$; and (d) reciprocal of the Fourier transform blows up, leading to instabilities.

In principle, one could estimate the Fourier transform of $f(t)$ using the preceding and reconstruct $f(t)$ using the inverse Fourier transform. However, as shown in Figure 3.10c, the Fourier transform of $h(t)$ approaches near-zero values quickly, and hence evaluation of $1/\widehat{h}(\omega)$ results in the blowing up of the spectrum. This is shown in Figure 3.10d. This is the exact cause of instability that makes the problem ill posed.

2. Example of an ill-posed problem: motion deblurring

One of the most practical examples of an ill-posed problem is *motion deblurring*. This situation arises when an object moves while an image is being captured using an imaging sensor. Consider the setting shown in Figure 3.11. An imaging sensor observes two different objects; in the first case (see Figure 3.11a), the object is stationary, and in the other case (see Figure 3.11b), the object is in motion. The imaging sensor acquires a photograph, which is essentially the number of photons collected at a given pixel during the exposure time. Exposure time is the duration for which the light enters the sensor. The higher the exposure time, the greater is the amount of light collected at the sensor.

Mathematically, capturing an image boils down to integration. In the context of the stationary object shown in Figure 3.11, the photograph at a given pixel (around x_0) and for exposure time $t \in [t_0, t_1]$ is

$$g(x_0) = \int_{t_0}^{t_1} f_s(t, x_0) p(x_0) dt = f_s(x_0) p(x_0) (t_1 - t_0)$$

where we have used $f_s(t, x_0) = f_s(x_0)$ because the object $f_s(t, x_0)$ is stationary with respect to time, and $p(x)$ is the point spread function.

Moving to the case of moving object $f_m(t, x)$, from Figure 3.11b it is clear that

$$g(x_0) = \int_{t_0}^{t_1} f_m(t, x_0) p(x_0) dt = f_s(x_0) p(x_0) (t_1 - t_0)$$

because the object is stationary or $f_s(t, x_0) = f_s(x_0)$ during the time of exposure, that is, $t \in [t_0, t_1]$. However, in the case of a longer time exposure $t_2 \gg t_1$

$$g(x_0) = \int_{t_0}^{t_2} f_m(t, x_0) p(x_0) dt = \int_{t_0}^{t_2} f_m(x_0 - t) p(x_0) dt$$

$$= p(x_0) \int f_m(x_0 - t) \mathbb{1}_{[t_0, t_2]}(t) dt$$

$$= p(x_0) (f_m * h)(x_0)$$

where $h(t) = \mathbb{1}_{[t_0, t_2]}(t)$ is the indicator function defined

$$\mathbb{1}_{[t_a, t_b]}(t) = \begin{cases} 1 & t \in [t_a, t_b] \\ 0 & t \notin [t_a, t_b] \end{cases}.$$

Ignoring $p(x_0)$, we observe that the image is nothing but filtering with the box function

$$g(x_0) = (f_m * h)(x_0), \qquad h(t) = \mathbb{1}_{[t_0, t_2]}(t).$$

In this setting, the inverse problem is to recover the moving object $f_m(x_0)$ given the image $g(x_0)$. To see why this problem is ill posed, note that the Fourier transform of the box filter is

$$\mathbb{1}_{[t_a, t_b]}(t) \xrightarrow{\text{Fourier}} j \frac{e^{j\omega t_b} - e^{j\omega t_a}}{\omega} = \widehat{h}(\omega)$$

which is the difference between two sinusoids with frequencies proportional to the length of the box with envelope $1/\omega$. For simplicity, consider the case when the box filter is symmetric, or $t_a = -t_b = t_0/2$. The Fourier transform simplifies to the classical sinc-function,

$$\mathbb{1}_{|t| \leq \frac{t_0}{2}}(t) \xrightarrow{\text{Fourier}} \text{sinc}\left(\frac{\omega}{2\pi} t_0\right).$$

3.3 Model-Based Inversion

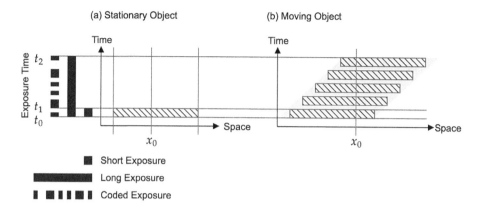

Figure 3.11 Motion deblurring is an ill-posed problem that can be made well posed by using computational imaging methods.

Hence, the measurements in the Fourier domain read

$$g(x) = (f_m * h)(x) \xrightarrow{\text{Fourier}} \widehat{g}(\omega) = \widehat{f_m}(\omega)\widehat{h}(\omega), \quad \widehat{h}(\omega) = \operatorname{sinc}\left(\frac{\omega}{2\pi}t_0\right).$$

This implies that the longer the exposure, the larger is t_0 and hence more frequent are the zeros of the sinc-function that annihilate the information in $\widehat{f_m}(\omega)$. Furthermore, the nulls in the high frequencies of $\operatorname{sinc}\left(\frac{\omega}{2\pi}t_0\right)$ zero out the information in $\widehat{f_m}(\omega)$, resulting in blurring or smearing. There is no way to recover the lost information in $\widehat{g}(\omega)$.

Because the zeros of the sinc-function lead to loss of information, making the problem ill posed, (Raskar et al., 2006) proposed a scheme to turn this setting into a well-posed problem. This scheme is known as *coded exposure photography*. Here, instead of simply opening and closing the camera shutter (yielding a temporal box filter), the idea is to use a sequence of on-off patterns leading to box filters of different widths, as shown in Figure 3.11. In this approach, the sequence is chosen such that zeros of one box filter are compensated by non-zeros of other box filters. When an optimal code is chosen, the Fourier transform is almost flat (or constant), thus preventing any loss of information and hence making $\widehat{h}(\omega)$ invertible.

Model-based inversion techniques provide a systematic solution to solving inverse problems. As we subsequently show in this section, there are many applications in which an unknown signal needs to be reconstructed from a set of measurements. Assuming that the image formation model is known, the inversion of the forward model depends on a set of system parameters intrinsic to the phenomenon being analyzed and a number of regularity conditions, which are characteristic to the inversion method. We consider three commonplace examples: (1) tomographic reconstruction in medical imaging, (2) image deconvolution or deblurring, and (3) subsurface imaging via seismic waveform inversion. Although the last example may seem somewhat unrelated, the main ideas are widely used in imaging examples such as subsurface imaging of tissues or non-line-of-sight imaging.

3.3.2 Tools and Techniques

Through preceding discussions, we have examined how in many cases an inverse problem can be modeled as a linear system of equations. Seeking solutions to an inverse problem boils down to inversion of the system that defines the relationship between the inputs and the outputs. In section 3.2.2, we studied basic examples of working with overdetermined and underdetermined systems. Here, we discuss additional ideas from least-squares and sparse optimization methods. The purpose of this chapter is to develop a familiarity with basic and frequently used regularization approaches. By no means is this discussion complete. For interested readers, we suggest the reference material on numerical methods (Björck, 1996), optimization theory (Boyd and Vandenberghe, 2004) and sparse recovery (Elad, 2010).

Least squares optimization Two least-squares methods are particularly useful: the regularized and the constrained methods.

1. Regularized least squares
   Section 3.2.2 explained that in the case of an overdetermined system, we aim at minimizing $\|\mathbf{g} - \mathbf{Hf}\|_2^2$, whereas in the case of an underdetermined system, our goal is to minimize $\|\mathbf{f}\|_2^2$. Furthermore, in either case it is required that $\mathbf{H}^\top \mathbf{H}$ is invertible.[6] Hence, in working with the general case in which $\mathbf{H}$ may be fat or tall, we can directly minimize the linear combination

$$\mu_1 \|\mathbf{f}\|_2^2 + \mu_2 \|\mathbf{g} - \mathbf{Hf}\|_2^2 = \mu_2 \left(\frac{\mu_1}{\mu_2} \|\mathbf{f}\|_2^2 + \|\mathbf{g} - \mathbf{Hf}\|_2^2 \right)$$

which depends on the ratio of linear coefficients, that is, $\lambda = \mu_1/\mu_2 > 0$. We can now define the cost function,

$$C_\lambda(\mathbf{f}) = \|\mathbf{g} - \mathbf{Hf}\|_2^2 + \lambda \|\mathbf{f}\|_2^2$$

and following the approach previously outlined, we obtain

$$\frac{\partial}{\partial \lambda} C_\lambda(\mathbf{f}) = 2 \langle \mathbf{H}, \mathbf{Hf} - \mathbf{g} \rangle + 2\lambda \mathbf{f} = 0 \Rightarrow \mathbf{f}^\star = \left(\mathbf{H}^\top \mathbf{H} + \lambda \mathbf{I} \right)^{-1} \mathbf{g}.$$

---

[6] Although the invertibility condition that appears in the context of underdetermined system is based on $\mathbf{HH}^\top$, the rank of this matrix is the same as that of $\mathbf{H}^\top \mathbf{H}$.

3.3 Model-Based Inversion

Here, the scalar $\lambda > 0$, also known as the regularization parameter, regularizes the solution. To see this in effect, we now work with

$$\left(\mathbf{H}^\top\mathbf{H} + \begin{bmatrix} \lambda & & \\ & \ddots & \\ & & \lambda \end{bmatrix}\right)^{-1} \quad \text{(Diagonal loading)}.$$

Even when $\mathbf{H}^\top\mathbf{H}$ (and hence $\mathbf{H}\mathbf{H}^\top$) is rank deficit, the regularization parameter ensures that $(\mathbf{H}^\top\mathbf{H} + \lambda\mathbf{I})^{-1}$ is invertible.

2. **Constrained least squares**

A strategy that is applicable to a broad variety of least-squares problems entails a solution to the system of linear equations with constraints. This is also known as the constrained least squares problem

$$\underbrace{\min_{\mathbf{f}} \|\mathbf{g} - \mathbf{H}\mathbf{f}\|_2^2}_{\text{Least squares}} \text{ such that } \underbrace{\mathbf{A}\mathbf{f} = \mathbf{b}}_{\text{Constraint}}. \tag{3.22}$$

To solve this optimization problem, we use the Lagrange multiplier-based strategy and define the cost function

$$C_\lambda(\mathbf{f}) = \|\mathbf{g} - \mathbf{H}\mathbf{f}\|_2^2 + \langle \lambda, \mathbf{A}\mathbf{f} - \mathbf{b}\rangle$$

where λ is a vector. Again, minimizing the cost function with respect to $\mathbf{f}$ and λ, we obtain

$$\frac{\partial}{\partial \mathbf{f}} C_\lambda(\mathbf{f}) = 2\mathbf{H}^\top(\mathbf{H}\mathbf{f} - \mathbf{g}) + \mathbf{A}^\top \lambda \quad \text{and} \quad \frac{\partial}{\partial \lambda} C_\lambda(\mathbf{f}) = \mathbf{A}\mathbf{f} - \mathbf{b},$$

respectively. Setting the first derivative to zero, we obtain

$$\frac{\partial}{\partial \mathbf{f}} C_\lambda(\mathbf{f}) = 0 \quad \Rightarrow \quad 2\mathbf{H}^\top(\mathbf{H}\mathbf{f} - \mathbf{g}) + \mathbf{A}^\top \lambda = 0$$

$$\mathbf{H}^\top\mathbf{H}\mathbf{f} - \mathbf{H}^\top\mathbf{g} = -\frac{1}{2}\mathbf{A}^\top \lambda$$

$$\mathbf{f} = (\mathbf{H}^\top\mathbf{H})^{-1}\left(\mathbf{H}^\top\mathbf{g} - \frac{1}{2}\mathbf{A}^\top \lambda\right).$$

The second derivative, when set to zero, trivially yields $\frac{\partial}{\partial \lambda} C_\lambda(\mathbf{f}) = 0 \Rightarrow \mathbf{A}\mathbf{f} = \mathbf{b}$, and we have the simultaneous equations

$$\mathbf{f} = (\mathbf{H}^\top\mathbf{H})^{-1}\left(\mathbf{H}^\top\mathbf{g} - \frac{1}{2}\mathbf{A}^\top \lambda\right) \tag{3.23}$$

$$\mathbf{A}\mathbf{f} = \mathbf{b}. \tag{3.24}$$

To solve for λ, we simplify (3.24) by multiplying (3.23) by $\mathbf{A}$ on the left-hand side. This yields

$$\mathbf{A}(\mathbf{H}^\top\mathbf{H})^{-1}\left(\mathbf{H}^\top\mathbf{g} - \frac{1}{2}\mathbf{A}^\top\lambda\right) = \mathbf{b}.$$

Solving for λ, we obtain

$$\lambda = 2\left(\mathbf{A}(\mathbf{H}^\top\mathbf{H})^{-1}\mathbf{A}^\top\right)^{-1}\left(\mathbf{A}(\mathbf{H}^\top\mathbf{H})^{-1}\mathbf{H}^\top\mathbf{g} - \mathbf{b}\right).$$

Substituting this value of λ in (3.23), we obtain the solution

$$\mathbf{f} = (\mathbf{H}^\top\mathbf{H})^{-1}\left(\mathbf{H}^\top\mathbf{g} - \mathbf{A}^\top\left(\mathbf{A}(\mathbf{H}^\top\mathbf{H})^{-1}\mathbf{A}^\top\right)^{-1}\left(\mathbf{A}(\mathbf{H}^\top\mathbf{H})^{-1}\mathbf{H}^\top\mathbf{g} - \mathbf{b}\right)\right). \qquad (3.25)$$

The generality of this solution is easily appreciated. For instance, let us set $\mathbf{g} = \mathbf{0}$, $\mathbf{H} = \mathbf{I}$. In this case, $\mathbf{f}$ simplifies to $\mathbf{f} = \mathbf{A}^\top(\mathbf{A}\mathbf{A}^\top)^{-1}\mathbf{b}$. This is the minimum norm solution to the underdetermined system of equations $\mathbf{A}\mathbf{f} = \mathbf{b}$ in (3.21).

Sparse regularization In the least-squares regularization discussion, we explained that the cost function is written as a sum of the data fidelity term (how close the estimated signal is to the measurements) and the term that defines the constraints or regularization. In practice, the constraints may arise from physical properties of the problem or may simply enforce an empirically desirable feature. Hence, in general, one may write the cost function

$$C_\lambda(\mathbf{f}) = \mathcal{D}(\mathbf{g}, \mathbf{H}\mathbf{f}) + \lambda\mathcal{R}(\mathbf{f}) \qquad (3.26)$$

where $\mathcal{D}$ measures distortion or distance between $\mathbf{g}$ and $\mathbf{H}\mathbf{f}$, and $\mathcal{R}$ is the regularization term that enforces desirable/undesirable features. The extent to which $\mathcal{R}$ affects the solution is controlled by the regularization parameter λ. As discussed previously, the conventional choice for $\mathcal{D}$ is the least-squares distance or the 2–norm, that is, $\mathcal{D}(\mathbf{g}, \mathbf{H}\mathbf{f}) = \|\mathbf{g} - \mathbf{H}\mathbf{f}\|_2^2$.

With regard to the choice of the regularization term, recently the notion of *sparsity* has been preferred over least-squares regularization. *Sparsity* simply implies that when solving for $\mathbf{H}\mathbf{f} = \mathbf{g}$, we also look for $\mathbf{f}$ that has few non-zero entries. In particular, we say that $\mathbf{f} \in \mathbb{R}^N$ signal is a K–sparse vector when any K out of N entries are non-zero. To instill an appreciation for the notion of sparsity, we present a toy example in Figure 3.12 that demonstrates the advantage in the case of denoising. In Figure 3.12a, we show a data vector (oracle) containing a sum of two sinusoids in the time domain. This data is corrupted by additive white Gaussian noise with 0 dB signal-to-noise-ratio (SNR). Figure 3.12b shows a plot of the Fourier domain representation of the data vector and noisy vector. As previously shown (cf. (3.9)), the Fourier transform sparsifies a sinusoid. Due to the linearity of the Fourier transform, a sum of K sinusoids results in a K-sparse signal in the Fourier domain. When working with noisy measurements, we can leverage this information and remove all but $K = 2$ largest Fourier components. Then, a simple inverse Fourier transform yields a nearly perfect reconstruction, as shown in Figure 3.12c. Sparsity regularized solutions are

3.3 Model-Based Inversion

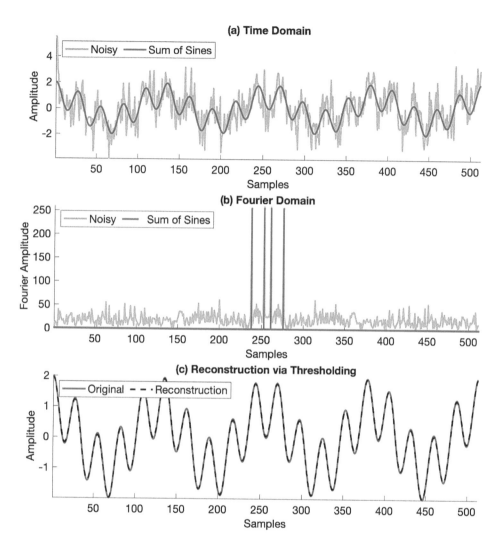

Figure 3.12 Denoising by leveraging sparsity: (a) Time-domain samples of a sum of two sinusoids and its noisy measurements with 0 dB signal-to-noise-ratio (SNR). (b) Fourier domain representation. Because the data comprises a sum of two sinusoids, its Fourier domain representation is a 2-sparse signal comprising two spikes. Adding noise changes this, leading to a number of spurious spikes. (c) Reconstruction via sparsity. In the Fourier domain, we remove all but the two largest coefficients because we know that the data comprises a 2-sparse signal. Time domain reconstruction shows the effect of denoising and results in a nearly perfect reconstruction.

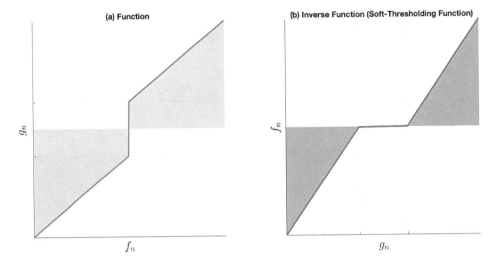

Figure 3.13 Soft-thresholding function as an inverse function. (a) Graph of the function $g_n = f_n + \lambda/2\,\mathrm{sgn}(f_n)$. (b) To evaluate f_n given g_n, we invert the graph in (a), which yields the definition of the soft-thresholding function. The grid lines in gray represent $\pm\lambda/2$.

leveraged in several applications presented in this book. Examples include (1) compressive imaging (section 4.4.1), (2) light-in-flight imaging in (section 5.5.2), (3) coded spectral imaging (section 8.3.4), and (4) subsurface and skin imaging (section 10.5.2).

To enforce sparsity as a regularization prior, a typical approach is to use the ℓ_1–norm. This is mathematically defined by $\mathcal{R}(\mathbf{f}) = \|\mathbf{f}\|_{\ell_1}$. For simplicity of notation, we simply write $\|\mathbf{f}\|_1$

$$\|\mathbf{f}\|_1 = |f_1| + |f_2| + \cdots |f_N|.$$

Clearly, the ℓ_1–norm of $\mathbf{f}$ is small when many of its entries are zero.

Sparse recovery by soft thresholding In the example presented in Figure 3.12, the key idea behind enforcing sparsity (see Figure 3.12b) was a nonlinear decision process: *keep K largest Fourier coefficients*. This entails removing many of the coefficients, an operation known as *thresholding*. The idea is used in the mathematical procedure for solving the sparse regularization problem. To this end, consider a special case of the cost function in (3.26) in which we set $\mathbf{H} = \mathbf{I}$ (identity operation) and $\mathcal{R}(\mathbf{f}) = \|\mathbf{f}\|_1$ is used to enforce sparsity. Our cost function takes the form of

$$C_\lambda(\mathbf{f}) = \|\mathbf{g} - \mathbf{f}\|_2^2 + \lambda\|\mathbf{f}\|_{\ell_1} = \sum_{n=1}^{N}(g_n - f_n)^2 + \lambda|f_n|. \qquad (3.27)$$

3.3 Model-Based Inversion

To minimize this cost function, we differentiate it and obtain

$$\frac{\partial}{\partial \mathbf{f}} C_\lambda(\mathbf{f}) = (g_n - f_n) + \lambda \operatorname{sgn}(f_n)$$

where sgn is the sign (signum) function. To obtain the optimal value of **f**, which minimizes (3.27), we set the derivative of the cost function to zero

$$\frac{\partial}{\partial \mathbf{f}} C_\lambda(\mathbf{f}) = 0 \quad \Rightarrow \quad g_n = \left(f_n + \frac{\lambda}{2}\operatorname{sgn}(f_n)\right) \quad \text{(element-wise)}$$

This gives a nonlinear relation between f_n and g_n. To define f_n in terms of g_n, we use the inverse function method. Figure 3.13a shows a plot $g_n = \left(f_n + \frac{\lambda}{2}\operatorname{sgn}(f_n)\right)$. Interchanging the axes shown in Figure 3.13b defines f_n in terms of g_n, which for an arbitrary **x** is analytically written

$$\operatorname{soft}_\lambda(\mathbf{x}) = \operatorname{sgn}(\mathbf{x})\left(|\mathbf{x}| - \frac{\lambda}{2}\right) \mathbb{1}_{|x| > \frac{\lambda}{2}}(\mathbf{x}).$$

This is the soft-thresholding function. The minimizer of (3.27) is

$$\mathbf{f}^\star = \operatorname{soft}_\lambda(\mathbf{g}).$$

The basic strategy for the cost function generalizes to the case in which the measurements **g** are explained by the forward model **H**, and hence one seeks to minimize

$$C_\lambda(\mathbf{f}) = \|\mathbf{g} - \mathbf{H}\mathbf{f}\|_2^2 + \lambda \|\mathbf{f}\|_{\ell_1}. \tag{3.28}$$

There are standard approaches to solve this problem, categorized into the following broad themes

1. Pursuit algorithms (Elad, 2010) such as orthogonal matching pursuit (OMP), basis pursuit (BP), and basis pursuit denoise (BPDN).
2. Thresholding-based algorithms such as the iterated soft-thresholding algorithm (ISTA) (Daubechies et al., 2004) and its accelerated version, the fast iterated soft-thresholding algorithm (FISTA) (Beck and Teboulle, 2009).
3. Majorization-minimization (Figueiredo et al., 2007) in which one seeks to break down a cost function in terms of simpler, typically, quadratic minimization problems.

Furthermore, for readers interested in more general optimization approaches around the theme of proximal splitting methods and alternating-direction method of multipliers (ADMM) based optimization, we suggest the reference materials (Parikh and Boyd, 2014; Combettes and Pesquet, 2011; Bach et al., 2011).

3.3.3 Examples of Model-Based Reconstruction
1. **Tomography** Model-based inversion techniques are required to recover data from measurements such as computer tomography (CT) scans or positron emission tomography (PET) scans. As opposed to localized imaging, which probes single points in the target,

in tomographic imaging the measurement contains contributions from larger regions, which leads to a more complex but interesting inversion problem. Those are both examples of ray tomography, which performs scans along given lines and then sums the distribution of the target object along those lines. Typically, the ray tomography scan results in measurements of the form

$$m = \exp\left(-\int_{\text{ray l}} \alpha(l)\,dl\right)$$

where $\alpha(l)$ denotes the absorption coefficient of the object along line l. The sum ray, or projection, is given by $\int_{\text{ray l}} \alpha(l)\,dl = -\ln(m)$. The sum ray can be expressed more rigorously using the Radon transform

$$p(s,\phi) = \int_{-\infty}^{\infty} d(s\cos\phi - \tau\sin\phi,\ s\sin\phi + \tau\cos\phi)\,d\tau$$

where $d(x,y)$ denotes the 2D distribution of the object, and $p(s,\phi)$ denotes the projection. The inverse problem in this case is recovering $d(x,y)$ given enough measurements of $p(s,\phi)$.

An important mathematical result, known as the *projection slice theorem*, shows an intuitive way to solve the inversion of the Radon transform. It states that each Radon projection $p(s,\phi)$ represents the 1D inverse Fourier transform of a slice of the object distribution $d(x,y)$. The corresponding equation is

$$\widehat{p}_\phi(k_s) = \widehat{d}(k_s\cos\phi,\ k_s\sin\phi).$$

This means that we can recover any slice of $\widehat{d}$ by computing the 1D Fourier transform of the Radon transform along variable s.

Let us take a very popular example of ray tomography, computer axial tomography (CAT), also known as computed tomography (CT). This system is designed to compute systematically as many projections from the 3D object as possible, which should generate enough slices of function D. The CAT scanner emits a series of parallel beams (organized in a fan shape for medical scanners) that are attenuated by the target object and then measured by a sensor array placed on the opposite side, as depicted in Figure 3.14. The ray emitters and detectors then rotate around the object to acquire samples of the object from all orientations.

2. **Image deconvolution** Imaging aims to recover the intrinsic properties of a scene. However, these properties are not directly accessible in most cases, and the cameras capture images that represent the result of filtering the image describing the scene with a kernel function that depends on the capturing device used. Model-based inversion methods reverse the filtering operation to uncover the underlying scene properties.

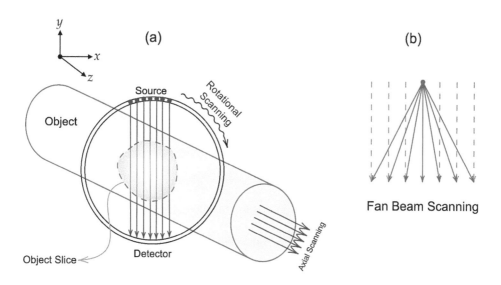

Figure 3.14 Computed axial tomography scanning: (a) scanner using parallel rays measured with sensor arrays, and (b) rays organized in a fan shape used by medical scanners.

An interesting example of model-based inversion can be found in 3D microscopy. The problem that occurs here is that typically the specimens analyzed with a microscope have several layers of depth, and only one can be in focus at a time (Vonesch and Unser, 2008). Can we recover the layers that are out of focus? In fact, an out-of-focus, blurry image is simply the original image filtered with a microscope specific kernel, known as the point spread function (PSF).

Generally speaking, a camera focused at different depth captures can be described by a 3D PSF that generates a 3D stack of blurry images apart from one in-focus image

$$\mathbf{I}_f(x, y, z) = (\mathbf{I} * \text{PSF})(x, y, z),$$

where $\mathbf{I}$ is the original 3D image stack, and $\mathbf{I}_f$ is the filtered stack. It is possible to compute the in-focus image stack $\mathbf{I}$ with a simple and fast deblurring algorithm known as nearest neighbor. This effectively computes the out-of-focus contribution of a blurry image as the average of directly adjacent images in the stack. This method is very imprecise.

The convolution operation can be expressed in the simple algebraic form

$$\mathbf{I}_f = \text{PSF} \cdot \mathbf{I}$$

where $\mathbf{I}_f$ and $\mathbf{I}$ are the original and filtered images in matrix form, and PSF is the matrix form of the point spread function. Then if we measure an out-of-focus 3D stack $\widetilde{\mathbf{I}}_f$, we can estimate the in-focus stack as $\min_{\mathbf{I}} \left\| \widetilde{\mathbf{I}}_f - \text{PSF} \cdot \mathbf{I} \right\|_2^2$ which is the classic least squares problem, where $\|\cdot\|_2^2$ is the squared norm. This is also known as the *inverse filter* method.

However this approach does not work well when the data is corrupted by noise. There are two types of noise in an imaging system. The *shot noise*, caused by the irregularity of the photon arrival times, is prevalent mainly at low levels of lighting. The *read noise* is determined by the imperfection of the imaging sensor, affected by temperature and gain (ISO value). The issue is addressed by adding a regularization term of the form

$$\widehat{\mathbf{I}} = \arg \min_{\mathbf{I}} \left\| \widetilde{\mathbf{I}}_f - \text{PSF} \cdot \mathbf{I} \right\|_2^2 + \lambda \|\mathbf{I}\|_1$$

where λ is the regularization parameter, and $\|\mathbf{I}\|_1 = \sum_{i,j,k} \|I_{i,j,k}\|$ denotes the ℓ_1 norm. This essentially creates a trade-off between minimizing the error (left-hand term) and generating images with a small number of components. This type of regularization assumes that the original image is sparse, in the sense that it has many zero entries. In microscopy, for example, this may be true in the case of a specimen with a dark background, but it does not apply in most cases. What if all the entries of the 3D stack are nonzero? A research direction called wavelet regularization shows that it is possible to select the right basis function of wavelets, such that the coefficients of I in that basis are sparse. We can then write the optimization problem

$$\widehat{\mathbf{I}} = \arg \min_{\mathbf{I}} \left\| \widetilde{\mathbf{I}}_f - \text{PSF} \cdot \mathbf{I} \right\|_2^2 + \lambda \|\mathbf{WI}\|_1$$

where λ is the regularization parameter, and **WI** denote the sparse set of wavelet coefficients of the in-focus image stack **I**.

Another problem is that the PSF function may be unknown, which makes it impossible to directly apply the inverse filter. This case can be addressed by implementing an iterative inversion method, which reconstructs both the PSF and the best image solution. Iterative methods are very precise and noise robust, but these come at the cost of a high computational complexity.

3. **Seismic imaging** In seismic inversion, the surface of the Earth is probed with a seismic vibrator or a dynamite explosion in order to quantify the geophysical properties of the underground layers of rock and fluid. One of the most widespread applications is determining the properties of underground petroleum reservoirs.

Seismic inversion methods are based on a *forward model*, which can consist of classical wave equations predicting particle displacement or fluid pressure variation during seismic propagation. The forward model is thus described by an equation of the form

$$s(t) = (w * r)(t)$$

where $s(t)$ is the synthetic seismic data, $r(t)$ is the reflectivity function to be estimated, and $w(t)$ is the source wavelet (also called the source signature), which is the shape of the pressure pulse created by the source. One should not confuse the meaning of the wavelet (referring to the waveform) in seismic inversion to the wavelet from signal processing used in the presented image deconvolution example in which *wavelet* refers to the wavelet transform.

An iterative method is proposed to derive the reflectivity function in a robust and precise way (Cooke and Cant, 2010). Initially, the reflectivity function is unknown, and an estimate is provided by the user, which is expected to lead to poor results. Knowing the waveform of the source wavelet, the model generates an initial estimate of the synthetic seismic data $s(t)$. By implementing partial derivatives of the forward model to each of the model parameters, the proposed algorithm is able to compute a new iteration of the reflectivity function $r(t)$. The process continues until the root mean squared error between the observed and synthetic seismic data is smaller than a predefined tolerance:

$$\text{err} = \frac{1}{N}\sqrt{\sum_{n=0}^{N-1} |s(nT) - s_{\text{true}}(nT)|^2} < \text{tol}$$

where T is the sampling time, and N is the data size.

3.4 Data-Driven Inversion Techniques

Take a scene property f and a set of camera measurements g. These two quantities are related by

$$g = \mathcal{H}[f] + \epsilon$$

where $\mathcal{H}[\cdot]$ is a nonlinear operator, and ϵ is additive noise. Typical imaging problems require us to solve the inverse problem, where we are given y and need to solve for x via the inverse model $\mathcal{H}^{-1}[\cdot]$. In this section, we study data-driven methods that help us find an inverse mapping from $y \to x$. These data-driven models search for patterns and structure in large amounts of data. The patterns learned from this data are then generalized to data that the model has never seen. In this section, we study how neural networks specifically are used to model this function mapping.

3.4.1 Machine Learning

In 1959, Arthur Samuel defined *machine learning (ML)* as the "field of study that gives computers the ability to learn without being explicitly programmed." He wrote the first self-learning program, which played checkers by learning from experience (i.e., data). In just sixty years, ML has evolved and become an integral part of our lives. ML has been

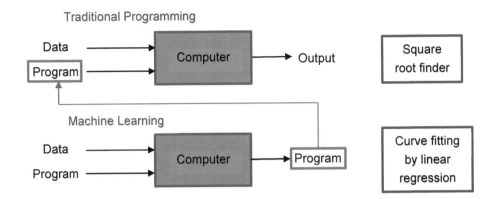

Figure 3.15 Traditional programming vs. machine learning.

deployed in cancer diagnosis, drug discovery, face recognition, recommendation systems (e.g., Netflix), linguistics, and more.

In traditional programming, we explicitly provide the computer with data and an algorithm that directly yields an output. In machine learning, we allow the computer to learn patterns and structure directly from the data itself. There are two major types of learning: *supervised* and *unsupervised learning*. A supervised learning algorithm is fed both input data (X) and output data (Y), that is, labels. From this, it is able to directly learn a function mapping $X \rightarrow Y$. Unsupervised learning algorithms, on the other hand, are given only input data. From this input data, unsupervised learning algorithms identify certain patterns and groupings within the data. These groupings are then used to classify new unseen data. In machine learning, there are three types of data: *training* data, *testing* data, and *validation* data. Training data is used to train the ML model and learn a function mapping. Testing data is used to evaluate the performance of the model, and validation data is used for hyperparameter tuning.

There are six major classes of machine learning: (1) clustering, (2) classification, (3) regression, (4) deep learning, (5) dimensionality reduction, and (6) reinforcement learning. Our discussions are confined to classes 1 through 4, and the interested reader is directed to (Ghodsi, 2006) and (Sutton and Barto, 2018) for details on classes 5 and 6, respectively.

- **Clustering** Clustering is an unsupervised learning algorithm. Each data point can be represented as a point in some representation (or feature) space. The clustering algorithm then groups data points on the basis of their proximity in this representation space. A classic example is k-means clustering (Likas et al., 2003).
- **Classification** Classification aims to label some input data as one of n classes. For example, a computer vision algorithm may classify an input image as either a cat or a dog. Classification algorithms typically work by searching for boundaries in feature

3.4 Data-Driven Inversion Techniques

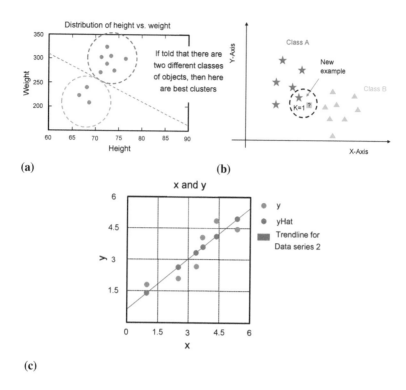

Figure 3.16 Examples of (a) clustering, (b) classification (KNN), and (c) linear regression. In (a) data is clustered into two groups; in (b) a new data point can be classified in either into class A or class B; and in (c) dots correspond to data points and the line corresponds to the linear fit.

space that separate different classes, as depicted in Figure 3.16b. One trivial example of a classification algorithm is k-nearest neighbors (Sutton, 2012).

- **Regression** Regression is a fundamental supervised learning algorithm. Given some unseen data, regression aims to predict some output variable given the value(s) of the input variable(s). A key distinction between classification and regression is that regression can take continuous values, whereas classification allows for only discrete outputs. The most basic example of this is first-order linear regression, in which we aim to relate the input and output as a linear function

$$\mathbf{y} = \mathbf{B_0} + \mathbf{B_1}\mathbf{x}$$

where $\mathbf{B_0}$ is the bias, and $\mathbf{B_1}$ is the slope. Regression would solve for the values of $\mathbf{B_0}$ and $\mathbf{B_1}$.

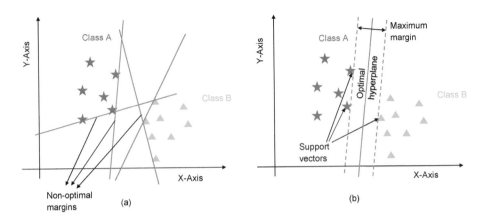

Figure 3.17 Support vector machine model classification: (a) examples of nonoptimal hyperplane margins, and (b) optimal hyperplane separation.

- **Support vector machine (SVM)** SVMs are a popular example of unsupervised learning. They can be used to solve both classification and regression problems. Given labeled training data, SVMs find decision boundaries between different classes. For example, suppose that we want to classify fruits as either apples and oranges. Each data point will be represented in an n-dimensional feature space, where features could be shape and color, for example.

The various types of margins that can be drawn to separate the two classes are shown in Figure 3.17a; these margins may not always classify a new data point correctly. In Figure 3.17b the support vectors and optimal hyperplane are displayed. *Support vectors* are data points that are closer to the hyperplane and influence the position and orientation of the hyperplane. Using these support vectors, we maximize the margin of the classifier. Deleting the support vectors will change the position of the hyperplane.

3.4.2 Neural Networks and Deep Learning

Neural networks are a machine learning models inspired by the pattern of neurons firing in the human brain. Neural networks attempt to process information and data similarly to our brains. As a simplified model, a neural network contains several *layers* of processing. Each layer contains several *nodes*, which are information processing units that simulate the neurons. Nodes are densely interconnected between adjacent layers, as shown in Figure 3.21. These artificial neural networks contain between dozens and millions of artificial neurons. This subset of machine learning is known as *deep learning*, in which deep layers of neurons are used to compute complex nonlinear inverse function mappings.

Model of a neuron A neuron is an information processing unit that is fundamental to the operation of a neural network. Figure 3.18 shows the model of a neuron that forms the

3.4 Data-Driven Inversion Techniques

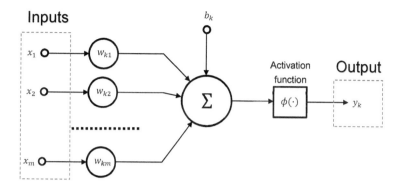

Figure 3.18 A nonlinear model of a neuron.

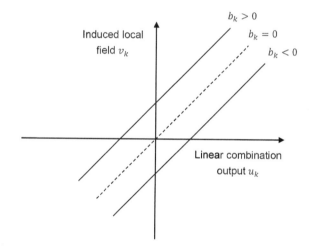

Figure 3.19 Affine transformation produced by the presence of bias.

basis for designing neural networks. Following are the three basic elements of the neuronal model:

1. The connecting links (synapses) characterize the weights. Specifically, a signal x_j at the input of synapse of j connected to a neuron k is multiplied by the synaptic weight $w_{k,j}$.
2. An adder for summing the input signals, weighted by their respective synapses of neurons.
3. An activation function (represented by σ) for limiting the amplitude of the output of a neuron.

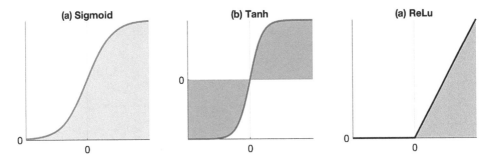

Figure 3.20 Various activation functions: (a) sigmoid activation function, (b) tanh activation function, and (c) ReLU activation function.

The model also includes an external bias b_k. The bias is used to shift the boundary line, that is, it has the effect of increasing or lowering the net input of the activation function, depending on whether it is positive or negative, as shown in Figure 3.19.

The output of the neuron is calculated

$$v_k = x_0 b_k + x_1 w_{k1} + x_2 w_{k2} + \cdots + x_n w_{kn}$$

where v_k is the local field, and $x_1 w_{k1}, \cdots, x_n w_{kn}$ constitute the weighted sum of input

$$v_k = x_0 b_k + \sum_{j=1}^{n} x_j w_{kj}.$$

To incorporate the bias value, the zeroth neuron is held constant ($x_0 = 1$), and the weight of the zeroth neuron is the bias ($w_{k0} = b_k$)

$$v_k = \sum_{j=0}^{n} x_j w_{kj}.$$

v_k is then input to a nonlinear activation function. The final output is defined

$$y_k = \sigma(v_k).$$

Types of activation functions The activation function decides whether a neuron should be activated or not by calculating the weighted sum and further adding bias with it. The purpose of the activation function is to introduce nonlinearity into the output of a neuron.

The most commonly used activation functions are the following:

1. **Sigmoid** The sigmoid function is defined

$$\sigma(x) = \frac{1}{1 + e^{-x}}.$$

It condenses the numbers into the range [0, 1]. Figure 3.20a depicts the sigmoid function. As shown, the sigmoid function is not zerocentric, and the saturated neurons kill the gradient.

2. **Tanh** The tanh function is defined

$$\tanh(x) = 2\sigma(x) - 1.$$

This function, unlike the sigmoid function, is zerocentric. It condenses the numbers into the range $[-1, 1]$. It is represented in Figure 3.20b.

3. **ReLU** The ReLU function is the most used activation function because it is computationally very efficient and it converges faster than tanh and sigmoid. The ReLU function is defined

$$\sigma(x) = \max(0, x).$$

It is represented in Figure 3.20c.

Loss function A loss function is the evaluation metric used to train a neural network. It is a scalar value indicating the quality of the network's output. In supervised learning, this constitutes an output consistent with the ground truth output. The choice of loss function is directly related to the activation function used in the output layer of the neural network. Following are a few loss functions:

1. **Mean squared error loss** This loss is calculated as the average of the squared differences between the predicted and actual values. The result is always positive regardless of the sign of the predicted and actual values, and a perfect value is 0.0.
2. **Mean absolute error loss** Absolute error for each training example is the distance between the predicted and the actual values, irrespective of the sign. Absolute error is also known as the ℓ_1 loss.
3. **Cross-entropy or log loss** Each predicted probability is compared to the actual class output value (0 or 1), and a score is calculated that penalizes the probability on the basis of the distance from the expected value. The penalty is logarithmic, offering a small score for small differences (0.1 or 0.2) and enormous score for a large difference (0.9 or 1.0).

Regularization The goal of machine learning is to have a model learn to generalize unseen data. When dealing with small data sets, however, these models may *overfit* to the data set. To prevent this overfitting, we can impose some restriction on the structure of the network. Specifically, regularization is a modification made to the learning algorithm to reduce its generalization error, even if it increases training error. One straightforward way to regularize is to constantly evaluate the training and validation loss on each training iteration and then return the model with the lowest validation error.

A simple-to-implement type of regularization is to modify the cost function with a parameter norm penalty. This penalty is usually denoted $\Omega(\theta)$. A common type of

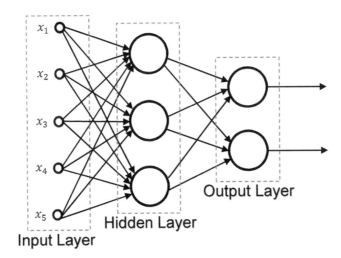

Figure 3.21 Fully connected feedforward network with one hidden layer and one output layer.

parameter norm regularization is to penalize the size of weights. We describe the following two parameter regularizations:

1. ℓ_2 **regularization** If **w** are the model parameters to be regularized, then ℓ_2 regularization penalty can be defined

$$\Omega(\theta) = \|\mathbf{w}\|_2^2 = \frac{1}{2}\mathbf{w}^T\mathbf{w}.$$

   ℓ_2 regularization causes the weights **w** to have a small norm. Large weights are often undesirable in that case small changes in the input would cause large changes in output, resulting in numerical instability.

2. ℓ_1 **regularization** ℓ_1 regularization defines the parameter norm penalty

$$\Omega(\theta) = \|\mathbf{w}\|_1 = \sum_i |w_i|.$$

   Intuitively, this form of regularization encourages the weights to be sparse, that is, only a few weights will be nonzero.

3. **Dropout** is a regularization method in which the output of certain nodes are randomly ignored, that is, set to zero. This helps the network from overtraining certain nodes relative to others. Dropout could also be considered a way to train different network architectures simultaneously, because the number of nodes and layers will change each iteration.

3.4 Data-Driven Inversion Techniques

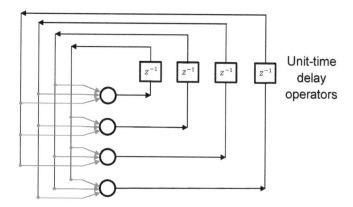

Figure 3.22 Recurrent network with no hidden neurons and no self-feedback loops.

Types of neural networks Figure 3.18 provides a functional description of the various elements that constitute the model of an artificial neuron. We may simplify the appearance of the model by using the idea of signal flow graphs. Here we discuss the various architectures used in neural networks.

1. **Multilayer feedforward network** In a single-layer network the input layer projects the input on the output layer of neurons. The output layer of neurons is not counted because no computation is performed. The difference between a single-layer and multilayer network is the existence of hidden layers. The added depth of hidden layers enables deeper processing of the data. One interpretation of these hidden layers is that they act as feature extractors before yielding an output. Figure 3.21 illustrates a multilayer fully connected feedforward network with one hidden layer and one output layer.

2. **Recurrent networks** A recurrent neural network (RNN) distinguishes itself from a feedforward network by having at least one feedback loop. The presence of a feedback loop has a profound impact on the learning capability of the network and its performance. In a traditional neural network we assume that all inputs (and outputs) are independent of each other, but sometimes this is an ill-posed assumption. For example, consecutive frames in a video are highly correlated due to their temporal relationship. RNNs are called recurrent because they perform the same task for every element of a sequence, with the output dependent on the previous computations. Another way to think about RNNs is that they have a memory that captures information about what has been calculated so far. Figure 3.22 shows a recurrent network with no hidden layers.

Perceptron The concept of a perceptron was proposed by Frank Rosenblatt in 1943 and was refined by Minsky and Papert in 1969. A perceptron is simply a binary linear classifier. The goal of a perceptron is to correctly classify the set of inputs into two classes. The perceptron algorithm automatically learns the optimal weight coefficients for the input

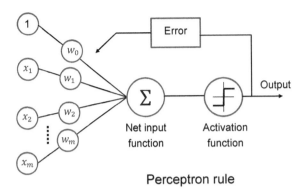

Figure 3.23 A simple model of a perceptron.

signals. The input features are then multiplied with these weights to determine if a neuron fires or not. The perceptron receives multiple input signals, and if the sum of the input signals exceeds a certain threshold, it either outputs a signal or does not return an output. In the context of supervised learning and classification, this can then be used to predict the class of a sample. Figure 3.23 illustrates the simple architecture of the perceptron.

The perceptron function is defined

$$f(\mathbf{x}) = \sum_i x_i w_i$$

where $f(\mathbf{x})$ is the output, x_i is the input, and w_i is the weight for the particular instance of node i.

Backpropagation algorithm Backpropagation is the most fundamental building block of neural network training. Backpropogation operates on the chain rule principle. The basic idea is to efficiently calculate the partial derivative of an approximating function $F(\mathbf{w}, \mathbf{x})$ realized by the network with respect to all the elements of the adjustable weight vector $\mathbf{w}$ for a given value of input vector $\mathbf{x}$. In simple terms, after each forward pass through a network, backpropagation performs a backward pass while adjusting the model's parameters (weights and biases). This is best explained by example. Let us say $x = -2$, $y = 5$, and $z = -4$ for the following equation

$$f(x, y, z) = (x + y) \cdot z.$$

3.4 Data-Driven Inversion Techniques

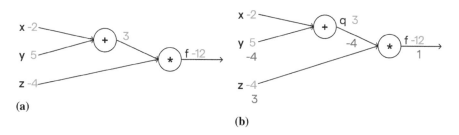

Figure 3.24 A simple example to demonstrate the backpropagation algorithm: (a) the function represented using signal flow graph rules, and (b) the signal flow graph with backpropagation values updated. (The values in green are the input values, and the values in red are the values calculated using the backpropagation algorithm).

Figure 3.24a shows the computational graph for this, which breaks the equation into intermediate variables.

$$q = x + y, \qquad \frac{\partial q}{\partial x} = 1, \frac{\partial q}{\partial y} = 1,$$

$$f = qz, \qquad \frac{\partial f}{\partial q} = z, \frac{\partial f}{\partial z} = q.$$

We want to calculate the below partial differentials

$$\frac{\partial f}{\partial x}, \frac{\partial f}{\partial y}, \frac{\partial f}{\partial z}.$$

We can make the following deductions from Figure 3.24a

$$\frac{\partial f}{\partial x} = 1, \frac{\partial f}{\partial y} = 3, \frac{\partial f}{\partial z} = -4.$$

Using the chain rule to find the partial derivative of f with respect to y

$$\frac{\partial f}{\partial y} = \frac{\partial f}{\partial q}\frac{\partial q}{\partial y}, \qquad \frac{\partial f}{\partial y} = -4.$$

Using the chain rule again to calculate the partial derivative of f with respect to x

$$\frac{\partial f}{\partial x} = \frac{\partial f}{\partial q}\frac{\partial q}{\partial x},$$

$$\frac{\partial f}{\partial x} = -4.$$

Figure 3.24b represents the output after the first round of backpropagation. Using these calculated derivatives (gradients, in practice), we can update the weights to reduce the value of the loss function. These weights are updated repeatedly over multiple iterations of training until a stopping criterion is met (usually number of iterations or performance threshold).

3.4.3 Convolutional Neural Networks and Computer Vision

Computer vision is currently a growing field with applications in autonomous vehicles, industrial automation, digital pathology, and more. Classic tasks in computer vision include object detection, segmentation, and tracking (Comaniciu et al., 2001). Computer vision techniques have also been applied to tasks such as monocular depth estimation (Laina et al., 2016; Liu et al., 2015; Eigen and Fergus, 2015), phase recovery (Rivenson et al., 2018), and even optical neural networks (Lin et al., 2018; Chang et al., 2018). State-of-the-art methods in these tasks typically incorporate a convolutional neural network (CNN). CNNs are particularly important in computer vision because they leverage 2D spatial features of images. These networks typically contain convolution layers, which convolve images with 2D or 3D convolution filters. CNNs have also been used in Natural language processing (NLP) and speech recognition. We now go through a few building blocks of a CNN.

Types of layers The various types of layers used to build CNNs include the following:

1. CONV layer (convolution) performs a 2D convolution of the input image with a filter. All entries in the filter are learnable parameters. The filter height and width are small relative to size of the image but extend through the full depth of the input volume. The convolution operator accounts for interdependencies between adjacent pixels.
2. POOL layer performs a downsampling operation along the spatial dimensions (width and height). It is common to periodically insert a pooling layer between successive CONV layers in a ConvNet architecture. Its function is to progressively reduce the spatial dimensions of the representation in order to reduce the number of parameters and computation in the network and hence to control overfitting. The pooling operator also introduces spatial invariance into the network.
3. FC (fully-connected) layer is typically the last layer of a network performing classification. It has n nodes, where n is the number of classes. This serves as the output layer, and each node contains a score for each class. The class with the highest score is chosen by the network to be the classified object.

CNN architecture There are many popular CNN architectures, many of which have gained recognition by achieving good results. A few of them include the following:

1. **LeNet-5** This 7-layer CNN classifies digits and digitizes 32×32 pixel grayscale input images (see Figure 3.25). It was used by several banks to recognize the hand-written numbers on checks. LeNet-5 architecture (Lecun et al., 1998) consists of two sets of convolutional and average pooling layers, followed by a flattening convolutional layer, then two fully-connected layers, and finally a softmax classifier.

3.4 Data-Driven Inversion Techniques

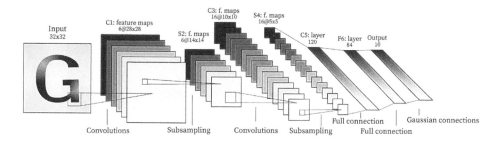

Figure 3.25 LeNet architecture.

In the first layer, the input for LeNet-5 is a 32×32 pixel grayscale image that passes through the first convolutional layer with six feature maps or filters of size 5×5 and a stride of 1. The image dimensions change from $32 \times 32 \times 1$ to $28 \times 28 \times 6$. Then the LeNet-5 applies an average pooling layer or subsampling layer with a filter size 2×2 and a stride of 2. The resulting image dimensions will be reduced to $14 \times 14 \times 6$. Next, there is a second convolutional layer with sixteen feature maps of size 5×5 and a stride of 1. In this layer, only ten out of sixteen feature maps are connected to six feature maps of the previous layer. The main reason is to break the symmetry in the network and keep the number of connections within reasonable bounds. That is why the number of training parameters in these layers is $1,516$ instead of $2,400$ and similarly the number of connections is $151,600$ instead of $240,000$. The fourth layer (S4) is again an average pooling layer with filter size 2×2 and a stride of 2. This layer is the same as the second layer (S2) except it has sixteen feature maps so the output will be reduced to $5 \times 5 \times 16$. The fifth layer (C5) is a fully connected convolutional layer with 120 feature maps, each of size 1×1. Each of the 120 units in C5 is connected to all 400 nodes ($5 \times 5 \times 16$) in the fourth layer S4. The sixth layer is a fully connected layer (F6) with eighty-four units. Finally, there is a fully connected softmax output layer $\hat{\mathbf{y}}$ with ten possible values corresponding to the digits 0 to 9.

2. **VGG16** As the name suggests, VGG16 has sixteen layers (see Figure 3.26). This architecture is from the Visual Geometry Group at Oxford (Simonyan and Zisserman, 2014). The VGG-16 network is characterized by 3×3 convolutional layers stacked on top of each other in increasing depth. Reducing the volume is handled by max pooling. Two fully connected layers, each with $4,096$ nodes, are followed by another fully connected layer of $1,000$ nodes. Then this is followed by a softmax classifier. In VGG-16 the blocks are of the same filter size and are applied multiple times to extract more complex and representative features. This concept of blocks became common in the networks developed after VGG.

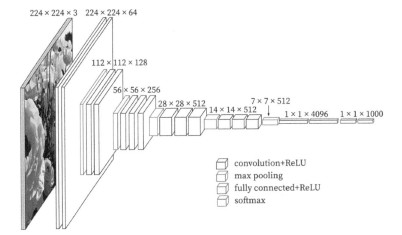

Figure 3.26 VGG16 architecture.

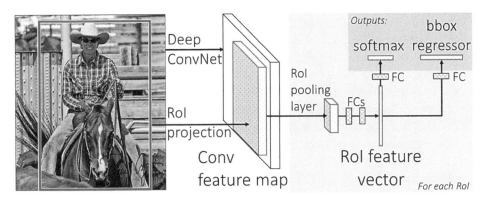

Figure 3.27 Fast R-CNN architecture (Girshick, 2015).

3. **Fast region-based convolutional network** A fast R-CNN network (Girshick, 2015) takes as input an entire image and a set of object proposals (see Figure 3.27). The network first processes the whole image with several convolutional and max pooling layers to produce a conv feature map. Then for each object proposal a region of interest (RoI) pooling layer extracts a fixed-length feature vector from the feature map. Each feature vector is fed into a sequence of fully connected layers that finally branch into two sibling output layers: one that produces softmax probability estimates over K object classes plus a catch-all background class and another layer that outputs four real-valued numbers for each of the K object classes (per-class bounding-box regression offsets). Training all network weights with backpropagation is an important capability of fast

3.4 Data-Driven Inversion Techniques

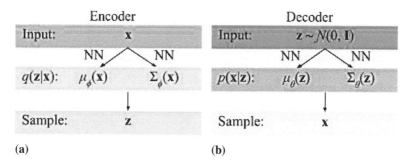

Figure 3.28 Architecture of variational autoencoder: (a) the encoder, and (b) the decoder.

R-CNN. In fast R-CNN training, stochastic gradient descent minibatches are sampled hierarchically, first by sampling N images and then by sampling R/N RoIs from each image. Critically, RoIs from the same image share computation and memory in the forward and backward passes. Making N small decreases minibatch computation.

4. **Variational autoencoders (VAE)** Many techniques in machine learning try to compress the dimensionality of the data into a smaller space. Autoencoders work on that principle (see Figure 3.28). The high-dimensional input is passed through a neural network to obtain a compressed output. It achieves this with two principal components. The first component, also known as an encoder, consists of a number of layers that can be fully connected or convolutional layers that take the input and try to compress it to a smaller representation. The smaller representation is known as a bottleneck. The second component consists of reconstructing the input from the bottleneck. The last function of training the autoencoder is to look at the reconstructed version at the end of the decoder and compute the reconstruction loss with respect to the input. This method can be used for denoising images, and neural inpainting (removing a small part of the image and asking the neural network to reconstruct the complete input). In autoencoders the input is mapped to a fixed vector, but in variational autoencoders the input is mapped to a distribution (Kingma and Welling, 2019). Hence the normal bottleneck vector is replaced by two separate vectors, one representing the mean of the distribution and the other representing the standard deviation of the distribution. (Hammernik et al., 2018) have used a VAE to learn the reconstruction of MRI data.

The loss function for this architecture consists of two terms:

$$\text{Loss} = \text{Reconstructionloss} + \text{KLdivergence}$$
$$\log p_\theta(x) = \mathbb{E}_z \log p(x \mid z) - D_{\text{KL}}(q(z \mid x) \| p(x)).$$

We note that there is a sampling operation between the encoder and the decoder. The node has to take a sample from a distribution and feed it through the decoder. The problem in VAE is that either we cannot run backpropagation or we cannot push

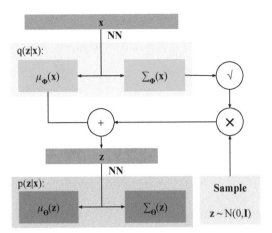

Figure 3.29 The reparameterization trick.

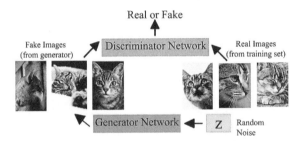

Figure 3.30 The generative adversarial network.

gradients through the sampling node. To run the gradients through the entire network, we use a reparameterization trick, shown in Figure 3.29. If we look at the latent vector that we are sampling, the vector can be a sum of fixed μ (learning parameter) and γ (learning parameter) multiplied by ε where $\varepsilon \sim \mathcal{N}(0, 1)$

$$z = \mu + \gamma \odot \varepsilon.$$

With the reparameterization trick, we can backpropagate to calculate the gradient with respect to all parameters.

5. **Generative adversarial network (GAN)** A GAN does not require any probabilistic learning, unlike the other methods. A GAN has two networks, a generator network and a discriminator, that compete with each other (see Figure 3.30). Conceptually, the GAN (Goodfellow et al., 2014) can be thought of as a game between two players, the generator and discriminator. The generator is a generative model and performs mapping

3.5 Hybrid Inversion Techniques (Data Driven and Model Based)

$\widehat{\mathbf{x}} = \mathfrak{G}(\mathbf{z})$, where $\mathbf{z}$ is some random noise. Its goal is to produce samples, $\widehat{\mathbf{x}}$ from the distribution of the training data $p(\mathbf{x})$. The discriminator is the generator's opponent and performs a mapping $\mathfrak{D}(\mathbf{x}) \in (0, 1)$. Its goal is to look at samples and determine whether they are real samples or synthetic samples from the generator. The generator is trained to fool the discriminator, and thus the two can be viewed as adversaries. Let us define the discriminator's loss function. We let $p_{\text{data}}(\mathbf{x})$ denote the data distribution and $p_{\text{model}}(\mathbf{x})$ denote the distribution of samples from the generator. Then the discriminator loss is defined

$$\mathfrak{L}^{(\mathfrak{D})} = -\frac{1}{2}\mathbb{E}_{\mathbf{x} \sim p_{\text{data}}} \log \mathfrak{D}(\mathbf{x}) - \frac{1}{2}\mathbb{E}_{\mathbf{z}} \log(1 - \mathfrak{D}(\mathfrak{G}(\mathbf{z}))).$$

The goal of the discriminator is to minimize the loss. The loss will be zero $\mathfrak{D}(\mathbf{x}) = 1$ for all $\mathbf{x} \sim p_{\text{data}}$ and $\mathfrak{D}(\widehat{\mathbf{x}}) = 0$ for all $\widehat{\mathbf{x}} \sim p_{\text{model}}$. A remarkable example of a GAN can be seen at thispersondoesnotexist.com, where a GAN generates a highly realistic image of a face of a person that does not actually exist.

Transfer learning and fine tuning Researchers in this field often make pretrained networks available open source to the computer vision community. The availability of these pretrained networks is particularly useful for other researchers, as training millions of parameters can be time consuming and ill posed with insufficient data and limited computing power. In practice, one can instead use the pretrained network weights as a starting point for their application. Following are the two major transfer learning scenarios:

- **Feature extraction** This is done by taking a pretrained network, removing the last fully connected layer, and treating the remainder of the network as a feature extractor for the new data set. A linear classifier (e.g., linear SVM or softmax classifier) is then trained, with the extracted features as input.
- **Fine tuning** The second strategy is to fine tune the pretrained network by using the network as a starting point for the training. The network can be fine tuned by continuing training on the new data set. It is possible to either fine tune all the layers of the network or keep some layers constant and fine tune the others.

3.5 Hybrid Inversion Techniques (Data Driven and Model Based)

At this point, let us revisit the benefits and limitations of inversion using data-driven approaches versus physical model-based approaches. Model-based inversion is highly interpretable and constrains the solution to be physically plausible. However, these models are limited by human bias (introduced by a partial understanding of the physical phenomena) and are not robust to noise. On the other hand, data-driven models are able to learn arbitrary function mappings $X \to Y$. However, these models lead to uninterpretable results with an unbounded solution space encompassing physically improbable solutions. They also require large amounts of data to meaningfully extract patterns and structure from the physical phenomena. By systematically integrating these two approaches, we are able to overcome deficiencies in both methods. (Willard et al., 2020) have provided an excellent survey of techniques to meaningfully integrate physics with data-driven approaches, which we summarize here.

3.5.1 Physics-Based Regularization

As we discussed previously, the goal of regularization in iterative models is to discourage solutions that do not satisfy some physical or mathematical property. In the context of data-driven models, regularization entails two terms in the loss function: (1) a data fidelity term and (2) a physics-based regularizer. A physical regularizer here aims to mitigate reconstruction errors by incorporating even just a partial understanding of the underlying physical phenomena

$$\widehat{f} = \arg \min_f \|g - \mathcal{H}[f]\|^2 + \alpha \Phi(f)$$

where $\mathcal{H}$ is the forward operator, f is the unknown property, Φ is some prior physical knowledge acting as a regularizer, and α is a hyperparameter balancing the trade-off between data fidelity and physical plausibility. For example, (Goy et al., 2018) solved the phase retrieval problem in low-light conditions (where shot noise is prevalent) by incorporating a regularization term on the possible values for phase, on the condition that the object modulates only the phase, and not the magnitude. Figure 3.31 shows the performance improvement of a physics-based neural network over standard deep learning methods (end-to-end), iterative methods (Gerchberg-Saxton), and inverse model projections (approximant).

3.5.2 Physics-Guided Network Initialization

The weights of a neural network are typically randomly initialized. Because neural networks learn a nonlinear function mapping, the initialization of the weights often plays a critical role in helping the network converge to an optimal loss minima rather than a local minima. One way to improve network convergence is to train the network to initialize based on the results from a physics-based model. This operates on the basis of a *transfer learning* principle. To do this, first synthetic data is developed using a physics-based forward model. A neural network can then learn from this physics-based simulation data. After training, this is used as a pretrained model. The network trained on simulation data is then used as an initialization point for the network to be trained on real data. Such a technique is also useful when the amount of experimental data available is limited.

3.5.3 Physics-Based Network Architectures

Neural networks often learn weights and biases in an uninterpretable manner, making it difficult to know whether the network is actually learning the same physics that humans

3.5 Hybrid Inversion Techniques (Data Driven and Model Based)

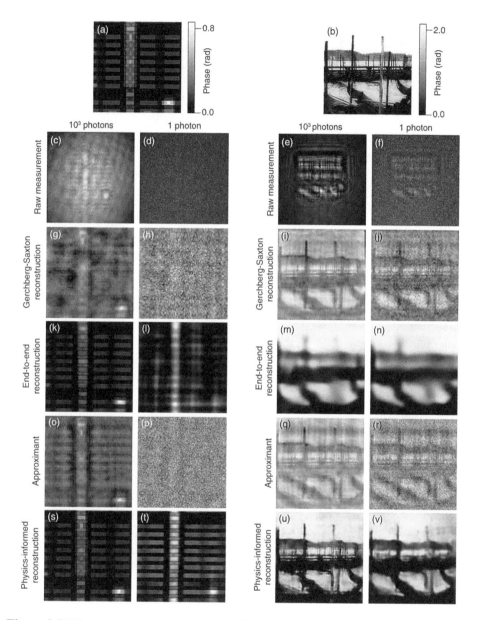

Figure 3.31 Phase retrieval under dominant effects of shot noise is shown to be more effective under a physics-based network, compared to standard deep learning, iterative, or model-based approaches (Goy et al., 2018).

understand. One approach to circumvent this is by directly manipulating layer connections on the basis of the variable dependencies between different parameters. For example, the use of a CNN inherently allows the assumption that objects in an image are scale, translation, and rotation invariant. Meanwhile, the use of a RNN encodes a time-invariant structure. For example, (Sturmfels et al., 2018) inserted a layer at the beginning of a CNN to spatially discretize different regions of the brain. This enabled the network to learn different parameters for different parts of the brain. This is particularly important for predicting age from neuroimages, because different regions of the brain behave differently at different stages of life.

3.5.4 Hybrid Models

An easy way to combine information from physics and data is to feed the output of a physics-based model as input to a deep learning model. One could think of this as feeding additional features of the data to the network. In chapter 7, we look at how polarization cues can be leveraged to obtain the shape of an object. In the context of a hybrid model, we can see how the shape estimates from a polarization-based model can be fed as input to a CNN, where ambiguities from the physics model can be corrected. In such a situation, the model-based shape estimate and the polarization images can be fed as input to the network. Another such hybrid model is a *residual model*, in which the network learns the errors, or residuals, between the physics model and the observed data. Such a model is able to learn from the deficiencies of the physical model and make corrections to it accordingly. The disadvantage of such a model lies in its inability to enforce any physical constraint as a physics-based architecture or loss function would.

3.5.5 Optical Neural Networks

There has been some interesting work on optical implementations of deep learning modalities. The layers of the neural network are physically constructed using diffractive materials, in contrast to the digital layers used in traditional neural networks. Each point in the layer can be interpreted as a neuron. This follows from Huygen-Fresnel's principle, in which every point in a diffractive material can be treated as a point source. Neurons in subsequent layers interact with each other, and modulate the phase and amplitude of the light based on their complex transmission/reflectance coefficients. The phase and amplitude of each neuron within a layer can be trained digitally via backpropagation. Once these parameters have been determined for a given task (e.g., MNIST digit classification), the individual layers can be fabricated. This type of network is termed a deep diffractive neural network and is illustrated in Figure 3.32 (Lin et al., 2018). Such diffractive networks are still being explored in the imaging community, with research underway into understanding the information capacity of such networks (Kulce et al., 2021) as well as improving their performance via methods such as diffractive ensemble learning (Rahman et al., 2021).

3.5 Hybrid Inversion Techniques (Data Driven and Model Based)

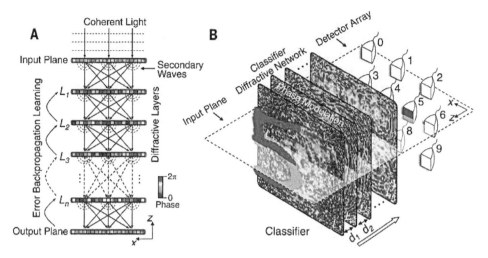

Figure 3.32 Deep diffractive neural network. (a) Each point on the diffractive layers behaves as a point source, in accordance with Huygen-Fresnel's principle. (b) Light emanating from the input plane (the digit '5') is propagated through a classifier diffractive network. The classifier network focuses light into one of ten detectors, each corresponding to a digit from 0–9 (Lin et al., 2018).

Optical neural networks can also be used to reduce the burden on digital computation by performing some computations optically. For example, one can insert a phase mask in front of the camera to act as an optical correlator that performs template matching. This layer acts as an optical preprocessor, making it easier for a digital CNN to achieve high classification accuracy with reduced training time (Chang et al., 2018).

Chapter Appendix: Notation

Symbol	Description
$f(t)$	Continuous-time function
$f[m]$	Discrete-time function
$\mathcal{L}$	Linear system
$f_\tau(t)$	Signal $f(t)$ delayed by τ
$\delta(t)$	Dirac delta distribution/impulse
ω	Angular frequency
$f_\omega(t)$	Eigenfunction of linear time-invariant (LTI) system
λ_ω	Eigenvalue of an LTI system

$\widehat{f}(\omega)$	Fourier transform of function $f(t)$
$g^*(t)$	Complex conjugate of function $g(t)$
T	Sampling interval
T_p	Period of a function in time
$\text{III}_{T_p}(t)$	Dirac comb with intertooth distance T_p
B_Ω	Set of bandlimited functions with maximum frequency Ω
Ω_0	Period of a function in frequency
$\|\mathbf{x}\|_2$	ℓ_2-norm of a vector
$\|f(t)\|_2$	L_2-norm of a function
$\langle \mathbf{x}, \mathbf{y} \rangle$	Inner product of two vectors $\mathbf{x}$ and $\mathbf{y}$
$\mathbb{1}_{[a,b]}$	Vector of all ones between $[a, b]$
$p(s, \phi)$	Radon projection
$d(x, y)$	Object distribution
$\mathbf{I}$	Image
$\mathbf{I}_f$	Filtered image
$\mathbf{W}$	Wavelet basis
$\|\mathbf{x}\|_1$	ℓ_1-norm of a vector $\mathbf{x}$
$D_{\text{KL}}(q\|p)$	Kullback–Leibler divergence between probability densities p and q
$\mathbb{E}_z$	Expectation with respect to random variable z
$\mathfrak{G}$	Generator network
$\mathfrak{D}$	Discriminator network

Exercises

1. Computing Fourier transforms
   Many of the imaging problems require computation of the Fourier domain representation of a function. The goal of this warm-up exercise is to compute basic Fourier transforms that are used subsequently in the book.

   a) Sparse signals
   Compute the Fourier transform of the continuous-time sparse signal
   $$s(t) = \sum_{k=0}^{K-1} \Gamma_k \delta(t - t_k).$$
   As discussed in chapter 5, here Γ_k and t_k are attributed to reflectivity and time delays in the context of time-resolved imaging.

   b) Exponential functions
   Compute the Fourier transform of the transfer function
   $$s(t) = \rho e^{-\frac{t-\tau}{\lambda}} \mathbb{1}_{t \geqslant \tau}(t).$$
   In the context of fluorescence lifetime imaging, which is an established imaging technique in the life sciences, ρ and λ take the meanings of emission coefficient and lifetime, respectively, and τ refers to the delay.

   c) Tensor spline functions
   In the areas of computer graphics and signal/image processing, splines are frequently used for interpolation-related tasks. A B-spline of order zero is defined by a box function or the following:
   $$\beta^0(t) = \begin{cases} 1 & |t| < \frac{1}{2} \\ \frac{1}{2} & t = \frac{1}{2} \\ 0 & t > \frac{1}{2} \end{cases}.$$
   Higher order splines are polynomial functions defined by recursive convolution of the basic spline
   $$\beta^N(t) = \underbrace{\beta^0 * \beta^0 * \cdots \beta^0}_{N+1 \text{ convolutions}}(t).$$
   Show that in closed form one can directly write
   $$\beta^N(t) = \frac{1}{N!} \sum_{n=0}^{N+1} \binom{N+1}{n} (-1)^n \left(t - n + \frac{N+1}{2}\right)_+^n$$

where $(t)_+^n = t^n \mathbb{1}_{t \geqslant 0}(t)$.

Multidimensional (tensor) splines in dimension M can be written as a separable basis function of the form
$$\beta^N(\mathbf{t}) = \prod_{m=1}^{M} \beta^N(t_m).$$
Show that its Fourier transform is
$$\widehat{\beta}^N(\omega) = \prod_{m=1}^{M} \left(\frac{\sin\left(\frac{\omega_m}{2}\right)}{\left(\frac{\omega_m}{2}\right)}\right)^{N+1}.$$

2. Weighted least-squares inversion

   When working with tall and fat matrices, we have seen that inversion amounts to solving a least-squares optimization problem. In practice, it may be of interest to minimize a weighted version of the cost function. Here, the weights refer to a diagonal matrix of the form
   $$\mathbf{W} = \begin{bmatrix} w_1 & & & \\ & w_2 & & \\ & & \ddots & \\ & & & w_N \end{bmatrix}.$$

   a) Overdetermined system of equations

   In the case of overdetermined system, we have shown that the cost function is given by (3.18), that is, $C(\mathbf{f}) = \|\mathbf{g} - \mathbf{Hf}\|_2^2$. For the weighted least-squares problem, minimize the cost function,
   $$C_{\mathbf{W}}(\mathbf{f}) = \left\|\sqrt{\mathbf{W}}(\mathbf{g} - \mathbf{Hf})\right\|_2^2, \quad \text{where} \quad \sqrt{\mathbf{W}} = \begin{bmatrix} \sqrt{w_1} & & & \\ & \sqrt{w_2} & & \\ & & \ddots & \\ & & & \sqrt{w_N} \end{bmatrix}$$
   and show that the optimal solution is
   $$\mathbf{f}^\star = \left(\mathbf{H}^\top \mathbf{W} \mathbf{H}\right)^{-1} \mathbf{H}^\top \mathbf{W} \mathbf{g}.$$

   b) Underdetermined system of equations

   Extending the preceding example for the case of an underdetermined system, minimize the weighted energy corresponding to (3.20) and show that the optimal solution

Exercises

to the problem
$$\min_{\mathbf{f}} \left\| \sqrt{\mathbf{W}} \mathbf{f} \right\|_2^2 \text{ such that } \mathbf{g} = \mathbf{H}\mathbf{f}$$
is
$$\mathbf{f}^\star = \mathbf{W}^{-1} \mathbf{H}^\top \left(\mathbf{H} \mathbf{W}^{-1} \mathbf{H}^\top \right)^{-1} \mathbf{g}.$$

3. **Sparse recovery beyond soft thresholding**
   In the context of sparse recovery, the cost function in (3.27) can be generalized in that the ℓ_1-norm can be replaced by a pointwise nonlinearity defined by function ξ, leading to
   $$C_\lambda (\mathbf{f}) = \|\mathbf{g} - \mathbf{f}\|_2^2 + \lambda \xi (\mathbf{f})$$
   provided that ξ is differentiable. The core idea is to use a ξ that mimics the ℓ_1-norm.

   a) Explain why we need ξ to be differentiable.

   b) Suppose that we define
   $$\xi (\mathbf{f}) = \frac{1}{\alpha} \log (1 + \alpha |\mathbf{f}|), \quad \alpha > 0.$$
   For one-dimensional and two-dimensional $\mathbf{f}$ plot $\xi (\mathbf{f})$ and compare it with the ℓ_1-norm of $\mathbf{f}$.
   In analogy to the minimizer of (3.27) given by, $\mathbf{f}^\star = \text{soft}_\lambda(\mathbf{g})$, what is the minimizer of the cost function with $\xi (\mathbf{f})$ defined as previously?

4. **Smoothing and trend filtering**
   Data smoothing is one of the most commonplace tasks in signal processing and computational imaging. Smoothing-related tasks work with discrete derivatives of the data, and for this purpose let us define the first-order forward difference matrix
   $$\mathbf{D}_N^1 = \begin{bmatrix} -1 & +1 & & & \\ & -1 & +1 & & \\ & & \ddots & \ddots & \\ & & & -1 & +1 \end{bmatrix} \in \mathbb{R}^{(N-1) \times N}$$
   which when acting on a vector $\mathbf{f} \in \mathbb{R}^N$ produces a vector $(\mathbf{D}_N^1 \mathbf{f}) \in \mathbb{R}^{N-1}$ where $\left[\mathbf{D}_N^1 \mathbf{f}\right]_n = f[n+1] - f[n]$.

   a) Verify that the Kth-order finite difference is given by
   $$\mathbf{D}_N^K = \mathbf{D}_{N-K+1}^1 \mathbf{D}_N^{K-1} \in \mathbb{R}^{(N-K) \times N}, \quad K > 1.$$

b) Smoothing filter
   For any $\mathbf{f} \in \mathbb{R}^N$ by minimizing the cost function
   $$C_\lambda(\mathbf{f}) = \|\mathbf{g} - \mathbf{H}\mathbf{f}\|_2^2 + \lambda \|\mathbf{D}_N^2 \mathbf{f}\|_2^2$$
   obtain the solution for optimal $\mathbf{f}$. Generally, this problem is known as the Tikhonov regularization problem when the difference matrix is replaced by a generic matrix.

   Show that $\mathbf{D}_N^K \mathbf{f}$ can be written as a convolution filter. On the basis of this, devise a fast Fourier domain filtering algorithm.

   For a vector $\mathbf{f}$ arising from a smooth function, suppose that we replace $\mathbf{f}_{\text{noise}} = \mathbf{f} + \mathbf{z}$ where $\mathbf{z}$ is a vector drawn from a independent and identically distributed (i.i.d) Gaussian distribution with variance parameter σ. What is the effect of changing λ for a given σ?

c) Total variation minimization
   For piecewise constant data, one usually minimizes the constant function
   $$C_\lambda(\mathbf{f}) = \|\mathbf{g} - \mathbf{f}\|_2^2 + \lambda \|\mathbf{D}_N^1 \mathbf{f}\|_{\ell_1}.$$
   Here, the term
   $$\|\mathbf{D}_N^1 \mathbf{f}\|_{\ell_1} = \sum_{n=1}^{N-1} |f[n+1] - f[n]|$$
   is known as the total variation of $\mathbf{f}$. By resorting to one of the references on thresholding or majorization-minimization approaches, design an algorithm to minimize $C_\lambda(\mathbf{f})$.

II PLENOPTIC IMAGING

4 Spatially Coded Imaging

Consider the case in which objects in the scene we want to capture are placed at different distances from the camera, so that it is not possible to have them all in focus. Alternatively, various applications might require estimating the depths at various points of the scene. Solving these tasks with conventional imaging setups is often challenging.

In this chapter, our goal is to study the key ideas at the heart of spatially coded imaging (SCI). As a flexible alternative to the conventional imaging setup, SCI allows spatial imaging parameters such as the aperture, sensor, and illumination to be engineered to enhance the quality of the imaging system. For example, conventional imaging uses a spherical aperture of various sizes. Allowing the aperture to have more general shapes endows the imaging device with additional capabilities such as measuring the scene depth and extending the depth of field. The sensors can be placed in various configurations depending on their sensitivity to different parts of the light spectrum. For applications that impose restrictions on the amount of data transmitted, compressive sensing techniques can be used in conjunction with sensors of reduced size, down to single pixel sensors.

From the computational imaging perspective, *coding* refers to applying specific tailoring of the spatial degrees-of-freedom that lead to spatially encoded measurements. From the *encoded* measurements, the image is then recovered using mathematical algorithms. In the last few years, several compelling applications have emerged in which the co-design of engineered imaging parameters and recovery algorithms have led to new imaging capabilities, for example, depth imaging and light field capture from a single, spatially coded image.

This chapter covers three well-known spatial coding modalities: aperture, sensors, and illumination. For insight into the general functions and interactions of these modalities, see the diagram shown in Figure 4.1.

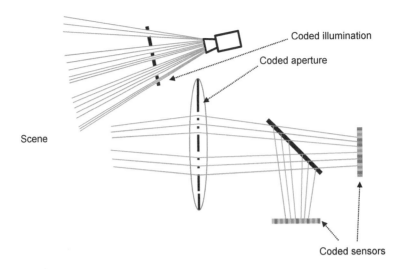

Figure 4.1 Three main categories of spatially coded imaging modalities. The illumination is usually coded by obstructing partially or completely the light from a projector in a predefined pattern. A traditional lens integrates all light from a point in the scene. A coded aperture selects light arriving from a number of angles. The sensors can be coded by arranging pixels sensitive to certain wavelengths. A beam splitter can be used to project the incoming light beam onto several sensors with modified parameters.

4.1 Coding the Aperture

In chapter 2, we studied the foundational principles of the image formation model of a basic imaging system. In this context, each imaging system has a point spread function (PSF) that characterizes the performance of the imaging method. The simplest form of the PSF is attributed to the pinhole camera. However, the pinhole imaging model suffers from low throughput of light and hence low signal-to-noise ratio as all but a pinhole-sized cavity allows for passage of light. From both physical and mathematical perspectives, we can think of the pinhole model to be a limiting case of the finite aperture model. In this analogy, coded aperture imaging can be seen as a trade-off between the two extremes, the pinhole model on one hand and the finite aperture model on the other. Before discussing the various formats of coded aperture imaging, we explain the intuition behind what makes the coded aperture imaging a flexible and desirable imaging alternative.

4.1.1 Physical Perspective

Let us begin by considering the physical instantiation of an aperture. An aperture is a hole that lets light pass through. A finite aperture is a hole of some finite radius, which collapses

4.1 Coding the Aperture

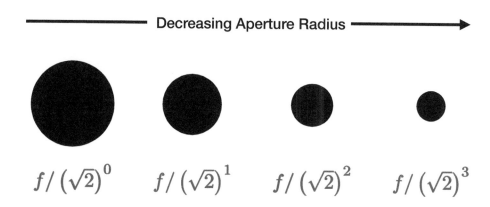

Figure 4.2 Four aperture sizes and their corresponding f-stop values.

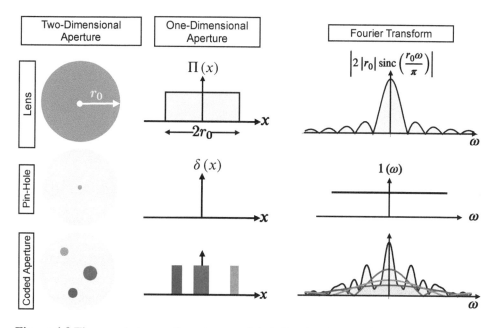

Figure 4.3 Three main types of apertures and their Fourier domain characteristics.

to a pinhole as the lens radius shrinks to an infinitesimally small quantity. This limiting behavior is best explained in observing the depth-of-field effect with varying lens apertures. Figure 4.2 shows this in terms of the f-stop.

4.1.2 Mathematical Perspective

In designing computational imaging centric systems, algorithms play a key role in the process of image formation and recovery. If we consider a one-dimensional lens as an abstraction of the optical system, then the pinhole PSF maps to the Dirac's delta function whereas a finite aperture lens with aperture radius r_0 maps to a box function of length $2r_0$. This is shown in Figure 4.3. A *coded aperture* is a more general setup consisting of multiple aperture openings that map to box functions with different widths and delays in the time domain, as shown in Figure 4.3. As a consequence, we obtain sinc functions of different widths and modulations. Coded aperture design is neither a fully open lens nor a pinhole but something between the two. For example, it may consist of multiple finite apertures at different locations carefully chosen to optimize a certain design criterion. As shown in Figure 4.3, the advantage of the coding scheme is best explained in the Fourier domain.

4.1.3 Noncoded Aperture

In this case, the aperture maps to a one-dimensional box function. The Fourier transform of this one-dimensional box function is the sinc function. The figure shows that the sinc function periodically touches zeros, which leads to a permanent loss of information at these frequencies that results in blurred images.

4.1.4 Pinhole

As discussed in a previous chapter, the pinhole camera is a box with a tiny hole on one side allowing light to be projected on the opposite wall, thereby creating an image on the projection plane. Despite having a measurable size, the pinhole can be modeled as a Dirac delta function due to its small radius.

The Fourier transform of this function is a constant function. Hence, there is no loss of information in the case of a pinhole camera. However, this setting suffers from a low throughput of light. Additionally, as discussed in chapter 2, if the pinhole diameter is close to the light wavelength, the effect of diffraction causes distortions in the image.

To avoid diffraction and increase the amount of light passing through the aperture, one could consider simply increasing the diameter of the pinhole. As shown in Figure 4.4, this leads to several points in the scene projected onto the same point on the sensor, causing a blurry effect.

In conclusion, a pinhole camera is not able to avoid all these problems at once. A lens integrates the light rays and bends them to converge onto one point. Therefore, by keeping the object in focus, the lens produces sharp and bright images, that are not distorted by diffraction. As explained in the next subsection, a lens allows choosing a wide range of aperture shapes with significantly less blur than pinhole cameras.

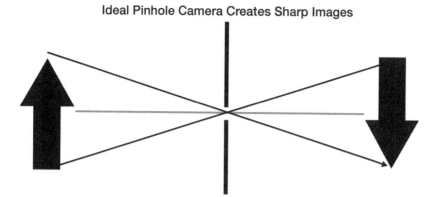

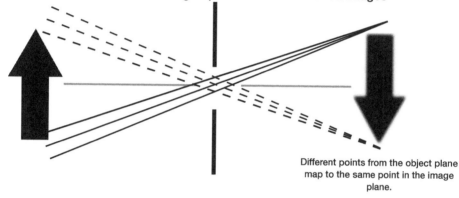

Figure 4.4 Pinhole camera with a large aperture. When the aperture size is increased, it can no longer be approximated with a Dirac delta function. As a consequence, different points in the scene are projected onto the same point in the projection plane, leading to a blurry image.

4.1.5 Coded Aperture

The distinct advantage of a coded aperture, which consists of multiple aperture openings, is that the zeros of one sinc function are compensated by the nonzero values of a sinc function with a different width. Hence, for a large range of frequencies (compared with the range for a noncoded aperture), we have no loss of information. However, in recovering information or images from such a coded aperture, the exact form of the Fourier transform is required to be known so that its effect can be undone in the recovery phase.

Using coded apertures for image and depth capture Although the idea of coding aperture dates to the pioneering efforts in X-ray astronomy (circa 1965), in recent years,

(Levin et al., 2007) were the first to demonstrate that both a conventional photograph and the corresponding *depth map* can be recovered by coding the aperture of a consumer-grade camera. The depth map is an image that contains at every pixel the distance to the scene from the view point. Their approach combined the idea of *depth from defocus* (DFD) with aperture coding. DFD-based methods for estimating 3D scene geometry exploit the optical nature of the image formation process. In particular, points that lie on the focal plane map to the sensor, creating a sharp image. Points away from the focal plane create a defocused image for which the amount of defocus depends on how far a point is from the focal plane. Hence, by knowing the amount of blur produced by a scene point in the image, one can estimate the depth of the same point. Clearly, the blur introduced in the process is a function of the aperture or the PSF of the optical system. To this end, (Levin et al., 2007) introduced coded aperture imaging in which the coded aperture was carefully designed so that it is sensitive to different depths, and hence one can discern depth information by solving the deblurring or the deconvolution problem. (Shedligeri et al., 2017) used a data-driven approach to design the coded aperture for depth recovery.

Using coded exposure (coding the aperture in time) Capturing objects in motion is a challenging task, often leading to images affected by motion blur. This effect can be addressed using deblurring algorithms, but it should be noted that deblurring in itself is an ill-posed problem for images captured with traditional architectures. As discussed in section 3.2.1, this process can be described mathematically by a convolution between the original image and a boxcar function, which in the frequency domain is translated to a multiplication of the original image spectrum with a sinc function. The information at every frequency where the sinc crosses 0 is lost, and high frequencies are also dampened as the sinc amplitude decreases.

To make the problem well-posed, a change in hardware architecture is necessary. Figure 4.3 shows how the aperture can be coded with multiple openings at once. But what if we generated multiple openings at different moments in time? This question led to the idea of coded exposure, also called *fluttered shutter* (Raskar et al., 2006). As the name suggests, this concept involves *fluttering* the shutter of the camera in a predefined sequence, which is often chosen to be a pseudorandom binary pattern. This preserves the high-frequency spatial details in the image and leads to an improved result, as depicted in Figure 4.5. Mathematically, the fluttered shutter approach allows modeling the aperture as several boxcar functions, which guarantee that their zeros in the frequency domain do not coincide. This means that the frequencies at which one sinc cancels are covered by other sinc functions, and no frequencies are lost.

The fluttered shutter is a very promising technique that has proved to be useful in other fields such as microscopy, in which fluorescence imaging of moving cells has a limited time resolution. By opening the shutter several times, it is possible to capture cells in motion with good resolution (see Figure 4.6). (Martel et al., 2020) built on this work to

4.1 Coding the Aperture

Figure 4.5 Coded exposure for objects in motion: (a) original blurred image, (b) rectification applied after estimating the vanishing point of motion lines, and (c) image deblurred using a camera with a fluttered shutter.

optimize the per-pixel shutter function in an end-to-end deep learning framework, in what was referred to as *neural sensors*. Coded exposure, however, has other applications. For example, using deep learning, (Okawara et al., 2020) jointly optimized the coded exposure and a classification model in a convolutional neural network to classify human actions in a single coded image.

Using multiple coded apertures Estimating the depth from defocus is an ill-posed inverse problem, because for this approach to work it is necessary to estimate the size of the defocus blur from a single image. A typical strategy to convert an ill-posed problem to a well-posed one requires the introduction of diversity in the measurements. The same applies to the case of DFD. In the literature, a common approach entails using multiple measurements with different defocus blurs. For example, this approach can arise from changing the focus setting by axially translating the sensor. Another method to create diversity in measurements uses different apertures, for instance, using two images, one with a large aperture introducing greater amount of defocus and another one with smaller aperture producing an image with large depth-of-field (compared with the former setting). However, it should be noted that in either case, the defocus is intimately linked with the aperture pattern, which is often chosen to be a circular disk. To this end, the use of multiple coded apertures has been proposed in the literature. For instance, (Farid and Simoncelli, 1998) used two images obtained with two different aperture patterns, that is, the Gaussian function and its derivative. (Hiura and Matsuyama, 1998) used a pair of pinhole apertures for depth measurement. When working with a pair of images, (Zhou et al., 2010) proposed a method for aperture pattern optimization for high-fidelity depth map recovery together

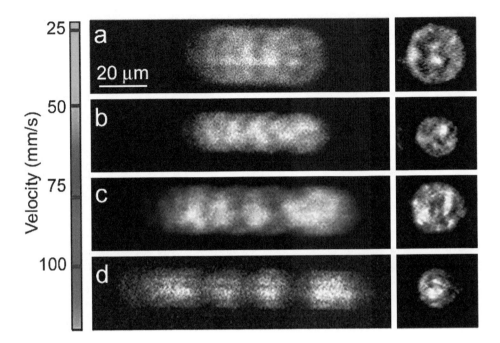

Figure 4.6 Fluorescence imaging for the nuclei of leukemia cells with different velocities. The motion blur of the cells, which is more pronounced with higher speeds, can be reversed with the fluttered shutter approach. Reprinted from (Gorthi et al., 2013).

with sharp, conventional image. Other imaging methods making use of coded aperture can be found in (Gottesman and Fenimore, 1989) and (Fenimore and Cannon, 1978).

Next we discuss various techniques to extend the depth of field of an image, either by capturing the full light field or by integrating the optical setup and computation via rayspace analysis. A simple method to extend the depth of field is to decrease the aperture size. However, this leads to a significant amount of noise. An alternative uses the concept of *focal stack*, which represents a series of images captured with the camera focused at different depths. This technique then requires combining the focal stack images using a technique similar to photomontage (Curless et al., 2004). As shown in Figure 4.7, this leads to a much clearer image.

The image projected by a scene point on the sensor is called the point spread function (PSF). When the point is in focus, the PSF is defined by a disk of infinitesimal size. When out of focus, the disk increases in size with measurable diameter. Therefore, the depth is a function of the PSF size. It acts as a convolution filter on the data to produce the

4.1 Coding the Aperture

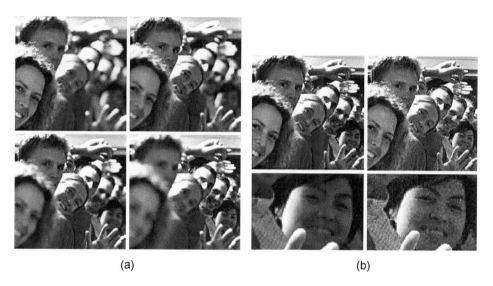

Figure 4.7 Extending the depth of field: (a) focal stack measured with different focus points, and (b) extended depth of field by reducing aperture size. Reprinted from (Ng et al., 2005).

measurements

$$\mathbf{m}(x, y) = (\text{PSF} * \mathbf{I})(x, y)$$

where $\mathbf{I}(x, y)$ represents the irradiance at location (x, y), and $\mathbf{m}(x, y)$ are the sensor measurements at the same coordinates. The PSF, which is disk-shaped in space, acts as a low-pass filter in frequency. This leads to high frequencies being canceled out, and these cannot be recovered via deconvolution. Moreover, the PSF function is generally unknown as a function of distance. All these reasons make recovering $\mathbf{I}(x, y)$ challenging in practice.

To recover the scene image $\mathbf{I}(x, y)$ while allowing efficient deblurring, there have been attempts to engineer the optics of the camera to allow it to generate a distance-independent PSF. One such attempt is to use an amplitude mask on the aperture of the lens (Levin et al., 2007; Veeraraghavan et al., 2007), which attenuates partially or fully the light rays intersecting the lens. Another option is to translate the sensor relative to the lens, which ensures that a large range of depths in the image are in focus (Nagahara et al., 2008). Using an image captured with such a modified PSF, one can then apply techniques based on deconvolving to recover the image in the blurred regions. Essentially, this produces a final image with an extended depth of field.

Conversely, some applications require reducing the depth of field. In photography a reduced depth of field may be considered more pleasant, artistically enhancing certain parts of the image. To this end, it is possible to use large lenses for their small depth of field

Figure 4.8 Reduced depth of field via blur estimation: (*left*) the original image; (*center*) the estimated blur; and (*right*) the image processed for depth reduction. Reprinted from (Bae and Durand, 2007).

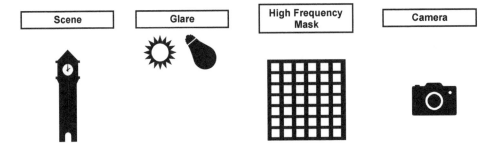

Figure 4.9 Removing glare from the scene with a high frequency mask outside the camera.

characteristics. However, they tend to be bulky and difficult to move around. A more convenient solution is using software to estimate and subsequently magnify the existing blur in the image, as depicted in Figure 4.8 (Bae and Durand, 2007). To alleviate the potential incorrect estimates due to noise sensitivity, it is also possible to use several lenses with different apertures to achieve the desired effect (Hasinoff and Kutulakos, 2007). This principle is currently used in many consumer-grade cameras and mobile phones.

As expected, it is more convenient to introduce the desired effects without updating the hardware, which increases the product affordability. To address this, the concept of *lens in time* translates the lens parallel to the sensor during the explosion time of a single capture, which allows using cheap hardware to achieve the desired effect with high accuracy (Mohan et al., 2009).

Glare is another important problem with many cameras. The glare is often caused by light reflected by the lens or by the lens diffraction effect. The reflection reduces the light reaching the sensor, but it also bounces back and causes unwanted glare effects. One way to address this is by coating the lens with multiple layers to cancel out reflections at multiple frequencies, as we discussed in chapter 2.

4.1 Coding the Aperture

Figure 4.10 Glare reduction using a high frequency mask near the sensor. The glare effect can be enhanced (*left*) or eliminated (*right*) by separating the light into global and direct components in the original image (*center*).

Figure 4.11 Prototype of a light field camera for glare reduction.

As mentioned briefly in chapter 2, the image captured by a camera can be separated into global components (reflectance from objects in the scene) and direct components (illumination by the source). Using a high frequency mask placed in various locations, it is possible to separate the two kinds of components (Talvala et al., 2007). Because glare is a global effect, it is convenient to eliminate it this way. The setup proposed in (Talvala et al., 2007) is depicted in Figure 4.9.

However, this technique requires a long capturing process that can take up to an hour. Furthermore, the technique works only when the mask is in focus, thus limiting the technique to indoor conditions.

Another interpretation of glare can be examined by looking at the frequency domain representation of the light field. It is situated mainly at high frequencies, therefore formulating the glare reduction operation as a filtering method (Raskar et al., 2008). This time, the setup involves a mask placed inside the camera on the sensor. For a Lambertian surface in the scene, characterized by diffusing light evenly in all directions, the angular component in the Fourier domain contains no information. The glare component, however, is present only in certain angular directions, therefore appearing as an outlier in the Fourier domain, so that it can be eliminated via relatively simple algorithms (Raskar et al., 2008). This requires measuring the light field, which is known to be demanding. However, using the fact that the glare is present only in the angular domain, the light field camera in this case is significantly simpler than a traditional one. A light field camera based on dappled photography (Veeraraghavan et al., 2007) is depicted in Figure 4.11.

The light field information is extracted using a mask placed directly in front of the sensor. The prototype is capable of separating glare into multiple categories without the need to capture several photographs. If the glare wavelength is known, it can be filtered out using an appropriate mask. This has the advantage that it does not dim the light from nearby objects of different colors (Mohan et al., 2008).

4.2 Coding the Sensor

An imaging sensor's capability can be enhanced by using the idea of sensor coding. That is, the concept of coded aperture imaging that was applied at the level of optics can also be brought to the imaging sensor. These ideas are used in innovations even today. (Sun et al., 2020) developed a probabilistic strategy to determine the optimal sensor sampling distribution for the optimal sensor design. (Chakrabarti, 2016) used a deep learning approach to backpropogate through the network parameters as well as the sensor parameters. These works were built on the principles discussed here.

4.2.1 Coded Sensors for Color Imaging

One of the earliest examples of coded sensor imaging led to the advent of colored digital imaging. Electronic imaging sensors cannot record the color of incident light as they record only the varying attenuation of light intensity levels, which makes them monochromatic sensors. To this end, Bryce Bayer, working at Eastman Kodak, used the idea of sensor coding (circa 1975) in which the sensor was made light sensitive by using the color filter array (CFA).

As shown in Figure 4.12, it takes four monochromic pixels to produce one colored pixel. One may wonder why four pixels are needed instead of three, one for each of the colors red (R), green (G), and blue (B). The use of twice as many green pixels is based on the physiological aspects of human visual apparatus. The human retina has higher sensitivity

4.2 Coding the Sensor

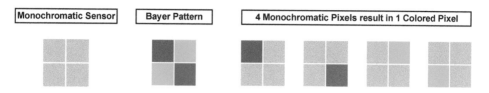

Figure 4.12 Three sensor architectures for color imaging.

to green light during the daytime. This is intricately tied to the cone cells in the retina which lead to a weighted sensitivity of luminance perception. So far, we have described the idea of sensor coding via CFAs. What about decoding? Clearly, in achieving a colored image from a monochromatic sensor, the essential resolution is downsized by a factor of four. To do this, we need to convert individual RGB pixels into colored pixels. Hence, decoding in the case of CFA entails recovering a full resolution colored image from the sensor-coded image. This is known *demosaicking*. As the name implies, when using the demosaicking approach, one demosaics the RGB tiles and combines them to produce a full resolution colored image. One of the simplest methods for demosaicking uses interpolation of color values of the pixels of the same color in their neighborhood. For any interpolation approach to work, a smoothness prior has to be assumed on the values to be interpolated. This is akin to the Shannon-Nyquist principle. Hence, demosaicking by interpolation is well suited to images with constant color regions and smooth gradients. Any abrupt jump in color or brightness levels would result in artifacts. This is typically the case with edges, and a well understood solution to this problem requires interpolation along the edge instead of interpolation across the edge.

Demosaicking represents only one of the steps in a processing pipeline that transforms the raw sensor image into the final product, such as a JPEG image. The pipeline generally introduces cumulative errors, but the work in (Heide et al., 2014b) overcame this problem by proposing a flexible image signal processor (ISP) called FlexISP, which optimizes the final image on the basis of assumptions on the final result, called image priors. For video capturing, the three-chip camera is a popular device, consisting of a prism, that splits the light into the RGB wavelengths, with each wavelength measured by a different sensor. The RGB video is subsequently synthesized by combining the three components. This mechanism is subsequently generalized to single-axis multiparameter (SAMP) cameras, which use beam splitters to send the light rays to multiple cameras, each with different sensor parameters (McGuire et al., 2007). This allows great flexibility. For example, by adding color filters to different cameras, the setup is equivalent to the three-chip camera. The possibilities are a lot more diverse. For example, each picture can be captured with a different exposure time, and the resulting images can be processed into a high-dynamic-range (HDR) image. A diagram and picture of this setup are depicted in Figure 4.13.

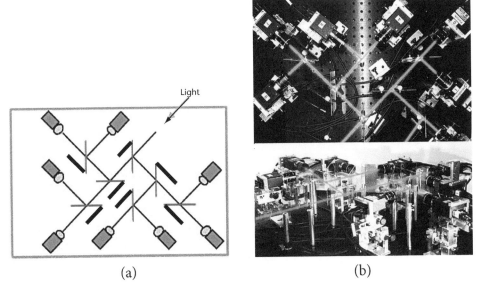

(a)   (b)

Figure 4.13 A single-axis multiparameter (SAMP) camera; (a) diagram depicting the incoming light beam split sequentially into eight beams, each captured by different cameras with different settings; and (b) the SAMP camera setup. Reprinted from (McGuire et al., 2007).

The SAMP camera is a self-explanatory setup, and therefore simple to implement on a theoretical level. However, the large amount of hardware required comes with greater cost and a bulkier size.

4.2.2 Coded Sensors for High Dynamic Range Imaging

Natural scenes are often composed of significant intensity ranges that may be far beyond what can be captured using a digital imaging sensor. As an example, consider the case of portrait photography against the sun. When such a situation arises, the amplitude or intensity to be captured is larger than the maximum recordable threshold of the digital sensor, and hence the measurements are clipped. This results in a permanent loss of information and is known as the sensor saturation problem. Human eyes are sensitive to subtle variations in contrast and can handle scenes with large range of intensities or dynamic range (Blackwell, 1946). On the other hand, a digital image capture device such as a digital camera or a video camera can handle only a finite dynamic range that is set by its bit-budget. For instance, an 8-bit digital system can handle 256 levels of illumination or brightness. The dynamic range of a system can be enhanced by simultaneously sampling across the spatial dimension and the exposure. This requires obtaining multiple exposures

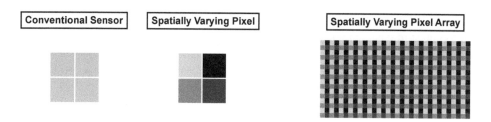

Figure 4.14 Three sensor architectures for high dynamic range imaging.

with varying intensities. The resultant set of images is then fused together algorithmically to produce a single high dynamic range image. This method is known as high dynamic range (HDR) imaging (Debevec and Malik, 1997). Although this may seem the ideal workaround because it requires no modification to the imaging setup, in several cases obtaining multiple exposures may not be feasible, specifically in scenes with fast motion between the exposures.

To circumvent the problem of using multiple exposures, (Nayar and Mitsunaga, 2000) proposed a coded sensor strategy. What was done with the colored filter array (CFA) in the previous case was extended to the idea of brightness. In their approach, the authors created a macropixel from four pixels (as before), each pixel attenuating the incoming light with different predesigned factors.

As shown in Figure 4.14, the brightness level linked with each pixel marked with a certain gray level represents its sensitivity. Consequently, light-shaded pixels will saturate faster for a given irradiance of light while their dark-shaded counterparts will record an attenuated exposure. As a result, even if one of the pixels is saturated, the illumination information can be recorded in one of the neighboring pixels provided that none of the other neighboring pixels saturate simultaneously. In this way, coded sensor imaging allows for simultaneous sampling of information along both the spatial and the exposure dimensions of the natural scene. Previously, in the case of Bayer's CFA, simultaneous sampling of information was performed along the spatial and the color dimensions of the scene. For the decoding step, two different approaches may be used for HDR image recovery. The first method is known as *Image Reconstruction by Aggregation*. The idea of this method is to average the macropixel consisting of 2×2 blocks. This averaging can be performed using a two-dimensional box filter. The second method is similar to the case of colored imaging and involves *image reconstruction by interpolation*. In this case, the interpixel values are interpolated using cubic interpolation. This approach works well in practice because sensor coding results in oversampled measurements. Hence, saturated and noisy pixels from each macropixel can be discarded and the unknown values can be interpolated. Even so, careful normalization must be performed to achieve realistic estimates of the brightness levels.

4.2.3 Modulo Sensors for HDR Imaging

In the context of computational sensing and imaging, a general purpose strategy for HDR capture and recovery has been introduced recently. Conventional digital systems acquire pointwise measurements that may potentially saturate, thus resulting in permanent loss of information. In contrast, in the *unlimited sensing* framework (Bhandari et al., 2017, 2020, 2021), one obtains folded measurements by injecting modulo nonlinearities into the sensing pipeline. This leads to a different kind of information loss in which the signal is folded into the dynamic range of the sensor. To see this in action, let us define the centered modulo operation using the mapping

$$\mathscr{M}_{V_{\max}} : f \mapsto 2V_{\max} \left(\left[\!\left[\frac{f}{2V_{\max}} + \frac{1}{2} \right]\!\right] - \frac{1}{2} \right), \quad [\![f]\!] \stackrel{\text{def}}{=} f - \lfloor f \rfloor \qquad (4.1)$$

where $[\![f]\!]$ and $\lfloor f \rfloor$ define the fractional part and floor function, respectively. Clearly, the modulo measurements defined

$$y[k] = \mathscr{M}_{V_{\max}}(g(kT))$$

are always smaller than the sensing threshold $V_{\max} > 0$. For a one-dimensional signal, the effect of modulo nonlinearity is shown in Figure 4.15a. The recorded measurements are orders of magnitude smaller than the original, HDR signal. Modulo samples for an image are shown in Figure 4.15b. Conceptually, analog-to-digital converters and imaging systems that implement signal folding were presented in circuit design literature (Rhee and Joo, 2003) and have also been implemented in recent years (Sasagawa et al., 2016; Zhao et al., 2010). Inversion of the modulo operator is a difficult problem in general (Bhandari, 2018; Bhandari et al., 2020). Akin to the Nyquist-Shannon recovery criterion (cf. theorem 3.2), the unlimited sampling theorem shows that a constant-factor oversampling suffices to recover any bandlimited signal from its low dynamic range, modulo samples.

Images are nonbandlimited objects, largely because image features such as corners and edges contain high frequency information. Mathematical guarantees for image recovery from modulo measurements were presented in (Bhandari and Krahmer, 2020) in which images are modeled in terms of shifts of spline functions (Unser, 1999). Let $\mathbf{x} \in \mathbb{R}^d$ be the spatial coordinates of a d-dimensional image $g(\mathbf{x})$. For multidimensional functions such as images, the shift-invariant model

$$V_{\mathbf{h}}^N = \left\{ g(\mathbf{x}) = \sum_{\mathbf{m} \in \mathbb{Z}^d} c[\mathbf{k}] \, B_N \left(\mathbf{H}^{-1}\mathbf{x} - \mathbf{m} \right) : \mathbf{c} \in \ell_2\left(\mathbb{Z}^d \right) \right\} \qquad (4.2)$$

is a flexible choice. In (4.2),

- **H** is a $d \times d$ diagonal matrix, with dilations $h_1, h_2, \ldots, h_d$ on the diagonal;

4.2 Coding the Sensor

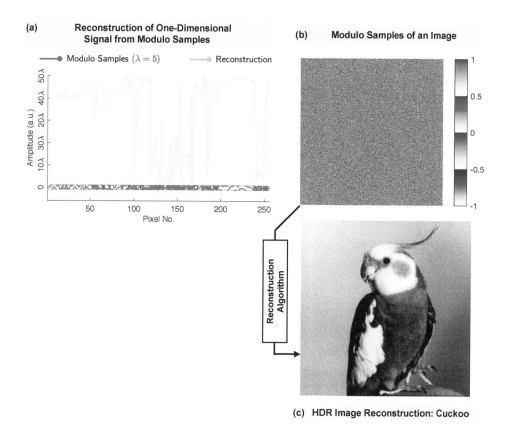

Figure 4.15 HDR imaging from modulo samples. In (a), $\lambda = V_{\max}$.

- $\mathbf{c} \in \ell_2\left(\mathbb{Z}^d\right)$ are the coefficients, and the expansion coefficients $c[k]$, can be evaluated using different strategies (Blu and Unser, 1999), namely, *interpolation*, *orthogonal projection*, and *quasi-interpolation*; and
- B_N is a B-spline of order N, and because the tensor product representation holds, $\mathsf{B}_N(\mathbf{x}) = \prod_{m=1}^{d} \mathsf{B}_N(x_m)$.

When modulo measurements of an image $g(x) \in \mathbb{R}$ are given

$$y[k] = \mathscr{M}_{V_{\max}}(g(kT)), \quad T > 0$$

the sampling interval (Bhandari and Krahmer, 2020),

$$T < \frac{h}{\pi e}\left(\frac{V_{\max}}{\max|g(x)|\widetilde{C}_{n,N}}\right)^{1/n}, \quad n \leqslant N, \quad \widetilde{C}_{n,N} = \left(\frac{\mathcal{K}_{N-n}}{\mathcal{K}_N}\right)$$

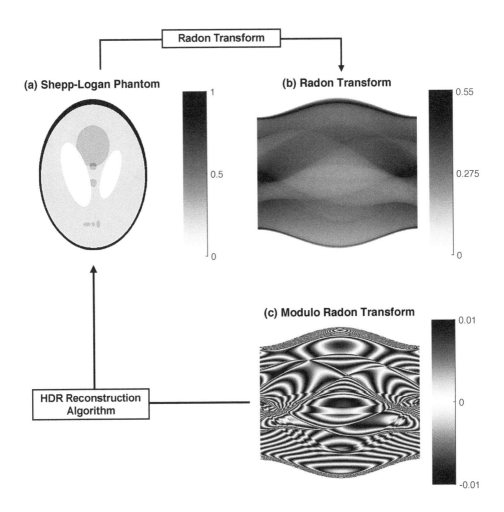

Figure 4.16 HDR tomography using modulo Radon transform.

guarantees recovery of the HDR image. In the preceding, $\mathcal{K}_N$ is the Bohr–Favard constant and e is the Euler number. Using the measurements presented in Figure 4.15b, an exemplary reconstruction is shown in Figure 4.15c.

Dynamic range constraint is a natural barrier in imaging systems that go beyond consumer photography. For instance, to overcome the dynamic range limitation of detectors used in computed tomography (CT), (Chen et al., 2015) extended the applicability of multiexposure fusion in consumer photography. In particular, they showed that high-dynamic-range CT reconstruction is possible by recording multiple exposures by varying the tube voltage. The principle of dynamic range compression using modulo nonlinearities can be applied to a wider class of computational imaging problems. For instance, the *modulo Radon transform* (MRT) (Bhandari et al., 2020a) allows for HDR reconstruction of Radon transform projections. Figure 4.16 shows conventional Radon transform projections together with the *low dynamic range* modulo Radon transform. The reconstruction algorithm (Beckmann et al., 2020) allows for recovery of the input image. The MRT has the advantage that it is a single-shot approach (avoids the drawbacks of the multiexposure fusion method) and is backed by mathematical guarantees.

4.2.4 Tone Mapping

The applications of *tone mapping*, which is a technique to create a mapping between two sets of colors, range from producing aesthetically pleasing images, to enhancing details to produce a higher contrast photograph. Many of today's displays do not support HDR content, so without tone mapping, the details in the image would be greatly reduced. Therefore, to enhance the applicability of HDR imagery we need to deal with the following two problems:

1. Displaying HDR content on low dynamic range (LDR) displays, such as computer monitors.
2. Displaying LDR content on HDR displays.

Those two problems can be addressed via tone mapping, by converting the content from one format to the other while taking into account human perception as a key factor. One approach is to compress the extended dynamic range of the available footage into a range that can be displayed on a LDR device. There are two ways to compress the footage:

1. Global compression acts on all pixels simultaneously, or
2. Local compression selectively converts a region of the image.

Global compression is performed by evaluating the luminance and other global variables, to compute the optimal transformation of each pixel independently from the values of the neighboring pixels. Examples of global compression methods include *reducing the image contrast*, *adjusting the brightness*, and *using gamma correction*. However, an image could have certain regions with different dynamic ranges from others, which would lead to decreased contrast in the resulting processed image.

Local compression extracts features in various image regions, and computes a region-specific transformation. This is why this process is heavier computationally and sometimes leads to small artifacts. However, local compression leads to much better results, given that the mammalian visual system perceives contrast mostly locally. Examples of local compression methods include *bilateral filtering*, *gradient domain computation*, and *constraint propagation*. We explain each of these as follows.

Bilateral filtering methods split the image to be processed into two layers: the detail layer, with a filter preserving the edges, and the base layer (Durand and Dorsey, 2002). The main

Figure 4.17 Dynamic range compression with the bilateral filter method. The base and detail layers are computed on grayscale images. Color is treated separately by reducing the contrast on each of its components, and then recomposing into the new color.

Figure 4.18 Gradient-based local compression. Starting with five images taken with different exposure values (*top row*), a radiance map is computed with pixel gradient calculations. The gradient attenuations (*lower left*) indicate the attenuation at that pixel corresponding to the gradient value. The final result contains adequate detail in both the dark and bright parts of the scene (*lower right*). Reprinted from (Fattal et al., 2002).

idea is that during dynamic range compression, the detail layer contains the information we want to keep unaltered, whereas the base layer information can be reduced. An example of an image processed with this method is depicted in Figure 4.17. This methodology was shown to work well in real time (Chen et al., 2007), and has also been used to generate creative and novel photograph processing.

4.2 Coding the Sensor

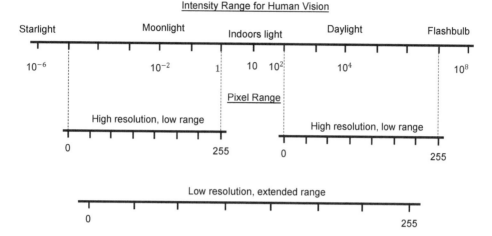

Figure 4.19 Mapping the high dynamic range intensities onto pixel values.

Gradient-based methods calculate a gradient field of the original image and reduce the values of the large gradients while simultaneously ensuring that the local contrast is not significantly modified (Fattal et al., 2002). Therefore, this method reduces the contrast in regions with no detail, and enhances contrast in dark regions. Figure 4.18 shows an example in which this method has been applied.

Alternatively, rather than using gradients, it is possible to propagate predefined constraints in an image. For example, the edges in the original image can remain unaltered while the dynamic range is compressed in local regions between the edges (Lischinski et al., 2006).

After summarizing some of the methods used for compressing the dynamic range, we can now look at displaying LDR content on the ever-increasing range of commercially available HDR displays. This method is known as reverse tone mapping. The problem in this case is unfortunately underdetermined, so that the resulting processed image contains more information than the original image.

In order to accurately store images of real scenes, we need to map the high dynamic range of intensities in the outside world on a value that will be assigned to the pixel. As depicted in Figure 4.19, to cover the extended range of intensities with a fixed number of bits per pixel, one needs a trade-off between the range covered and the resolution. Initially, the number of bits per pixel was very large. The current widely accepted format was devised in 2003 by Industrial Light and Magic in collaboration with independent partners, which included flexible bit depth, backward compatibility, computing platform independence, and open-source licensing (http://www.openexr.com). The EXR format is widely used in cases in which accuracy has a high priority, such as photorealistic rendering and texturing.

The display of HDR content was addressed by placing two screens parallel to each other: a low-resolution display behind an LCD screen (Seetzen et al., 2004). The image viewed on the opposite side of the LCD screen has a high contrast, which is equal to the product of the contrasts of the two screens. The contrast of a display is evaluated using the contrast ratio (CR) measure, which is the ratio of the luminance of the brightest color to the luminance of the darkest color that the display can achieve. A consumer-grade display may have a CR on the order of thousands to one. The two display setting in (Seetzen et al., 2004) allowed visualizing HDR content achieving CRs as high as 50,000:1.

The classic representation of color is based on three primaries: red, green, and blue (RGB). However, this simplified representation does not always encode the color faithfully. For instance, *metamerism* is the phenomenon in which objects for which the spectra of their reflected light are significantly different but the tricolor representation makes them look very similar. Interestingly, each set of RGB primaries, or mapping between colors and wavelengths, defines a continuous set of human-perceived colors, shaped as a convex hull. Therefore a device can reproduce only a subset of the human-perceived colors. Usually this subset is defined by the RGB filters in the Bayer pattern. The use of color filters precedes the advent of color photography. Even during the times of black-and-white photography, color filters were used in order to enhance the contrast in an image. For instance, without the filters the sensor would hardly distinguish between clouds and the sky, which is an example of imaging metamers in a scene. Some modern cameras allow filtering the image in real time at desired frequencies, which allows distinguishing between metamers in the scene.

4.2.5 Exposure Metering

In photography, there are rules of thumb on how to set the camera parameters for a good image exposure, for example, the aperture setting to f/16 and shutter speed to 1/ISO when there is a lot of ambient light, also known as the *sunny-16 rule*. We prefer that the intensity of the ambient light is measured automatically, which is done with an in-camera light meter. Light meters can be grouped in two categories:

- reflected-light meters and
- incident-light meters.

The reflected-light meters evaluate the light reflected by the scene and include all in-camera light meters. Examples are spot meters, which measure the light reflected by a small view angle (1 degree or less), and center-weighted meters, which average the light from a larger portion of the scene.

The reflected-light meters are not effective in the case of highly reflective scenes such as large areas covered in snow. To achieve good results in this case, incident-light meters are placed at the scene to measure the arriving light, which avoids the reflective properties of the scene.

Placing meters on the scene is difficult and often impossible. A popular system designed to produce accurate exposure values for a scene is called the zone system. This system allows human perception to be included in the choice of exposure value (Adams, 1980). It relies on the fact that the photographer can recognize which objects are more reflective and which are not, therefore making the appropriate choice in each case. It consists of ten zones, each representing an increase in exposure value by a factor of two over the preceding one.

This concept was later implemented commercially by Nikon, which included multizone metering in their Nikon FA camera. The concept is extended in the sense that the sensors are organized in a matrix and measure the brightness in several locations in the scene, ranging from 5 to several thousand. The camera then automatically computes the optimal settings for that capture.

4.2.6 Improving the Resolution

The resolution of an image is in direct connection with the ISO parameter measuring the light sensitivity. Dating back to the film-based analog cameras, decreasing the ISO would lead to a finer film granularity, and thus higher resolution, and large ISO values lead to a higher granularity and thus lower resolution. In digital cameras, the equivalent of granularity is the number/size of the discrete light-sensitive sensors called pixels. The ISO sensitivity in this case adjusts the gain of the analog-to-digital (A/D) converter, leading to brighter images.

Unfortunately, the higher gain also amplifies the noise. One may say that this does not change the resolution of the sensor, which is fixed. However, the usable resolution drops in this case: fewer distinct pixels capture useful details from the scene. This problem as well as other causes of decreased resolution can be alleviated with the methodology known as *superresolution*. In the case of noise images, several exposures with low effective resolution can be used to compute a high resolution image by fusing the information from all the image captures. Specifically, the superresolution algorithms first estimate the relative movement between the camera and the scene, and then find a common coordinate system. This is then used to filter each image and fuse them together. Examples include the works of (Tsai and Huang, 1984; Kim et al., 1990; Irani and Peleg, 1990, 1991; Kim and Su, 1993); and (Elad and Feuer, 1997).

Just as in the case of the noisy image, the superresolution can be implemented by fusing images captured with low resolution images and varying camera parameters. For example, the sensor position can be changed in a noisy way to introduce a variation between images (Keren et al., 1988; Vandewalle et al., 2006). Alternatively, the aperture of each image capture can be altered (Komatsu et al., 1993), or the zoom level (Joshi et al., 2004), or the level of blur, shading and defocus characteristics (Rajan et al., 2003).

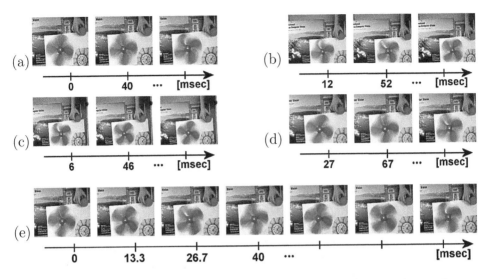

Figure 4.20 Space-time superresolution technique. Using four video recordings at low frame rate and low resolution, a new higher resolution video is generated with frames at times where there is no physical measurement. Reprinted from (Shechtman et al., 2005).

Interestingly, the resolution can also be increased by using a single photograph. By inducing motion blur through long exposure in capturing a moving object, the resulting image consists of an elongated capture of that object, therefore allowing more pixels to encode its details. The result is a high-resolution image of the object in question (Agrawal and Raskar, 2007).

4.2.7 Capturing Fast Phenomena

Imaging fast motions dates back to the 1870s, when the English photographer Eadweard Muybridge designed a multicamera setup to photograph a horse galloping. It consisted of a system synchronizing the shutters of several cameras using electromagnets. This allowed one of the first glimpses of the details in the movement of humans or animals.

Today modern cameras can reach frame rates as high as 1,000 frames per second, revealing details of the scene that would be impossible to see with the human eye. However, certain high speed phenomena still cannot be captured in this way. The concept of superresolution, presented in the previous section, can be used here to further enhance the frame rate. Just as it was used to recover the values of the light intensities in between the measured pixels, superresolution is used to recover frames located between the captured frames. In (Shechtman et al., 2005), the authors captured a scene with multiple cameras of relatively small resolution and frame rate. As the pixels in the cameras are not perfectly aligned, and the frames are not perfectly synchronized in time, the authors exploit this to infer additional information about the scene. The application of this methodology is depicted in Figure 4.20.

However, the superresolution cannot increase the performance arbitrarily high due to the reconstruction error. High-speed video was also achieved by using a set of 128 digital video cameras packed together, so that they all have the same field of view. The cameras all had a frame rate of 30 fps, but they were synchronized such that in each 1/30 s interval all cameras were capturing a frame at equidistant times. This allowed their setup to produce frame rates as high as 3,000 fps.

4.2.8 Using Coded Sensors for Light Field Capture

Another interesting application of coded sensor imaging is linked to recovery of four-dimensional light fields from two-dimensional images. Conventional methods for capturing a light field rely on trading off spatial resolution for angular differences, for example, using an array of lenses or using a large lens covering a microlens array. However, as demonstrated by (Veeraraghavan et al., 2007), a coded sensor approach allows for a solution that does not require any use of refractive optics. In this approach, coined *dappled photography*, the key idea is to place a mask on the sensor. By placing a high frequency sinusoidal mask between the sensor and the optical elements of a camera, spectral tiles of the light field in a four-dimensional Fourier domain can be created. Hence, the encoding measurements multiplex a four-dimensional light field onto a two-dimensional image. The light field is then decoded from the measurements in two steps. First, the Fourier transform of the two-dimensional image is computed and then reassembled in a way such that the two-dimensional tiles can be stacked into the four-dimensional plane. There on, a four-dimensional inverse Fourier transform results in the desired light field. A method for high-resolution imaging was presented in (Cossairt et al., 2011). Masks have also been used to reduce the blur in the out-of-focus regions of the image in a process called *defocus deblurring*, which consists of finding good coded apertures that allow recovering a sharp image from its blurred original version (Masia et al., 2012). The authors showed that the results are significantly improved by considering the human visual perception factor in designing the masks.

4.3 Coding the Illumination

The preceding examples showed how coded aperture and coded sensor imaging can enhance the capability of an imaging system. In these examples it was assumed that the ambient illumination was fixed by design. However, if illumination is used as a degree of freedom in designing an imaging system, coding illumination can lead to substantial advantages. A coded illumination strategy can be based on something as simple as a light flash used in consumer photography, or it may employ more sophisticated setups such as a projector or a laser. Next we present a few examples.

4.3.1 Coded Illumination Imaging with Flash

The scene lighting has a big impact on the end result of the imaging process. If a scene is dimly lit, some of the options for preserving the ambient light include increasing the exposure, using a larger aperture, or increasing the sensor ISO. However, each comes with its own drawback. Longer exposure times create motion blur, due either to moving objects or to moving camera position; larger apertures lead to smaller depths of view; and larger ISO values decrease the signal-to-noise ratio.

Flash photography addresses all these problems at the expense of images that do not reproduce the true ambient illumination. In addition, flash images lead to brighter closer objects and cause the red-eye effect and harsh shadows.

However, the flash and ambient images can be combined in order to access the benefits of both modalities. (Petschnigg et al., 2004) proposed a method of generating flash/ambient image pairs and combining them with algorithms that perform denoising of the ambient image, transfer detail from the flash image to the ambient image, perform white balancing on the ambient image, and allow a continuous adjustment between the information provided by the two images.

In this method, the two images are captured with the same aperture and focal length. The quality of the ambient image is optimized by tuning the exposure time and ISO values. The flash image is captured with low ISO and exposure time in order to minimize the noise and provide more high-frequency details.

The denoising algorithm is based on an existing technique that processes the image with an edge-preserving bilateral filter. However, the flash image contains more detail on the edges, which is used to design a joint bilateral filter that leads to more natural results with less noise.

The next step is transferring high-frequency detail from the flash image to the ambient image. To this end, the detail from the flash image is computed as the variation around the denoised image. However, the computed detail is not accurate in the shadow or specular regions, and a mask is used to avoid transferring any detail from the respective image areas (Petschnigg et al., 2004). An example of image denoising and detail transfer is depicted in Figure 4.21.

Using two images also improves white balancing, in which the known color of the flash light provides useful information for the image coloring. The authors additionally introduce a method to allow a user to generate an intermediate image between the flash and ambient images in real time, therefore essentially adjusting the flash intensity after the capture. Inspecting the differences between the two images allows correcting for undesirable artifacts such as red eye.

4.3 Coding the Illumination

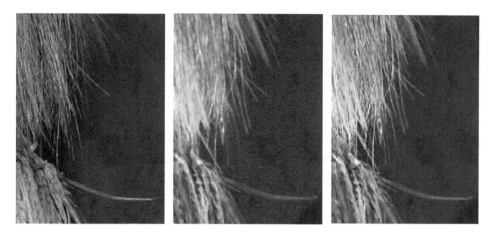

Figure 4.21 The procedure of transferring details from the flash image to the ambient image: (a) flash image, (b) ambient image, and (c) ambient image processed with denoising and detail transfer. Reprinted from (Petschnigg et al., 2004).

4.3.2 Coded Illumination Imaging with Lasers

The illumination of a scene for image capturing has two sources: the direct illumination by the light source and the global illumination from other points in the scene (Nayar et al., 2006). The separation of these two sources is desirable because each reveals different information about the scene. The *direct component* enhances the material properties of a given point, and the *global component* reveals the optical properties of the scene by indicating how a certain point is illuminated by other points in the scene.

The scene was divided in a number of patches, such that each visible patch corresponds to a pixel of the light source. The main observation that makes the separation of the two components possible is that using high frequency illumination lights up patches in the scene that have both global and direct components and leaves unlit the patches with only global components. For uniform coverage of the scene, the illumination was performed using checkerboard patterns. However, an off-the-shelf projector suffers from imperfections such as light leakages in its optics, which causes some unwanted brightness variation between the squares on the checkerboard. To compensate for this, the authors captured five times more images, shifting the checkerboard pattern slightly each time.

All these experiments include artificial illumination, indicating that they can be performed indoors only. However, it is possible to separate the global and direct illumination components outdoors, using *occluders*, which have the opposite effect of a light projector, casting shadow on various portions of the scene. Examples include the line occluder (a stick) and the mesh occluder.

The future of this line of research could see camera flashes with high-frequency components, tailored to allow an in-camera separation of the two components. This would allow the users to create novel images in which the appearance of objects could be edited using their global and direct components (Nayar et al., 2006). The depth of a scene can be inferred accurately in the presence of global illumination (Gupta et al., 2009). A theoretical lower bound was derived on the number of images required to separate the global and direct illumination components (Gu et al., 2011).

4.3.3 Coded Illumination Imaging with LEDs

Let us consider the context of microscopy. Changing the viewing angle, or moving the specimen in order to acquire an image from a different perspective can be problematic. Using illumination can help solve this problem. In microscopy there is a common trade-off between the resolution and the field of view, so that a specimen can be captured in high detail only in a relatively narrow region. To increase the field of view without affecting the accuracy, *Fourier ptychography* is a well known computational imaging technique. This technique increases the numerical aperture of the microscope, which is essentially the range of angles it can capture, by recording images illuminated from a range of angles. This results in increased resolution over that of a conventional microscope. The illumination is done with an array of LEDs. In the Fourier domain, changing the illumination angle corresponds to Fourier values computed in a shifted domain. The large numerical aperture image is then computed by stitching the Fourier slices. This requires a large overlap between the domains of each slice of around 60%.

The traditional methods capture one illumination angle at a time by turning the LEDs on in a sequential manner. (Tian et al., 2014) have improved on the traditional methods by turning on a larger number of LEDs, thereby saving a lot of acquisition time. Specifically, when K LEDs are turned on at the same time, the exposure time can be decreased K times, because there are K times more light rays illuminating the specimen. The number of images can also be decreased by K, leading to a total reduction by a factor of K^2 in acquisition time. The K LEDs are selected randomly, but it is ensured that different images do not use the same LEDs. Figure 4.22 gives an example for $K = 4$ for four randomly generated patterns of LEDs. The lateral resolution can be doubled by using a pair of images with asymmetric illumination patterns (Tian and Waller, 2015). (Kellman et al., 2019) have presented a method of optimizing the LED illumination pattern for phase retrieval in microscopy by combining the physics of the measurement scheme with the nonlinearity of deep learning.

Decreasing the exposure time allows capturing videos of samples to record dynamical phenomena among populations of cells such as division and migration (Tian et al., 2015).

Other works optimize illumination to increase the performance in depth estimation (Nayar et al., 1996), or to perform differential phase imaging, which recovers the optical path length of the sample (Tian et al., 2015). In conditions of poor visibility, controlling the light transport by, for example, using polarized light, leads to images of quality superior to that of images produced using the existing illumination (Gupta et al., 2008). A method

4.4 Further Research

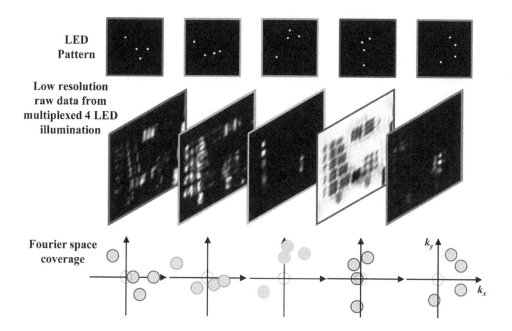

Figure 4.22 A Fourier ptychography technique with multiplexed illumination. The LED pattern (*top*) illuminates the target object, leading to a different image (*center*). For each illumination, the resulting image has a spectrum computed in four subsets of the two-dimensional frequency domain, corresponding to the four LEDs (*bottom*). Reprinted from (Tian et al., 2014).

generating wide field-of-view images using reflective surfaces was presented in (Hicks and Bajcsy, 2001).

4.4 Further Research

In this section we discuss three interesting applications of spatially coded imaging: compressive imaging, ghost imaging, and spectrometry. With the ever increasing resolution of 2D cameras and with the advent of plenoptic cameras for capturing light fields, the constraint on the bandwidth of the data transmission infrastructure becomes evident. Compressive imaging offers a solution to this problem by minimizing the measurements transmitted. It is possible to capture images of objects that are not in the line of sight of the camera, using a setup known as ghost imaging. Moreover, it is also possible to use it in conjunction with compressive sensing for increased efficiency. Last, most photographs taken capture the light intensity distributed in space. Analyzing the light in the frequency domain using a process called spectrometry can reveal a lot about the material properties.

4.4.1 Compressive Imaging

Compressive imaging is an image processing technique to efficiently capture an image with a reduced number of samples, which allows recovering a higher resolution image of the scene by finding solutions to underdetermined linear systems.

So far, we have examined how varying parameters of the imaging system help achieve improved results. Here, we discuss sampling methods and ways to minimize the number of samples needed to make a good decision a good representation of the original data. Shannon's sampling theorem proves that a signal can be perfectly recovered if sampled at a rate twice the maximum frequency in its spectrum, known as the Nyquist rate. What happens if we take fewer samples? In the general case, this would lead to aliasing, so that the high frequencies in the signal no longer can be recovered from the generated samples. However, in many practical applications, signals can be represented by samples taken at sub-Nyquist rates due to a property called *K-sparsity*, which we define as follows.

Note that although we discuss the one-variable scenario, this analysis can be easily generalized to two variables. Let $\mathbf{f} = [f_1, \cdots, f_N]$ be a vector of N samples, which can be represented in an orthonormal basis $\{\mathbf{g}_p\}_{p=1,\cdots,N}$ as $\mathbf{f} = \sum_{p=1}^{N} c_p \mathbf{g}_p$. In matrix form, this amounts to $\mathbf{f} = \mathbf{Gc}$, where the lines of $\mathbf{G}$ are given by vectors $\{\mathbf{g}_p\}_{p=1,\cdots,N}$. We say that $\mathbf{f}$ is K-sparse in this basis if $K < N$ and $\mathbf{f} = \sum_{i=1}^{K} c_{p_i} \mathbf{g}_{p_i}$, meaning that only K elements from the set $\{c_1, \ldots, c_N\}$ are different from 0.

One way to encode such a signal is to use *transform coding*. This involves generating the full set $\mathbf{f}$ of N samples; then estimating the coefficients c_p as $c_p = \langle \mathbf{f}, \mathbf{g}_p \rangle$, $p = 1, \ldots, N$; and finally computing the largest K coefficients. This process is unnecessarily complex, especially if $K \ll N$.

The *compressive sensing* problem aims to decrease the number of generated samples to $M \ll N$ from the start and then introduce methods to recover the signal from the samples. The new M samples, denoted $\{y_p\}_{p=1,\cdots,M}$ are generated by projecting signal $\mathbf{f}$ onto a new set of sampling kernels $\{\mathbf{d}_p\}_{p=1,\cdots,M}$, such that $y_p = \langle \mathbf{f}, \mathbf{d}_p \rangle$, $p = 1, \cdots, M$. In matrix form, we can define the *measurement matrix* $\mathbf{D}$ composed of M vectors $\mathbf{d}_p^\top$, each with dimensions $1 \times N$, resulting in a final matrix dimension $M \times N$. The resulting measurements are given by $\mathbf{y} = \mathbf{D}\mathbf{f}$, where $\mathbf{y}$ is a $M \times 1$ vector.

There are ways to generate sampling kernels to guarantee perfect recovery. However, in a practical setting, it is very convenient to generate matrix $\mathbf{D}$ randomly, which has been demonstrated to allow perfect recovery with a very high probability. Specifically, the elements in $\mathbf{D}$ are taken from the Gaussian distribution with zero mean and variance $1/N$.

Clearly, for $M \ll N$ the problem of recovering $\mathbf{f}$ from $\mathbf{y}$ is ill conditioned, because the system $\mathbf{y} = \mathbf{D}\mathbf{f}$ is underdetermined, that is, there are fewer equations than unknowns.

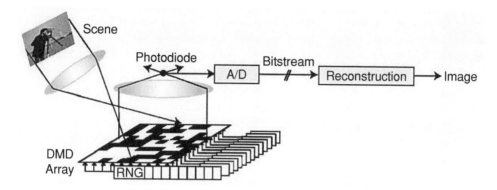

Figure 4.23 A single-pixel camera performing compressive sensing. Reprinted from (Baraniuk, 2007).

However, this problem can be approached assuming $\mathbf{f}$ is K-sparse in basis $\{\mathbf{g}_p\}_{p=1,\cdots,N}$ and $M \geqslant K$, which means that its vector of coefficients $\mathbf{c}$ in this basis has only K nonzero entries. In practice, K-sparsity is not always precise and the measurements are noisy, therefore a more reasonable assumption is that $\mathbf{c}$ has only K large elements. This is can be quantified using vector norms, which leads to the following recovery formulations:

$$\widehat{\mathbf{c}} = \arg \min_{\mathbf{c}} \| \mathbf{c} \|_2^2 \quad \text{subject to} \quad \mathbf{y} = \mathbf{DGc}, \tag{4.3}$$

$$\widehat{\mathbf{c}} = \arg \min_{\mathbf{c}} \| \mathbf{c} \|_0 \quad \text{subject to} \quad \mathbf{y} = \mathbf{DGc}, \tag{4.4}$$

$$\widehat{\mathbf{c}} = \arg \min_{\mathbf{c}} \| \mathbf{c} \|_1 \quad \text{subject to} \quad \mathbf{y} = \mathbf{DGc}. \tag{4.5}$$

The ℓ_2 norm from (4.3) is not very good at selecting the large coefficients. The number of nonzero entries, given by ℓ_0 norm, seems like the ideal choice (4.4). However, the routines implementing it are numerically unstable and complex. Moreover, they assume precise K-sparsity. The most widely used reconstruction is based on ℓ_1 norm, which recovers well signals that are precisely K-sparse, is less complex numerically, and leads to good approximations in the case of noisy measurements.

One may wonder how to generate compressive samples $\mathbf{y}$ without generating the N uncompressed samples $\mathbf{f}$. A method proposed in (Wakin et al., 2006b,a) uses a digital micromirror device (DMD), which is essentially an array of small mirrors, each representing a pixel that can switch between two angles. An image is projected onto this device, and each pixel in the DMD can reflect the corresponding rays from the image either toward a photodiode sensor, or away from it. The light reflected toward the sensor is integrated into a one-dimensional data stream. This device, depicted in Figure 4.23, effectively generates samples $y_p = \langle \mathbf{f}, \mathbf{d}_p \rangle$, where $\mathbf{d}_p$ is a vectorized version of the binary matrix defined by the orientations of each DMD pixel, and $\mathbf{f}$ is a vectorized version of the two-dimensional image.

This design enables choosing the kernels $\mathbf{d}_p$ and the number of samples M in a flexible and easy way. Increasing the number of samples leads to more precise reconstructions, and decreasing it boosts the compression and decreases the acquisition time. Another advantage of this device is that it avoids the use of a shutter. This is particularly significant for video encoding, in which a shutter typically opens and closes for every frame, whereas in this case samples can be captured in a continuous way (Duarte et al., 2008).

This method is also known as single-pixel imaging, because every measurement is acquired by mapping an image onto a single-pixel sensor represented by the photodiode. However, this involves a special setup with a DMD that takes up space and leads to increased power consumption. Additionally, because it acquires one pixel at a time, it needs many measurements for one image. A video captured with this setup would have a very poor temporal resolution.

An intermediate step between capturing high-resolution and single-pixel measurements was proposed in (Marcia et al., 2009), where the assumption is that the camera has a small sensor array. The authors introduce a method called *compressive coded aperture imaging*, which recovers the high-resolution data that is sparse on some basis from low-resolution sensor data. Their setup includes an aperture mask that processes the incoming video stream, followed by a downsampling operation, which reduces the resolution of the data to match the sensor. Their proposed aperture mask is compatible with nonlinear reconstruction. This allows them to superimpose frames originating from a wider field of view and then perform disambiguation, effectively recovering a much larger field of view than would be possible with other masks.

The trade-off between spatial and temporal video resolution has also been addressed by encoding the temporal information in a high-speed video in a single frame (Serrano et al., 2017; Liu et al., 2013; Hitomi et al., 2011). The work uses a technique called *sparse coding*, which identifies a representation of the data in a sparse dictionary of atoms. The benefit of sparse coding is also used to recover high quality HDR images using single coded camera exposures (Serrano et al., 2016). The image can also be represented using features extracted from the data in an unsupervised manner in a process called *convolutional sparse coding*. This approach leads to faster and better solutions than the state of the art (Heide et al., 2015).

Other examples of compressive imaging setups are given in (Wakin et al., 2006b; Pitsianis et al., 2006; Takhar et al., 2006; Romberg, 2008; Watts et al., 2014; Gan et al., 2008; Som and Schniter, 2012; Lohit et al., 2018); and (Keller and Heidrich, 2001).

4.4.2 Ghost Imaging

All the methods presented so far generate images from the recorded light rays arriving from a target scene. This is not the case for ghost imaging, which is based on a quite interesting principle. A light source generates a beam that is split into two: the first outgoing beam is

4.4 Further Research

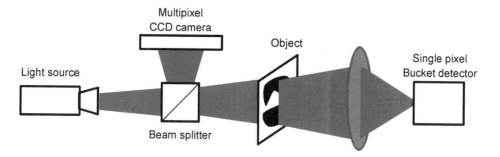

Figure 4.24 The ghost imaging paradigm.

captured by a pixel array, often based on a charge-coupled device (CCD) camera, and the second beam intersects an object of interest and then converges into a single pixel sensor, also known as *bucket detector*. Even though the light reaching the CCD did not intersect the object, its intensity in each pixel can be correlated with the intensity captured by the bucket detector to acquire the shape of the object in question (Shapiro, 2008).

The interpretation of these correlations ranges from classic intensity correlations to quantum correlations, and it is not a straightforward problem. A comparison between these correlations can be found in (Gatti et al., 2004). It was also shown that in order for the ghost imaging scheme to work, it is crucial to use incoherent light beams consisting of photons with different phase and frequency (Gatti et al., 2006). The classic setup for ghost imaging is depicted in Figure 4.24.

There have been several variations around the ghost imaging setup. For example, instead of measuring the transmitted photons, the bucket detector can capture light reflected and scattered by the object (Meyers et al., 2008). The two-detector setup can be replaced by a single-detector setup (Bromberg et al., 2009). This is done by calculating the field propagation for the reference beam, typically captured by the CCD camera, thus leaving the bucket detector as the only sensor. The transmission function of the object, which is the main result in ghost imaging, can be computed in absolute units, therefore revealing inherent properties about the object (Ferri et al., 2010).

The light source emits several intensity patterns, each corresponding to an object sample. A novel approach to ghost imaging looks at the required sample size for a good reconstruction and uses compressive sensing to decrease the acquisition time (Katz et al., 2009). The intensity captured by the bucket detector for each pattern is given by $B_r = \iint \mathbf{T}(x, y) \mathbf{I}_r(x, y) \, dx \, dy$, where $\mathbf{T}(x, y)$ is the transmission function, and $\mathbf{I}_r(x, y)$ is the field generated by the light source. Under this interpretation, $\mathbf{I}_r$ acts as a sampling kernel for $\mathbf{T}$. Because natural images are sparse in carefully selected bases, the authors showed that it is possible to achieve good image reconstructions with as little as 15% of the Nyquist sampling rate. Ghost imaging using deep learning has also improved reconstruction accuracy while reducing the number of overall number of needed measurements (Lyu et al., 2017).

4.4.3 Spectrometry

An interesting and well known application of imaging is to evaluate the light in the frequency domain, that is, to measure its spectrum. A straightforward approach is to pass the incoming light through a collimating lens, which generates parallel rays, and then interpose a prism in the path of the generated parallel beam. As we know, this will cause the light to separate into its frequency components. Then each wavelength interval can be measured with a separate sensor (James, 2007).

Spectrometry is the field of analyzing the spectra of point sources. *Imaging spectrometry*, also called *multispectral photography*, is a recent subfield of spetrometry based on analyzing the spectrum of an object in each of its points (8.1.3). If we look at a 2D object, the spectrometer generates a measurement modeled as function $\mathbf{m}(x, y, \lambda)$ where x, y are the spatial coordinate and λ is the wavelength. A spectrometer typically scans a 2D slice of the measurements, in which at each point one of the x, y, λ coordinates is kept constant. Each measurement is affected by noise and therefore can be evaluated using the signal-to-noise ratio (SNR). The SNR of some of the common spectrometers was calculated and reported in (Harvey et al., 2000).

From a mechanical perspective, spectrometers can be grouped into the following categories:

- pushbroom cameras, which do not scan the scene but have a camera attached to a platform that moves forward; and
- whiskbroom cameras, which use gimbals, or pivoted supports allow the rotation of the camera in a single axis to actively scan the scene.

The spectrometers make use of *interference filters*, which consist of optical filters that reflect a number of predefined spectral bands and transmit others. A key characteristic of the interference filter is that it absorbs almost none of the light in the wavelengths of interest. A linearly variable interference filter (LVIF) uses interference films varying in thickness along one dimension. A spectrometer with an associated mounted LVIF is known as a *wedge imaging spectrometer* (Demro et al., 1995). Such a device was used to capture several photos of a mosaic and then process them into a multispectral mosaic. A more systematic approach to spectrometry uses a narrow wavelength disperser, known as a monochromator, and then uses a moving slit that narrows it down to a single emitted wavelength. The multitude of wavelengths generated are then recombined with another monochromator (Li and Ma, 1991). To decrease the amount of information measured, compressive sensing techniques have been used in conjunction with spectrometry (Willett et al., 2007).

The reconstruction of the three-variable function $\mathbf{m}(x, y, \lambda)$ from slice measurements is called *chromatography*. It includes a tomography step that recovers the 3D data. For example, it is possible to recover the function from five 2D slices computed with transmission gratings and a consumer-grade camera (Okamoto and Yamaguchi, 1991). The 2D measurements can also be achieved with a 1D spectrometer and a rotation mechanism (Betremieux et al., 1993).

Another distinct line of work is focused on constructing multispectral projectors, which usually are made with lamps, diffraction devices, and color filters. A notable way to implement a multispectral projector involves separating white light into its wavelength components using a diffraction grating and then direct all the beams onto a digital micromirror device (DMD). As we explained previously, the DMD is a spatial modulator consisting of micromirrors that selectively direct specific rays toward a point of interest. In this case the DMD can select the desired wavelengths and direct them toward a prism to recombine them, which in principle allows generating light of any spectrum characteristic of interest (Wall et al., 2001; MacKinnon et al., 2005; Brown et al., 2006; Farup et al., 2007).

Chapter Appendix: Notations

Notation	Description
$\mathbf{m}(x, y)$	Measurements at location (x, y)
$\mathbf{I}(x, y)$	Irradiance at location (x, y)
f	Focal length
$\{\mathbf{g}_p\}$	Orthonormal basis
G	Orthonormal basis $\{\mathbf{g}_p\}$ stacked in a matrix columnwise
$\{\mathbf{d}_p\}$	Sampling kernels
D	Sampling kernels $\{\mathbf{d}_p\}$ stacked in a matrix rowwise
y	Measurements
f	Signal
c	Signal coefficients
$g(\mathbf{x})$	Image with spatial coordinates $\mathbf{x}$
$\mathcal{M}_{V_{max}}(x)$	Centered modulo operation
V_{max}	Maximum recordable sensor voltage
$\mathcal{K}_N$	Bohr–Favard constant
B_N	B-spline of order N
$\mathbf{m}(x, y, \lambda)$	Measurements at location (x, y) for wavelength λ

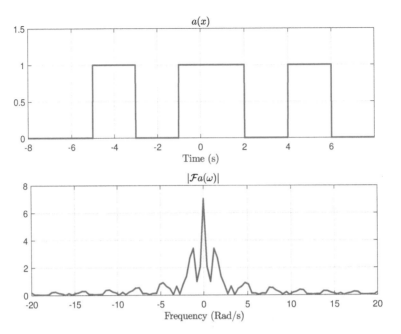

Figure E4.1 The aperture function (*top*) and the magnitude of its Fourier transform (*bottom*) in the case of aperture coded imaging.

Exercises

1. Aperture coding

    As explained in this chapter, by exposing selectively smaller regions within the aperture we can learn more about the light field captured. Projected in a single dimension, an aperture can be described by a function such as

    $$a(x) = 1_{[-5,-3]}(x) + 1_{[-1,2]}(x) + 1_{[4,6]}(x).$$

    Here, we denote by $1_{[-5,-3]}(x)$ a function that is 1 inside $[-5, -3]$ and 0 otherwise. Let us define its Fourier transform

    $$\mathcal{F}a(\omega) = \int_{\mathbb{R}} a(x)e^{-j\omega x}dx.$$

    The aperture function $a(x)$ and the absolute value of its Fourier transform are depicted in figure E4.1.

Exercises

a) Fourier transform

Using the properties of the Fourier transform, derive the expression of $\mathcal{F}a(\omega)$ analytically.

b) Modulation theorem

The aperture can be coded by placing a mask between the lens and the sensor, as in dappled photography. This involves an important result called the modulation theorem.

Prove that

$$\mathcal{F}\{\cos(\omega_0 x)s(x)\}(\omega) = \pi\left[\mathcal{F}s(\omega - \omega_0) + \mathcal{F}s(\omega + \omega_0)\right],$$

where $s(x)$ is an input signal.

c) Image blurring

In capturing a blurred image with coded aperture we can estimate the scene depth. Let us look at an example on how to blur and then deblur an image.

Feel free to capture your own image or use the one provided in this example. Reduce the image to a resolution of 220×220, and then convolve it with a blur kernel, which will be a matrix of ones with size 22×22. The kernel must be normalized a priori by dividing it by the sum of its elements, to ensure that the image values are in the same range. The result should be similar to the one shown in figure E4.2. Please plot your own images.

d) Image deblurring

Compute the naive deconvolution to recover the original image. The two-dimensional discrete Fourier transform can be used for this task.

2. Color coding

In this chapter, we learned that a sensor can be coded by placing together pixels sensitive to different colors in a certain pattern. Let us now simulate the image captured by a sensor coded with the Bayer pattern. For this example, we need a color image. Even though all colors have already been acquired at each pixel, the simulation cancels out the components corresponding to the other colors in order to generate a Bayer pattern.

Each color image is made up of three images, each containing the intensities of red, green, and blue, respectively. We then split each of the three images into groups of four adjacent pixels, as shown in figure E4.3. For the red image, we keep only pixel 4 and cancel out the other pixels. The same is repeated for the other two colors, to result in a Bayer pattern.

The resulting image, depicted in figure E4.4, is similar to an image captured by a digital camera with its sensor pixels arranged in a Bayer pattern.

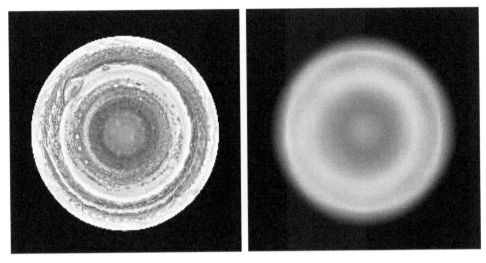

(a) Grayscale image of Jupiter (b) The image blurred with a matrix of ones

Figure E4.2 Simulating the out-of-focus blur effect by convolving with a kernel.

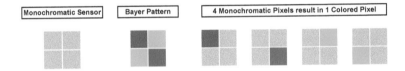

Figure E4.3 A monochromatic sensor (*left*), the pixels arranged in the Bayer pattern (*center*), and breaking down the Bayer pattern in three components for each RGB color (*right*).

a) Generating the Bayer pattern

   Now plot and display your own Bayer pattern image. To make sure that the pattern is visible, keep the resolution low, around 100×100.

b) Image demosaicking

   As mentioned previously in the chapter, we now need to recover the original image before mosaicking, in a process called demosaicking. Because we set a number of pixels to 0, we have lost $2/3$ of the information from the original image, as shown in figure E4.3.

   Assuming smoothness conditions for the captured image, we can still recover the original image with reasonable quality. Let us perform nearest neighbor and linear

Exercises

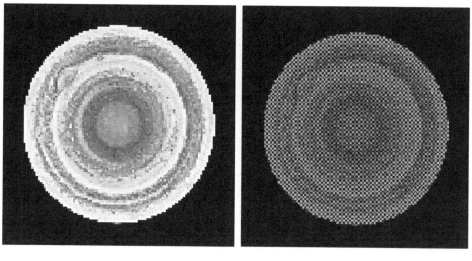

(a) Original image (b) Image mosaicked in the Bayer pattern

Figure E4.4 Simulating an image captured by a sensor in the Bayer pattern.

interpolation to compute the values of the missing pixels. The resulting images should be similar to the ones shown in figure E4.5. Please plot your own figures demonstrating demosaicking via interpolation.

Can you explain why each image looks in this particular way? Why does the linear interpolation have an effect similar to the blurring examined in the previous example?

3. Compressive imaging

   Here we analyze how the theory of compressive sensing can be used to generate compressed images. We use a grayscale image in this case, from which we extract a small patch of size 40×40 located on an edge, as shown in figure E4.6. To compress the samples we first turn the image $I(k,l)$ into a vector $I_v(k)$ of size $1,600$. The compressed samples satisfy

   $$y = D \cdot I_v, y \in \mathbb{R}^M$$

   where $M \ll N = 1,600$, and $D \in \mathbb{R}^{M \times N}$ is a random matrix with elements drawn from the Gaussian distribution with zero mean and standard deviation $1/N$.

   The recovery of the image is performed with one of the following:

   $$\widehat{I_v} = \arg\min \|I_v\|_2, \text{ s.t. } y = D \cdot I_v$$
   $$\widehat{I_v} = \arg\min \|I_v\|_1, \text{ s.t. } y = D \cdot I_v.$$

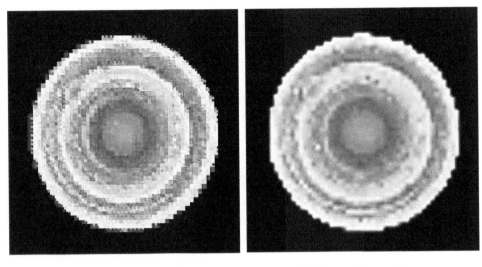

(a) Nearest neighbor interpolation   (b) Linear interpolation

Figure E4.5 Reconstructing the original full resolution image via demosaicking with nearest neighbor and linear interpolation.

a) Generating compressive samples
   Increase M until you achieve a good reconstruction using norm $\|\cdot\|_1$, and display the results for $\|\cdot\|_1$ and $\|\cdot\|_2$. The reconstructions using 70% of the samples for the proposed image ($M = 1,120$) are given in figure E4.6. Why does the result look so noisy when you use $\|\cdot\|_2$?

b) Selecting the number of compressive samples
   In the following, let us define the reconstruction error
   $$\text{Error} = 100 \cdot \frac{\left\|\widehat{I}_v - I_v\right\|_2}{\|I_v\|_2} \quad (\%).$$
   Plot the error function for reconstructing the image patch using $\|\cdot\|_1$ and $\|\cdot\|_2$ as a function of the number of compressed samples M. The result should look similar to figure E4.7. In order to get consistent results, make sure you generate a large random matrix D and then pick the first M lines for each iteration.

   How can it be explained that $\|\cdot\|_2$ leads to better results than $\|\cdot\|_1$ when M is small?

Exercises

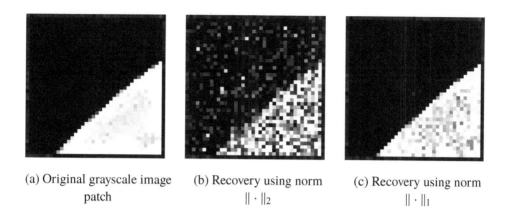

(a) Original grayscale image patch

(b) Recovery using norm $\|\cdot\|_2$

(c) Recovery using norm $\|\cdot\|_1$

Figure E4.6 Compressed image reconstruction using 30% fewer samples than the number of pixels in the original image.

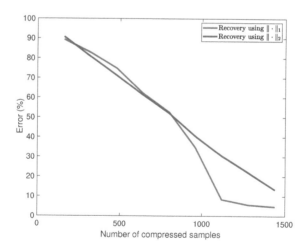

Figure E4.7 The reconstruction error as a function of the number of compressed samples M.

5 Temporally Coded Imaging

In recent years many approaches have been proposed to capture the high-dimensional plenoptic function by introducing diversity into its low-dimensional measurements. This idea is explained in Figure 5.1. For example, emerging from multiple views leads to viewpoint diversity. Similarly, illuminating a scene from different positions leads to illumination diversity. In both these examples, redundant information is then used to reconstruct the plenoptic function. Redundancy in measurements may also be introduced by spatial coding. As we have previously discussed in the case of dappled photography, coding the sensor leads to the recovery of the four-dimensional light field from a two-dimensional image.

In this chapter, we focus on a different aspect of the plenoptic function that helps us go beyond the steady scene assumption. In particular, if we consider the speed of light to be finite, then the information in the echoes of light can be harnessed in unconventional ways. The reason is that when the light interacts with a scene, the information about the scene is parametrically encoded in the time delay of light arriving at the sensor. This recently emerging field of research is known as time resolved imaging (TRI) or time-of-flight (ToF) imaging. TRI fundamentally combines time-stamped photos with computational methods to redefine the conventional camera.

This new way of reinterpreting the camera directly leads to applications such as 3D imaging as well as fluorescence lifetime imaging. Beyond conventional applications, (Velten et al., 2012a) demonstrated that the information in time delays can be used for non-line-of-sight imaging.

The goal of this current chapter is to develop an understanding about imaging a scene at different time scales. The technological challenge behind this idea is that capturing time information at the speed of light requires exorbitant sampling rates and very sophisticated apparatus. However, both of these restrictions can be relaxed by using computational imaging-centric approaches. The key idea here is to recover the temporal information using computational approaches.

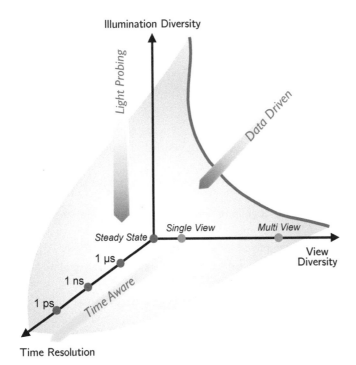

Figure 5.1 Conceptual diagram showing different mechanisms for creating diversity in measurements when capturing the high-dimensional plenoptic function.

5.1 A Brief History of the Time-of-Flight Revolution

The ToF principle exploits the idea that distance and time are proportional quantities. As the name suggests, ToF is the round-trip time between the source and the destination taken by a particle or a wave. Hence, knowing one entity is equivalent to knowing the other. Nature is replete with examples that rely on the ToF principle. For example, bats, dolphins (Au and Benoit-Bird, 2003), and visually impaired human beings use the ToF principle for navigational purposes.

From a chronological standpoint, the use of sound waves was known to human beings long before the notion of electromagnetic waves came to be known. Human beings have known to use stones for estimating the depth of wells for millennia. One of the earliest attempts using electromagnetic waves for ToF measurements can be traced back to an experiment conducted by the Italian scientist Galileo, who together with his assistant wanted to estimate the speed of light. It is well known now that the speed of light is

$$3 \times 10^8 \text{ m/s}.$$

5.1 A Brief History of the Time-of-Flight Revolution

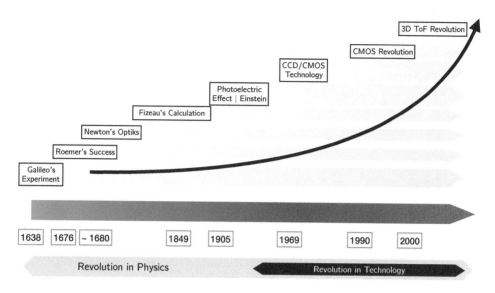

Figure 5.2 Brief history of the time-resolved imaging revolution.

It was unknown at the time of Galileo, who tried to estimate the speed of light by using two far apart hills to time the flicker of a lantern. Unfortunately, this chosen distance was not adequate given the speed of light, and hence his experiment was inconclusive.

Making progress on this front, the Danish astronomer Ole Römer worked with planetary distances and overcame the hurdle in Galileo's experiment. About two hundred years later (circa 1849), French physicist Hippolyte Fizeau was the first to estimate the speed of light. Following that, Albert Abraham Michelson, improving upon the previous experiments of Hippolyte Fizeau and Foucault, in 1879 used a laboratory setup to estimate the speed of light to be 2.99864×10^8 m/s.

Further revolutions, such as the discovery of the photoelectric effect by Albert Einstein in the early twentieth century followed by the development of the electronic imaging sensors (CCD/CMOS), led to the development of consumer-grade, mass producible optical ToF sensors. An example of such a device is the popular Microsoft Kinect XBox One released in 2013. A brief history of major scientific and technological revolutions culminating into the modern day ToF sensor technology is shown in Figure 5.2.

In contrast to the conventional image or a photograph produced by imaging sensors, the ToF sensors capture 3D images. This is made possible by recording two images per exposure: an amplitude image and a depth image. The amplitude image is the standard two-dimensional photograph. For the same exposure, at each pixel, the unique depth image represents the corresponding distance in the scene. This is based on the ToF information.

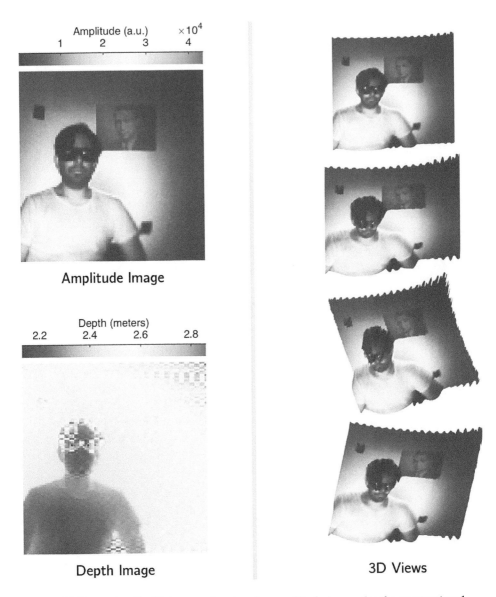

Figure 5.3 Example of a 3D image, showing the amplitude image (or the conventional digital image), the depth image, and 3D images seen from multiple viewpoints.

The combination of the amplitude and depth images produces the 3D image. We show the amplitude, depth, and resulting 3D images in Figure 5.3.

Early scientific instrumentation for computational imaging based on the ToF principle required high-quality equipment that was often fragile, prohibitively expensive, and constrained to controlled laboratory environments. The reason is that fundamental to the ToF principle is the fact that the speed of light is assumed to be finite. This, in practice, is achievable only when electro-optical elements of the imaging system are extremely precise. However, in the context of the gaming and entertainment industry, a number of consumer companies such as Mesa Imaging, Microsoft, and PMDtec have developed consumer-grade ToF sensors that are not only affordable but also alleviate all the issues associated with their expensive and sensitive counterparts—custom-designed scientific hardware.

While optical ToF sensors are a recent phenomenon, other ToF systems such as ultrasound, seismic, and radar technologies have been around for decades. More recently, terahertz imaging systems (also based on ToF principle) have become increasingly popular. The knowledge transfer between optical and other ToF systems is far from reality. Each ToF modality has its own idiosyncratic constraints that stem from the physics of the problem. However, there are commonalities that are shared by all these systems.

5.2 Optical Time-Resolved Imaging

In conventional imaging, each sensor pixel integrates the photo-generated carriers over a time interval, creating a low-dimensional projection of the plenoptic function. The time window of the exposure (or integration) is typically in the range of milliseconds. Hence, conventional digital sensors provide a count of photons reaching each pixel, producing a photograph. Such a photograph lacks any time-resolved information. On the other hand, during the time of exposure the photons travel distances that are much larger than the scale of the scene, and this detail can provide information about the scene that is not available with a conventional image. At the end of the imaging process, one can only estimate the number of photons that arrive at each pixel and not the time-of-arrival of each of those photons. Such an information loss is unavoidable with a conventional imaging sensor.

To overcome this barrier, time-resolved imaging sensors bring the time dimension to imaging, eventually becoming able to image the light in motion as it propagates through the scene (Velten et al., 2012a). In working time scales at the speed of light, a depth resolution on the order of millimeters translates into picosecond time resolution. It is not possible to build large arrays of conventional pixels that can control the integration windows of the order of picosecond or subpicosecond resolution. Consequently, there is a natural trade-off:

Spatial resolution		**Time resolution**
	versus	
Number of pixels in the imager		Time scale of exposure

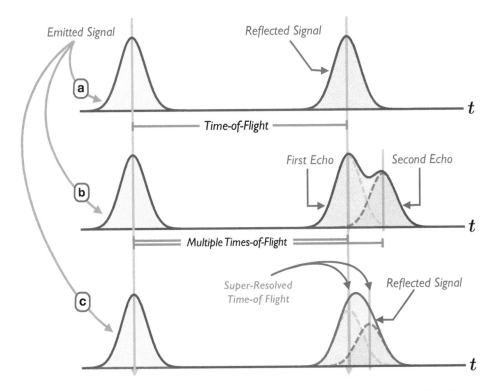

Figure 5.4 Time-resolved information at a single pixel in the following cases: (a) when a signal is backscattered from a single object; (b) when the signal is backscattered from two objects, such as in imaging through a window pane; and (c) similarly to (b) but in more challenging scenario, when reflections occur from closely spaced objects. Recovering individual light paths in (c) is known as superresolution.

Reconciling this fundamental trade-off, a variety of time-resolved imaging hardware is available, such as

- Gated cameras,
- single photo avalanche diode (SPAD) arrays,
- Ultrafast probing with a single detector,
- Ultrafast probing with streak cameras, and
- Lock-in sensors (coded and continuous wave).

Each of these imaging modalities is an active imaging system. In such systems, the scene is illuminated with a light source, and reflected information is captured as a temporal signature at each pixel. Hence the measurements can be written,

$$m(x, y, t)$$

5.3 Time-Resolved Image Formation Model

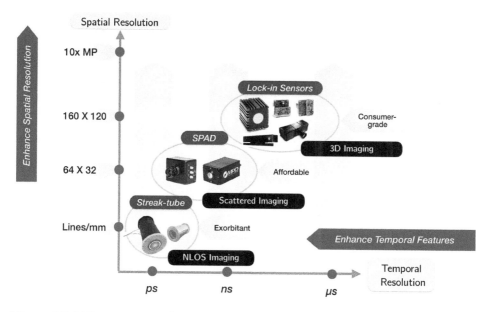

Figure 5.5 Different time-resolved imaging sensors with their spatiotemporal parameters. Subnanosecond and picosecond range illumination resembles a spike; hence SPAD and streak tube-based sensors are known as impulse imaging devices. Lock-in sensors use a periodic waveform with frequencies in the range of a few megahertz to a few hundreds of megahertz; such sensors are known as continuous-wave imaging sensors.

where (x, y) denotes the spatial coordinates and t denotes the time. For a single-pixel time-resolved sensor, we plot the measurement as shown in Figure 5.4.

As we have mentioned, time-resolved sensors are active sensors, and depending on the active mode of operation, they may be classified as *impulse-based* or *continuous-wave sensors*. This classification is based on the shape of the active illumination, which affects the temporal resolution. For subnanosecond and picosecond range illumination, the pulse shape resembles a spike in time, and hence this class of sensors is referred to as impulse-based systems. In working with wavelike illumination, the current sensors are able to generate temporal waveforms (typically periodic) in the range of a few megahertz to few hundred megahertz. Such sensors are known as continuous-wave sensors. Figure 5.5 shows different time-resolved imaging sensors together with their spatiotemporal parameters.

Next we present a general image formation model that is common to most of the time-resolved imaging systems.

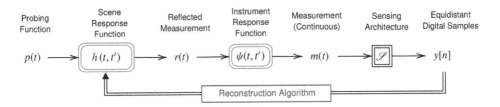

Figure 5.6 Time-resolved imaging pipeline.

5.3 Time-Resolved Image Formation Model

The time-resolved imaging systems follow the imaging pipeline shown in Figure 5.6. Its basic elements are as follows.

5.3.1 Probing Function

The probing function denoted by $p(t)$ represents the waveform emitted by the ToF sensor's illumination unit. The probing function may be a time-localized pulse, for example, a B-spline, Gaussian, or exponential-Gaussian mixture. Alternatively, this may be a continuous wave. This is decided by the time-resolved imaging apparatus being used.

- **Streak-tube** (Velten et al., 2012a; Wu et al., 2013) and **SPAD detectors** (Hernandez-Marin et al., 2007) offer picosecond range timing, and hence the illumination can be characterized as an impulse or the Dirac delta function $\delta(t)$.
- In the case of **continuous-wave imagers** that use the lock-in mechanism, one may use either sinusoidal illumination (this is the case with Microsoft Kinect XBox One (Bhandari et al., 2014a) or maximum length sequences (Kadambi et al., 2013) to achieve time localization.
- Other examples include the use of first- and second- order derivatives of Gaussian pulses, which are used in ultra wideband systems (Chen and Kiaei, 2002).

5.3.2 Scene Response Function

The scene response function (SRF), denoted by $h(t, t')$, models the transfer function of the scene. This may be a filter, a shift-invariant function such as

$$h(t, t') = h_{\text{SI}}(t - t')$$

or even a solution to a partial differential equation. This depends only on the scene that one is interested in imaging and not the modality of the time-resolved imaging system. The different classes of scene response functions are as follows.

5.3 Time-Resolved Image Formation Model

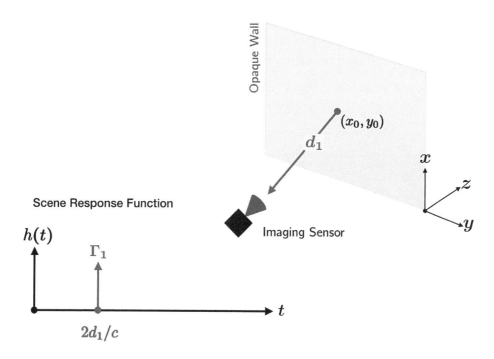

Figure 5.7 A scene with one light path.

- **Single depth imaging** In this setting, shown in Figure 5.7, there is a one-to-one mapping between the scene and the sensor. At any given pixel, the SRF in this case is written

$$h(t, t') = \Gamma_1 \delta\left(t - t' - \frac{2d_1}{c}\right) = \Gamma_1 \delta(t - t' - t_1) \tag{5.1}$$

where Γ_1 denotes the reflectivity and $t_1 = 2d_1/c$ is the time delay due to the object at a distance d_1 from the sensor.

- **Multiple depth imaging** As shown in Figure 5.8, there may be scenes in which more than a single light path contributes to a given pixel in the sensor. In this case, multiple delays need to be accounted for at the detector. Consequently, when accounting for K light paths, the SRF is written

$$h_K(t, t') = \sum_{k=0}^{K-1} \Gamma_k \delta\left(t - t' - \frac{2d_k}{c}\right) = \sum_{k=0}^{K-1} \Gamma_k \delta(t - t' - t_k) \tag{5.2}$$

where Γ_k and t_k for $k = 0, \cdots, K-1$ are the $2K$ unknown parameters.

- **Fluorescence lifetime imaging (FLI)** FLI is a significant research area spanning many engineering applications. Knowledge of a sample's fluorescence lifetime allows ap-

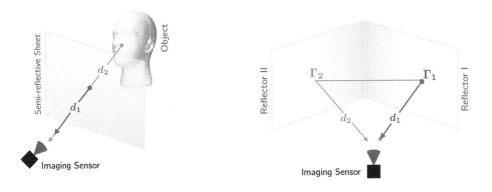

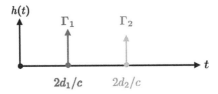

Figure 5.8 Examples of scenes that lead to two light paths.

plications such as DNA sequencing, tumor detection, fluorescence tomography, and high-resolution microscopy. In this case, temporal excitation is modeled by a first-order differential equation whose solution is the SRF given by (Bhandari et al., 2015),

$$h(t, t') = h^{\text{Depth}}(t, t') + h^{\text{Decay}}(t, t')$$

where $h^{\text{Depth}}(t, t')$ is defined in (5.2) and represents a delay of t_1 due to the fluorescence sample's placement at depth $2d_1/c$ meters from the sensor, and

$$h^{\text{Decay}}(t, t') = \mu \exp\left(-\frac{t - t' - \tau}{\lambda}\right) \Pi(t - t' - \tau)$$

where μ and λ are emission strength and the lifetime of the fluorescent sample respectively, and $\Pi(t)$ is the Heaviside function. A plot of the SRF is shown in Figure 5.9.

- **Transient imaging and subsurface scattering** Decomposing global light transport in terms of its temporal response leads to the study of transient phenomena. The first steps in this direction were set in (Wu et al., 2013). Using time-resolved information, a scene can be decomposed into its constituent light transport elements. In particular, a scene can be decomposed into the

5.3 Time-Resolved Image Formation Model

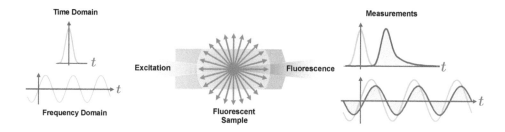

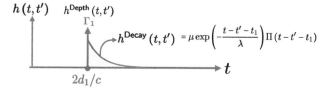

Figure 5.9 Fluorescence lifetime imaging.

- *Direct component*, corresponding to all the first bounce light that arrives at the sensor;
- *Subsurface component*, corresponding to short-range scattering;
- *Interreflection component*, corresponding to longer range scattering. Based on this model, the transient SRF is written

$$h_{Tr}(t, t') = \underbrace{h_D(t - t')}_{\text{Direct reflection}} + \underbrace{h_I(t - t')}_{\text{interreflections}} + \underbrace{h_S(t - t')}_{\text{Subsurface}},$$

where the individual contributions are as follows:

- Direct component

$$h_D(t) = \alpha_D \delta(t - 2d_D/c)$$

- Interreflections

$$h_I(t) = \sum_{k=0}^{K-1} \alpha_k \delta(t - 2d_k/c)$$

- Subsurface scattering

$$h_S = \delta(t - 2d_S/c) * \left(\alpha_S e^{-\beta_S t} \mathbb{1}_{t \geq 0}(t) \right)$$

5.3.3 Reflected Function

The reflected function is the backscattered signal arising from the interaction between the probing signal and the SRF. The reflected function is modeled as a Volterra/Fredholm integral

$$r(t) = \int_{\Omega_1} p(\tau) h(t,\tau) d\tau. \tag{5.3}$$

Whenever the SRF is a shift-invariant kernel, that is, $h(t,t') = h_{SI}(t-t')$, the reflected signal is simply a convolution/filtering operation between the probing function and the SRF

$$r(t) = (p * h_{SI})(t).$$

5.3.4 Instrument Response Function

The instrument response function or the IRF denoted by $\Psi(t,t')$ models the transfer function of the electro-optical elements of the ToF sensor. For example, in conventional digital cameras, the spatial IRF is the point spread function of the lens. The different classes of instrument response functions are as follows.

- **LiDAR systems** In the case of SPAD-based LiDAR systems, as presented in the literature (Buller and Wallace, 2007; Hernandez-Marin et al., 2007), the IRF due to SPAD detectors may be modeled as a parametric, shift-invariant kernel of form

$$\Psi_L(t) = \alpha \begin{cases} \theta\left(-\frac{(T_1-T_0)^2}{2\sigma^2}\right)\theta\left(\frac{t-T_1}{\lambda_1}\right), & t < T_1 \\ \theta\left(-\frac{(t-T_0)^2}{2\sigma^2}\right), & t \in [T_1, T_2) \\ \theta\left(-\frac{(T_2-T_0)^2}{2\sigma^2}\right)\theta\left(-\frac{t-T_2}{\lambda_2}\right), & t \in [T_2, T_3) \\ \theta\left(-\frac{(T_2-T_0)^2}{2\sigma^2}\right)\theta\left(-\frac{(T_3-T_2)^2}{\lambda_2}\right)\theta\left(-\frac{t-T_3}{\lambda_3}\right), & t \geqslant T_3 \end{cases}, \tag{5.4}$$

where $\theta(t) = e^{-t}$ and $\mathbf{L}$ is an unknown parameter vector

$$\mathbf{L} = [\alpha \; \sigma \; T_0 \; T_1 \; T_2 \; T_3 \; \lambda_1 \; \lambda_2 \; \lambda_3]^T. \tag{5.5}$$

- **Lock-in mechanism–based ToF sensors** This is one of the most widely used mechanisms for consumer-grade time-resolved imaging. In this case, IRF is used as the probing function,

$$\Psi(t,\tau) = p(t+\tau).$$

- **Streak tube** In this case, as pointed out in (Wu et al. 2013), the IRF may be modeled as a Gaussian profile

$$\Psi(t,t') = \exp\left(-\frac{(t-t')^2}{2\sigma^2}\right).$$

5.3.5 Continuous-Time Measurements

Measurements denoted by $m(t)$ are a result of sensing the reflected signal via the electro-optical elements of the ToF sensor. Continuous-time measurements are modeled as,

$$m(t) = \int_{\Omega_2} r(t') \Psi(t, t') \, dt'. \tag{5.6}$$

5.3.6 Discrete-Time Measurements

The sampling operator maps the continuous-time measurements to digital samples

$$\mathcal{S}: m(t) \to y[n] = m(nT), \quad n \in \mathbb{Z}, T > 0.$$

The ToF sensor stores discrete measurements by sampling continuous-time signal $m(t)$, which results in the discrete sequence $y[n] = m(nT), n \in \mathbb{Z}$ where $T > 0$ is the sampling interval.

This general model applies to modalities beyond the optical time-resolved imaging systems. Whereas the probing and the instrument response functions are characterized by the imaging modality at hand, the scene response function is entirely characterized by the scene that we are interested in imaging. The different variations of the probing functions, namely SRF and IRF, lead to different forms of computational imaging problems.

In many practical cases of interest, both the SRF and the IRF are shift invariant. In that case, the measurements can be written as a convolution product or $m(t) = (p * h * \Psi)(t)$. Whenever the IRF is a function of the form $\Psi(t, \tau) = \Psi(t + \tau)$, the measurements amount to $m(t) = (r \otimes \Psi)(t)$, where $\otimes$ denotes a cross-correlation operation. Lock-in sensors operate on this principle.

5.4 Lock-in Sensor–based 3D Imaging

5.4.1 Continuous Wave Imaging

ToF sensors such as the Microsoft Kinect XBox One use a continuous-wave-based probing function $p(t) = 1 + p_0 \cos(\omega t)$, $p_0 < 1$, where ω is the modulation frequency and p_0 is the modulation amplitude. With the SRF defined in (5.2), the reflected signal reads $r(t) = \Gamma_0 p(t - t_0)$ where $t_0 = 2d_0/c$. Again, the lock-in sensor acts as an electronic homodyne detector such that $\Psi(t, \tau) = \Psi(t + \tau)$ and the measurements result in (Bhandari et al., 2014b; Gupta et al., 2015)

$$m(t) = \Gamma_0 \left(1 + \frac{p_0^2}{2} \cos(\omega t + \theta_\omega) \right), \quad \theta_\omega \in [0, 2\pi) \tag{5.7}$$

where $\theta_\omega = \omega t_0 = 2d_0\omega/c$ is the frequency dependent phase. The ToF sensor records discrete measurements of form, $m_k = m(kT)$ with $T = \pi/2\omega$ and uses a phase estimation algorithm commonly known as the **four bucket method** (Foix et al., 2011) to estimate the

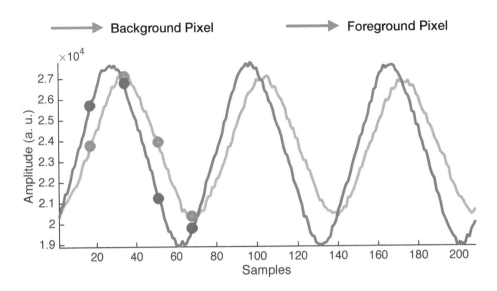

Figure 5.10 Raw data samples based on a continuous wave, time-of-flight imaging sensor.

unknown parameters Γ_0 and d_0. For a given modulation frequency, this method works with four discrete measurements

$$m_0 \quad m_1 \quad m_2 \quad m_3$$

that are used to form a complex number $z_\omega \in \mathbb{C}$

$$z_\omega = (m_0 - m_2) + j(m_3 - m_1)$$

where,

$$\begin{bmatrix} m_0 & m_1 \\ m_2 & m_3 \end{bmatrix} = \frac{\Gamma_0}{2} \begin{bmatrix} 2 + p_0^2 \cos(\omega t_0) & 2 - p_0^2 \sin(\omega t_0) \\ 2 - p_0^2 \cos(\omega t_0) & 2 + p_0^2 \sin(\omega t_0) \end{bmatrix}$$

The scene parameters are then estimated

$$\Gamma_0 = \frac{|z_\omega|}{p_0^2} \text{ and } d_0 = c\frac{\angle z_\omega}{2\omega}.$$

In fact, the depth images shown in Figure 5.3 are obtained using this method with $\omega = 2\pi f$ where $f = 50\,\text{MHz}$. The raw data samples corresponding to the experiment are shown in Figure 5.10.

5.4.2 Coded Time-of-Flight Imaging

In the previous section, sinusoidal illumination was studied for time-resolved imaging. As an alternative mode of imaging, time localized pulses may also be used to probe the scene. Even though the probing function is time-localized, it may be modeled as a periodic signal of form $p(t) = p(t + \Delta)$, $\Delta > 0$. Although specialized scientific instruments such as the streak tube, may be able to produce a pulse that mimics the Dirac's delta distribution δ (Velten et al., 2012a), this form of precision is impractical for consumer-grade instruments. In practice, either the pseudonoise (PN) sequences (Buttgen and Seitz, 2008) or the maximum-length sequences (MLS) (Kadambi et al., 2013) is an optimal choice of probing function in regard to time-localization. For further study on the choice of optimal codes, we suggest (Gupta et al., 2018).

In this case, given the SRF defined in (5.2), the reflected signal is,

$$r(t) = \Gamma_0 p(t - t_0),$$

with delay $t_0 = 2d_0/c$. Due to the lock-in sensor architecture (Foix et al., 2011), which constrains the IRF to the form

$$\Psi(t, \tau) = p(t + \tau)$$

the measurements simplify to

$$m(t) = (r \otimes p)(t) = (\bar{r} * p)(t)$$

where $\otimes$ denotes the cross-correlation operation and $\bar{r}(t) = r(-t)$. Due to the lock-in constraint, we can rewrite the measurements in terms of $p(t)$

$$\bar{m}(t) = \Gamma_0 (p * \bar{p})(t - t_0).$$

Consequently, we may write $\phi = p * \bar{p}$. The ToF is then estimated by estimating the delay using,

$$\widetilde{t_0} = \arg\max_{t_0} \bar{m}(t) = \arg\max_{t_0} \phi(t - t_0).$$

Whenever $p(t)$ is modeled to be some parametric waveform, such as a Gaussian function, B-spline, or a combination of parametric pulses, parameter estimation techniques may be used to estimate the ToF t_0, and the reflection coefficient Γ_0. However, this may not be the case in practice because of model mismatch or the physical aspects of light propagation. In such a setting, it is effective to use the property of *bandlimited approximation*: approximate the probing function, and hence ϕ, with the first few dominant Fourier components,

$$\widetilde{p}(t) = \sum_{|m| \leq M_0} \widehat{p}_m e^{jm\omega_0 t} \text{ with } \widehat{p}_m = \frac{1}{\Delta} \int_0^\Delta p(t) e^{-jm\omega_0 t} dt, \qquad (5.8)$$

where $\omega_0 = 2\pi/\Delta$ is the fundamental frequency, and Δ is the maximum operating range of the ToF sensor. The reasons behind this choice are that

- most electronic/optical instrumentation is approximately bandlimited due to physical constraints; and

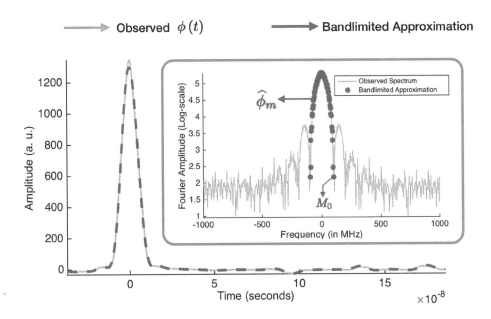

Figure 5.11 Bandlimited approximation of autocorrelated probing signal ($\phi = p * \overline{p}$) in time-domain ToF setup. The low-pass property is evident from its Fourier spectrum. This is a result of an experiment with $\Delta = 310$ ns and $M_0 = 30$.

- the probing function may not admit a parametric representation. Even if the probing function assumes a parametric representation, bandlimited approximation via Fourier series coefficients circumvents the estimation of parameters of the probing function.

The utility of the bandlimited approximation property is demonstrated via experiments shown in Figure 5.11. Starting with a maximum length sequence (Kadambi et al., 2013), we design a probing function. We plot $\phi = p * \overline{p}$ together with its *bandlimited approximation*

$$\widetilde{\phi}(t) = \sum_{|m| \leq M_0} \widehat{\phi}_m e^{\jmath m \omega_0 t} \text{ with } \widehat{\phi}_m = \frac{1}{\Delta} \int_0^\Delta \phi(t) e^{-\jmath m \omega_0 t} dt, \quad (5.9)$$

obtained by retaining the first M_0 Fourier series coefficients, $\widehat{\phi}_m = \widehat{p}_m^2$. We are thus able to rewrite the measurements

$$\overline{m}(t) = \Gamma_0 \phi(t - t_0) \equiv C_0 \sum_{|m| \leq M_0} \widehat{\phi}_m e^{\jmath m \omega_0 t}$$

where the complex-valued constant $C_0 = \Gamma_0 e^{-\jmath \omega_0 t_0}$ is the constant to be estimated.

5.5 Application Areas

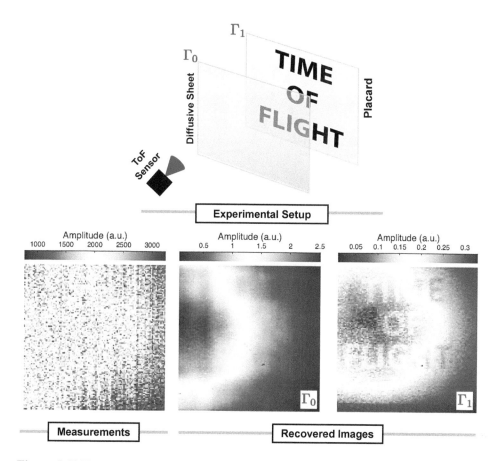

Figure 5.12 Experimental setup for diffuse imaging. The goal is to be able to read the placard that is covered by a diffusive surface. Although conventional measurements seem corrupted by noise (due to specular reflection), in working with time-of-flight sensors, it is possible to recover the hidden information.

5.5 Application Areas

In this section, we discuss the various applications of temporally coded imaging that have emerged in the last decade and cover several topics that include the areas of scientific and bioimaging, computer vision, and computer graphics.

5.5.1 Diffuse Imaging

In diffuse imaging, we are interested in recovering scene information that is covered by a diffusive object. The scenario is shown in Figure 5.12. In the context of our discussion,

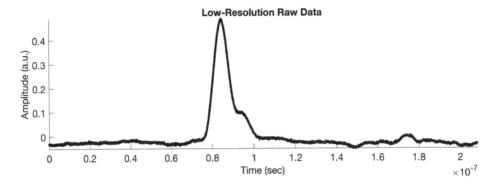

Figure 5.13 Coded time-of-flight measurements at a given pixel corresponding to the experimental setup in Figure 5.12. The probing signal used in this experiment is shown in Figure 5.11.

the scenario leads to a scene response function in (5.2) that arises due to the schematic shown in Figure 5.7. Following the discussion in section 5.4.2, the reflected signal reads, $r(t) = \Gamma_0 p(t - t_0) + \Gamma_1 p(t - t_1)$ where, as usual, $\{\Gamma_0, \Gamma_1\}$ are the unknown amplitudes and $\{t_0, t_1\}$ are the unknown delays due to the depth of the respective reflections. As shown in (Kadambi et al., 2013), in working with the coded time-of-flight system, the measurements are

$$m(t) = \Gamma_0 \phi(t - t_0) + \Gamma_1 \phi(t - t_1) \text{ where } \phi = p * \overline{p}.$$

For a given pixel, the plotted measurements are shown in Figure 5.13, and the probing function used in this experiment is shown in Figure 5.11. There are different methods for recovering the unknown SRF from the measurements, and one of the most common approaches leverages the idea of sparsity in that there are very few reflections in the scene in comparison to the dimensionality of the measurements. This essentially arises because that the SRF defined in (5.2) is a sparse signal that consists of spikes. To this end, the sampled measurements $m(nT)$, $n = 0, \cdots, N - 1$ can be stacked into a vector **m**. According to the measurement model, the measurements can be written

$$\mathbf{m} = \mathbf{Ts},$$

where

- **T** is a Toeplitz/convolution matrix comprising shifts of the pulse ϕ, that is, each element of the matrix **T** is of the form, $T_{n,k} = \phi((n-k)T)$ where T is the measurement sampling rate; and
- **s** is a K-sparse vector, that is, a vector of size N of which K elements are nonzero.

5.5 Application Areas

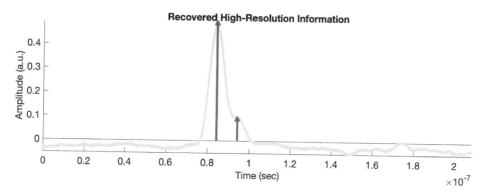

Figure 5.14 Recovering the unknown (sparse) scene parameters using the orthogonal matching pursuit algorithm.

According to this model, we are interested in recovering a sparse vector **s** from the measurements **m**, that is

$$\mathbf{s}^* = \arg\min_{\mathbf{s}} \|\mathbf{m} - \mathbf{Ts}\|_2^2 \text{ subject to } \|\mathbf{s}\|_0 \leqslant K$$

or, in other words, find a vector **Ts** that is closest to the measurements **m** in the least-squares sense such that the unknown vector **s** has at most K nonzero values. This problem can be solved using the *orthogonal matching pursuit* (OMP) algorithm (Elad, 2010) and this is the approach followed in (Kadambi et al., 2013). In our example, $K = 2$, and the result of running the OMP algorithm is that we are able to find the vector **s** whose elements are $\{\Gamma_0, \Gamma_1\}$. The corresponding nonzero locations in the vector provide the information about $\{t_0, t_1\}$. The result is shown in Figure 5.14. By processing the time-resolved information at each pixel, we are able to recover the hidden information that lies behind the diffuser. This is shown in Figure 5.12.

5.5.2 Light-in-Flight Imaging

Whereas ultrafast light-in-flight imaging was proposed in (Velten et al., 2012a,b, 2013), the apparatus used in this work consisted of the streak tube. Replacing this sophisticated and expensive imaging system by a simpler, consumer-grade time-of-flight sensor, in their work, (Heide et al., 2013) demonstrated an alternative approach to low-budget light-in-flight imaging. Following the insights developed in (Wu et al. 2013), Heide et al. modeled the scene information at pixel as a measurement of the form,

$$m(t) = \underbrace{\sum_{k=0}^{K-1} \Gamma_k g_\sigma (t - t_k)}_{\text{Reflections}} + \underbrace{s_k e^{-\lambda_k (t - t_k)}}_{\text{Scattering}}$$

where $\phi(t) = g_\sigma(t)$ is modeled as a Gaussian pulse, and $\{s_k, \lambda_k\}$ are the scattering coefficients. By resorting to an alternating minimization approach, the authors were able to recover the unknown scene parameters

$$\bigcup_{k=0}^{K-1} \{\Gamma_k, s_k, \lambda_k, t_k\}$$

which in turn allows for visualization of the light-in-flight phenomenon. This is shown in Figure 5.15a. By ignoring the scattering coefficients s_k, $k = 0, \cdots, K-1$, one is left with a K-sparse signal, and in that case the theory of section 5.5.1 applies. The corresponding light-in-flight phenomenon was presented in (Kadambi et al., 2013). We show the results in Figure 5.15b. A frequency domain analysis for recovering transient information corresponding to the light-in-flight imaging setup was considered in (Lin et al., 2017). An inexpensive time-of-flight camera was proposed to perform low-budget transient imaging (Heide et al., 2013). Light transport analysis of the time-resolved phenomenon was discussed in (O'Toole et al., 2014). A survey of the recent advances in transient imaging from a graphics and vision perspective is given in (Jarabo et al., 2017).

5.5.3 Multidepth Imaging

As mentioned previously, almost all consumer-grade sensors, such as the Microsoft Kinect Xbox One (Bhandari et al., 2014b) and the PMD sensor (Bhandari et al., 2014), are based on the continuous-wave model, that is, the probing signal is of the form $p(t) = 1 + p_0 \cos(\omega t)$, $p_0 < 1$, where ω is the modulation frequency and p_0 is the modulation amplitude (cf. section 5.4.1). Such models are designed to work with single depth only. In practice, however, the scene may be composed of multiple depths (see, for example, Figure 5.7) and in this case the scene response function (SRF) is given by (5.2). For the case of two depths, we show the SRF in Figure 5.16a. Consequently, the continuous-wave measurements take the form (assuming that $p_0 = 1$ in the definition of the probing signal)

$$m(t, \omega) = \frac{1}{2} e^{j\omega t} \sum_{k=0}^{K-1} \Gamma_k e^{j\omega t_k},$$

which is a sum of K complex exponentials or phasors. Because the probing function is a sinusoidal waveform, from basic linear systems theory we know that sinusoidal functions are eigen-functions of a linear system, and hence the measurements amount to observing the Fourier transform of the scene response function at the modulation frequency; this is shown in Figure 5.16b. Clearly, given measurements, $m_k = m(kT, \omega)$ with $T = \pi/2\omega$, it is impossible to discern the multiple depth information encoded in the unknowns $\{\Gamma_k, t_k\}$ where $k = 0, \cdots, K-1$ because the different phases add up for a given frequency. To overcome this problem, the idea used in the literature is to use multiple frequency

5.5 Application Areas

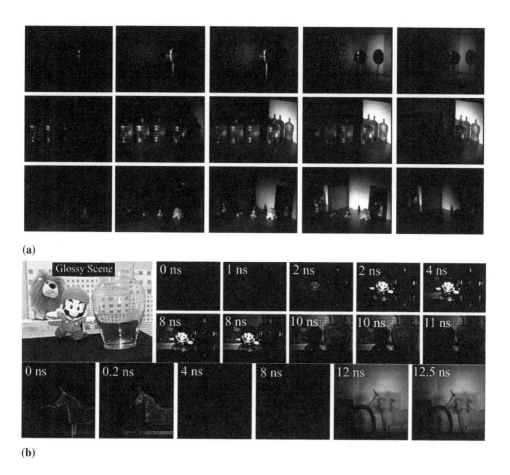

Figure 5.15 Time slices of different scenes demonstrating light-in-flight imaging. (a) Adapted from (Heide et al., 2013). (b) Adapted from (Kadambi et al., 2013).

measurements that trace a path parameterized by individual phasors, that is

$$m(t, \omega) \propto \underbrace{\Gamma_0 e^{J\omega t_0}}_{\text{Phasor 1}} + \underbrace{\Gamma_1 e^{J\omega t_1}}_{\text{Phasor 2}} + \cdots + \underbrace{\Gamma_{K-1} e^{J\omega t_{K-1}}}_{\text{Phasor } K},$$

and for the case with $K = 2$, we show the resulting measurements and the phasors in Figure 5.16c. Given multiple frequency measurements, the individual phasors can be estimated using the orthogonal matching pursuit (Bhandari et al., 2014) or sparse regularization (Freedman et al., 2014). This specific problem is also known as the spectral estimation method in signal processing literature (Bhandari and Raskar, 2016; Kirmani et al., 2013) and similar algorithms can be tailored for multidepth imaging. Examples of reconstructions

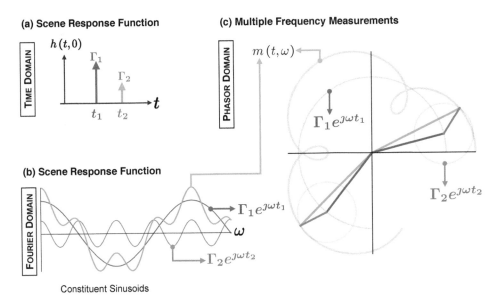

Figure 5.16 Continuous-wave imaging with two depths. (a) Scene response function in the time-domain. (b) Scene response function in the Fourier domain. (c) Multiple frequency measurements amount to a phasor addition, and the identification of the scene response function is equivalent to estimation of the phasor components.

for different cases are shown in Figure 5.17. When multidepth components are undesirable, the setting is known as multipath interference. This topic has been studied in detail.

An ultrafast imaging approach to capture space-time images that can separate different bounces of light based on path length was proposed in (Naik et al., 2011). A light transport model for mitigating multipath interference was discussed in (Naik et al., 2015), and the same work compares and contrasts the various approaches related to multipath interference.

5.5.4 Fluorescence Lifetime Imaging

Fluorescence lifetime imaging (FLI) is an established tool for estimating useful information that is otherwise unavailable from a conventional intensity image. Due to the high precision optics and electronics involved in the setup, the imaging system is quite expensive. This is a situation in which consumer-grade sensors such as the time-of-flight imaging system can be leveraged. When a fluorescent sample is exposed to a continuous-wave illumination at a given modulation frequency, the interaction results in a phase shift parameterized by the lifetime of the sample. Unlike depth imaging, in which phase shift is a linear function of the unknown depth parameter, in lifetime imaging the dependence of the phase on the lifetime is a nonlinear function (Bhandari et al., 2015). In particular, the observed phase is a sum of a line (with slope depending on depth) and arctangent function of the lifetime. The observed

5.5 Application Areas

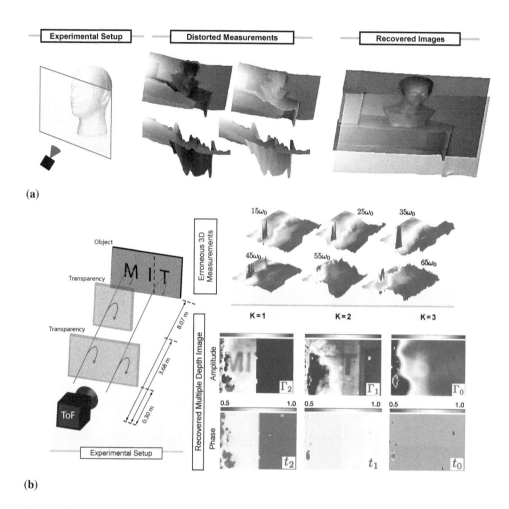

Figure 5.17 Example of imaging with multiple depths. (a) The case when $K = 2$. Measurements are based on the Microsoft Kinect XBox One, adapted from (Bhandari et al., 2014b). (b) The case when $K = 3$. Measurements are based on the PMD sensor, and the experiment was adapted from (Bhandari et al., 2014).

phase is shown in Figure 5.18. However, because the parametric form of this relationship is known, parameter estimation allows for lifetime imaging using consumer-grade depth sensors.

5.5.5 Non-Line-of-Sight Imaging

Seeing an object that is not in our line of sight is easy to do using mirrors. However, mirrors are characterized by *specular reflection*, which allows seeing the target easily. A wall, for

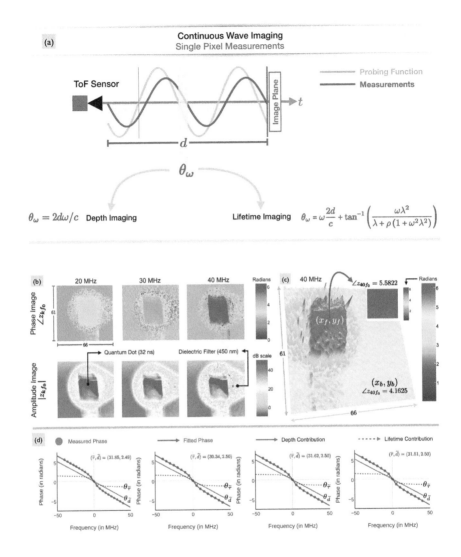

Figure 5.18 The fundamental difference between depth imaging and lifetime imaging. (a) In depth imaging, the phase of the measurements is linearly proportional to the modulation frequency of the probing signal, whereas in lifetime imaging, the phase is nonlinearly dependent on the depth (d) and the lifetime (λ) parameters that arise due to the modified scene reflection function. (b) Time-of-flight measurements at different modulation frequencies. (c) Phase image at 40 MHz. (d) Parametric curve fitting of observed phase for estimation of lifetime.

5.5 Application Areas

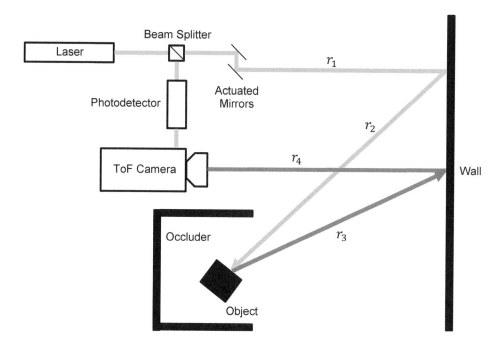

Figure 5.19 The NLOS imaging setup.

example, is characterized by *diffuse reflection*, making it difficult to visualize the target using traditional cameras.

This problem is addressed by a research field entitled non-line-of-sight (NLOS) imaging, which is also discussed in section 10.4. ToF techniques are more suitable in this case because the temporal structure of the light is the same for specular and diffuse reflections, and allows a reconstruction of the target (Velten et al., 2012a).

NLOS imaging has a multitude of applications, from sensing in hazardous environments, such as areas with radioactive or chemical leaks, to imaging inside machinery with moving parts. The authors propose using a pulse of light to illuminate the occluded object. One single point can be easily identified from its reflection. However, the reflections of multiple scene points may overlap at the detector, leading to a loss in correspondence. The result is that we do not know which point reflected which of the pulses received by the camera.

The imaging setup is depicted in Figure 5.19. A camera and a laser are pointed toward a wall, and a patch is positioned behind an occluder, so that neither the camera nor the laser has a direct line to the patch. The laser emits a pulse of light that is directed toward a point on the wall using actuated mirrors. Some of this light is reflected diffusively three times: first by the wall toward the object, second by the object back to the wall, and third by the wall to the ToF camera. The laser is oriented to ensure that no single-bounce light reaches the camera. The distances between the laser, the three reflection points, and the camera are denoted r_1, r_2, r_3, and r_4, respectively.

The problem proposed is to estimate the distances r_2 and r_3, which are the only unknowns in this setup. The algorithm addressing this estimation problem using the camera measurements is called *backprojection* (Velten et al., 2012a).

In the absence of noise, the location of an occluded patch can be derived using only measurements from three camera pixels. Specifically, for each pixel, all possible points in space that could have generated a set of measurements lie on an ellipse, as depicted in Figure 5.20b. Therefore, in the absence of noise, the patch location can be identified as the unique intersection of the three ellipses. In practice, the scene is split into 3D volume units, called voxels. For each voxel, the backprojection algorithm computes the likelihood that the patch is in that voxel on the basis of the time-of-flight $r_1 + r_2 + r_3 + r_4$. In Figure 5.20c the likelihood is displayed as a heatmap for one laser orientation, and it is displayed in Figure 5.20d for multiple laser orientations. The final reconstruction after filtering is depicted in Figure 5.20e.

The method proposed in (Heide et al., 2014c) can recover both the object shape and the albedo, which is the proportion of reflected incident light. Moreover, compared with the method presented in (Velten et al., 2012a), this method uses a cheaper hardware setup that has no moving parts. The capturing speed was greatly improved, and the method is more robust under ambient illumination.

However, the recovery problem is more complex in this case, and it is addressed by making assumptions about the model, that is, using sparsity priors. The reconstruction is formulated as an optimization problem, where the object volume is the unknown to be estimated from the camera measurements. The optimization problem is ill conditioned, and it cannot be solved in its raw form. The assumptions, such as single surface reflection and minimal object volume, are used as regularizers to generate a solvable optimization problem.

This line of work was continued in (Kadambi et al., 2016), where the illumination shape was changed from a pulse to a continuous wave. This change allows a custom resolution that depends on the modulation frequency and wall shininess. Moreover, this work is the first to provide bounds on the NLOS reconstruction.

The performance depends on the shininess of the wall and the camera modulation frequency (Kadambi et al., 2016). The authors characterize the relationship between the reflectivity and resolution as nonlinear. Small improvements in reflectivity lead to large improvements in resolution. Given the reflective properties of realistic scenes, commercial ToF cameras can be used in the proposed setup to achieve good results for NLOS imaging.

To decrease the power consumption, (Buttafava et al., 2015) proposed using a single SPAD detector, capturing light corresponding to a single pixel. This has additional advantages such as lower cost, smaller size, and reduced reconstruction time.

5.5 Application Areas

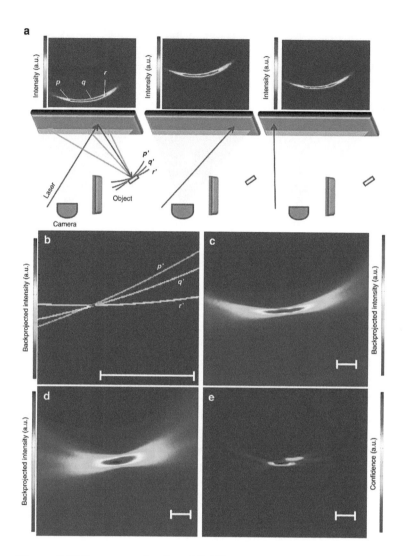

Figure 5.20 Reconstruction from NLOS imaging measurements. (a) Data collected for three different laser positions, where the object is a 2 × 2 cm white patch. Three of the pixels of the streak camera are denoted as p, q, and r. (b) The voxels that could have contributed to pixels p, q, and r are determined by the corresponding ellipses p', q', and r'. (c) The heatmap resulted from the backprojection algorithm, computed by super-imposing the elliptical curves corresponding to all pixels. (d) The heatmap resulted from 59 laser positions. (e) The final heatmap computed after filtering, representing the reconstruction of the patch.

The observation that the first reflected photons are characterize single points of the NLOS scene has enabled estimating the 3D NLOS scene using fewer measurements (Tsai et al., 2017). Moreover, assuming the scene can be approximated locally with planes, the proposed algorithm allows an efficient estimation of the surface normals. Overall, the work in (Tsai et al., 2017) avoids solving complex processing tasks from NLOS imaging, such as the elliptical tomography problem.

A new interpretation of the NLOS scene geometry allows speeding the reconstruction as much as 1,000 times (Arellano et al., 2017). Specifically, the authors interpret the hidden geometry probability map as the intersection of three space-time manifolds, defined by the generated and NLOS scene reflected light.

Instead of estimating the scene with voxels and recovering each voxel using the reflected light, (Pediredla et al., 2017) use a polygonal estimation composed of multiple planar walls. They demonstrate that estimating the planes instead of voxels leads to lower levels of noise and is compatible with larger spatial scales. Further examples and several other state-of-the-art implementations are discussed in the context of light transport in section 10.4.

5.6 Summary of Recent Advances and Further Applications

Looking at lower level reflection properties such as subsurface light scattering enables characterizing the surface material. For instance, it is possible to produce a two-dimensional image of optical scattering from internal tissue microstructures (Huang et al., 1991; Hee et al., 1993). This technique leads to capturing new low-level features of the scene (Su et al., 2016).

The material of an object can be identified through the way the light interacts with it via the scene response function $h(t, t')$, also called the temporal point spread function (TPSF), which describes the way light is scattered and reflected.

The setup used for material classification in (Su et al., 2016) is depicted in Figure 5.21. As for classic ToF imaging, a probing function is used to describe the waveform of the illumination. Here, it is chosen as a periodic function of period ω denoted $p_\omega(t)$. The light rays are scattered through the material and some are reflected toward the camera. This process is modeled by $h(t, t')$. The measurements are then correlated with a scene-independent function $f_\omega(t - \phi/\omega)$.

In order to account for the material properties, the scene response function incorporates the delays and attenuations introduced by each light ray

$$h(t, t') = h(t - t') = \int_{s \in S} h_s \delta\left(|s| + (t - t')\right) ds,$$

5.6 Summary of Recent Advances and Further Applications

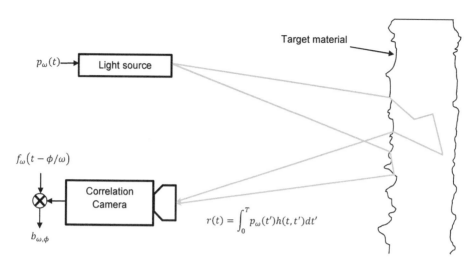

Figure 5.21 Material classification setup using time-of-flight imaging.

where s denotes the total length of all light rays reaching the camera, and h_s denotes the attenuation. Then the reflected measurement $r_\omega(t)$ amounts to

$$r_\omega(t) = \int_0^T p_\omega(t') h(t-t') dt' = \int_0^T h(t') p_\omega(t-t') dt'.$$

It follows that the final measurements $b_{\omega,\phi}$ satisfy

$$b_{\omega,\phi} = \int_0^T f_\omega(t-\phi/\omega) \int_0^T h(t') p_\omega(t-t') dt' dt,$$

$$= \int_0^T h(t') \int_0^T f_\omega(t-\phi/\omega) p_\omega(t-t') dt dt',$$

$$\implies b_{\omega,\phi} = \sum_{k=1}^\infty g_k \int_0^T h(t') \cos(k\omega(\phi/\omega+t')+\phi_k) dt', \quad (5.10)$$

where g_k are the Fourier coefficients of function

$$g_{\omega,\phi}(t') = \int_0^T f_\omega\left(t-\frac{\phi}{\omega}\right) p_\omega(t-t') dt,$$

which is periodic with period ω. Then it follows that $b_{\omega,\phi}$ is computed using the Fourier coefficients of $h(t')$, the scene response function.

This result in (5.10) is extremely useful, because it means that the setup shown in Figure 5.21 samples the spectrum of the scene response function, and this can be used to uniquely identify a material. Specifically, features are constructed by taking several measurements of $b_{\omega,\phi}$, which are then matched against a database of previously computed features (Tanaka et al., 2017).

5.6.1 Time-Resolved Imaging through Scattering Media

The data from correlation image sensors can be further analyzed to allow processing data in more complex environments such as scattering and turbid media. Imaging through scattering media has many applications such as deep-tissue imaging and artwork inspection. The work of (Heide et al., 2014c) allows recovering transient, or light-in-flight images through such environments using coding with significantly improved sparsity.

This type of imaging can be performed using a setup similar to the one shown in Figure 5.21. If we go back to the result in equation (5.10), we have

$$b_{\omega,\phi} = \int_0^T h(t') g_{\omega,\phi}(t') \, dt'. \tag{5.11}$$

In this new context, we are trying to estimate $h(t')$, where

$$g_{\omega,\phi}(t') = \int_0^T f_\omega \left(t - \frac{\phi}{\omega} \right) p_\omega (t - t') \, dt,$$

is known and is scene independent. To this end, the data is sampled in time, frequency, and phase, as follows. Let $\mathbf{G}$ be a $N \times M$ matrix with elements $[\mathbf{G}]_{i,j} = g_{\omega_i,\phi_i}(t_j)$; $\mathbf{b}$ a $N \times 1$ vector $[\mathbf{b}]_i = b_{\omega_i,\phi_i}$ and $\mathbf{h}$ a $M \times 1$ vector $[\mathbf{h}]_j = h(t_j)$. Then (5.11) results in matrix equation $\mathbf{b} = \mathbf{Gh}$, which is solved to estimate $\mathbf{h}$.

Recall that $h(t) = \int_{s \in S} h_s \delta(|s| + t) \, ds$, where s stands for the light path length. The solution of $\mathbf{b} = \mathbf{Gh}$ solves a rather complex problem. Besides identifying image intensities, computing $\mathbf{h}$ untangles path contributions with different path lengths, which is known as the multipath problem.

However, solving $\mathbf{b} = \mathbf{Gh}$ is not trivial, and in many cases results in an ill-posed problem. One way to address this is by assuming a sparse representation for $h(t)$ in the compressive sensing framework (Heide et al., 2014c). The solution of $\mathbf{b} = \mathbf{Gh}$ is formulated as the optimization problem

$$\mathbf{h}_{\text{opt}} = \arg\min_{\mathbf{h}} \|\mathbf{h}\|_1 \text{ such that } \|\mathbf{b} - \mathbf{Gh}\|_2^2 < \epsilon$$

which is stable numerically and leads to sparse representations. The prototype used in (Heide et al., 2014c) for imaging through scattering media is depicted in Figure 5.22, and the results of an experiment with scattering through milk are depicted in Figure 5.23.

5.6 Summary of Recent Advances and Further Applications

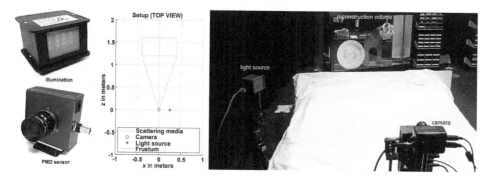

Figure 5.22 The imaging prototype used in (Heide et al., 2014c). The cameras are imaging through a tank field with scattering medium placed frontally. Shown are an array of laser diodes and imaging sensor (*left*), arrangement diagram (*center*), and experiment setup (*right*). Reprinted from (Heide et al., 2014c).

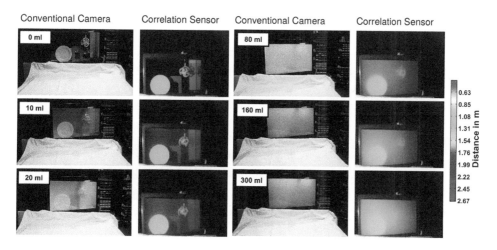

Figure 5.23 Depth estimation in a scattering medium. A tank filled with water (*top*) and then with a gradual increase in milk volume up to 300ml (*bottom*). Imaging with a conventional camera (*left*) and with ToF correlation image sensors (*right*). Reprinted from (Heide et al., 2014c).

We now review some of the other work done in imaging through scattering media. (Satat et al., 2018b) proposed a LiDAR-based system for scanning through fog. The system, based on SPAD sensors, uses a probabilistic model to distinguish between photons reflected by the fog and photons reflected by the target. Experiments demonstrate imaging with good accuracy beyond the visibility range of the fog.

The calibration parameters for imaging through scattering media, such as the field of view or illumination, can cause variability in measurements. However, it is possible to design machine learning imaging techniques that are insensitive to these parameters on a given range (Satat et al., 2017). The authors use a convolutional neural network that leads to results comparable to those of a time-resolved camera without the need of calibration on a given training range.

Many time-resolved imaging techniques for scattered media select a subset of the reflected photons on the basis of their arrival time. A method using the whole range of reflected photons can achieve a twofold improvement in imaging resolution (Satat et al., 2016). An analytical solution has been proposed to model a highly scattering medium based on its optical absorption and fluorescence source distribution using diffusive light (Schotland, 1997). Polarimetric approaches for imaging include the use of scattering section (7.4) as well as light transport decomposition section (10.5.3).

5.6.2 Time-Resolved Imaging Systems

ToF sensors are generally produced with low resolution due to practical limitations. In (Li et al., 2017a), the nonlinear relationship between the depth and the ToF measurements is represented as a linear imaging problem. Using this representation in conjunction with compressive sensing techniques, the authors managed to boost the resolution of the reconstructed image up to fourfold.

The problem of low temporal resolution of ToF cameras can be addressed using an array of light sources, which introduce subnanosecond time shifts in the captured waveform (Tadano et al., 2016). This is exploited to increase the temporal resolution of the ToF imaging system tenfold. In the case of out-of-focus, low-resolution ToF images, the resolution of the data can be enhanced via a technique called superresolution, while simultaneously performing deblurring (Xiao et al., 2015).

Increasing the temporal resolution of imaging devices to the picosecond scale reveals important information about the scene. This can be done using active illumination with picosecond temporal resolution, which allows solving the inverse multipath problem with higher accuracy in the presence of noise (O'Toole et al., 2017).

Most ToF imaging systems estimate the position of an object in the scene. For dynamical scenes it is then possible to approximate the velocity using several position instances. However, it has been shown that the instantaneous radial velocity, which is the velocity along the observer line of sight, can be measured directly. This can be done using the Doppler effect, which creates a frequency shift in the probe signal emitted by the ToF camera (Heide et al., 2015). Additionally, the authors measured the color, depth, and velocity of a pixel simultaneously. It is possible to increase the precision of ToF cameras from centimeter to micrometer by designing a form of cascaded ToF, by using a Hertzstyle intermediate frequency to encode a high-frequency path length (Kadambi and Raskar, 2017).

The problem of waveform interference can also be addressed using novel hardware approaches. A new hardware system has been introduced with multiple ToF cameras, which allows synchronizing exposure times and waveforms from three cameras. The system is also used to capture radial velocity from interfered waveforms using the Doppler frequency shift principle (Shrestha et al., 2016).

ToF cameras have applications in day-to-day activities such as web conferencing and gaming by providing enhanced gesture recognition capabilities (Kolb et al., 2010).

A novel result for ToF cameras significantly decreases the complexity of the encoded data and associated encoder implementation. Instead of using quantization on different amplitude levels, the work of (Bhandari et al., 2020b) encodes the data into a sequence of ± 1 values. Although this leads to an information loss, the underlying time-resolved information can still be recovered accurately in a noniterative fashion. The setup is compatible with both single and multiple paths of light.

5.7 Related Optical Imaging Techniques

5.7.1 Optical Coherence Tomography

Along with widespread scanning techniques such as X-ray computed tomography, magnetic resonance imaging, and ultrasound imaging, it is possible to image the internal structure of biological tissues with a noninvasive technique known as optical coherence tomography (OCT). OCT measures the optical reflections of various layers of biological tissues and returns their 3D shapes by using the ToF information at multiple locations (Huang et al., 1991).

As is the case for ToF, OCT can function with time-localized pulses or continuous-wave (CW) probing functions. The former approach can be achieved by time gating the detected light to separate direct transmission light from light obtained in the turbid tissue. The CW approach utilizes low coherence interferometry. At the heart of this setup is the Michelson interferometer. An interferometric setup is used due to the extremely high speed of light, which would otherwise require detectors with time resolution on the scale of 10^{-15}s. Along one arm of the interferometer lies the tissue sample, and the other arm is the reference arm, which usually contains a movable mirror to introduce an adjustable optical path length. The reference path is used to measure ToF information from each layer of tissue and the intensity of backscattered light. Because low coherence light is used, a fringe pattern will be observed only if the two path lengths along each arm are within the coherent length of the light source. Therefore, different reference arm positions correspond to measurements at different depths (i.e., different tissue layers). This ToF information gives spatial information about the tissue at each layer. By scanning laterally across the entire tissue sample, it is possible to obtain ToF information for all spatial locations in the tissue. The ToF information obtained from the fringe patterns can then be used for 3D reconstruction (Popescu et al., 2011).

For objects that are opaque at optical frequencies, imaging can be performed with radiation in the terahertz frequency range in real time (Lee and Hu, 2005). Moreover, imaging small scenes at very high spatial resolutions can be achieved using a method closely related to OCT, which decomposes light transport on the basis of properties such as path length and wavelength (Gkioulekas et al., 2015).

5.7.2 Digital Holography

A distinct example that uses the same concept as ToF is digital holography. In this imaging method, a coherent light source is split along two paths, namely, the reference beam and the object beam. The object beam is reflected by the object of interest, and the light reflected from and diffracted by the object will interfere with the reference beam at an electronic sensor array (Schnars et al., 2015). In this case, the depth is not directly encoded in the difference in arrival time, but rather in the difference in phase of the two beams. The phase difference is introduced by the differing optical path lengths, which therefore enable the encoding of ToF information. The sensor receives interferograms in the form of intensity measurements. Reconstructing the 3D object leads to a phase retrieval problem that can be addressed with analytical approaches (Waller et al., 2010) or machine learning (Rivenson et al., 2018; Peng et al., 2020).

5.7.3 Time-Stretched Optics

Thus far in the chapter, we have discussed ultrafast computational imaging methods which require specialized techniques to extract meaning from high-speed events. However, it is also possible to slow down a high-speed event to a lower speed so that it can be detected. For example, we often seek to obtain spectral images of ultrafast events in real time, but are limited by the available detector speed, computing power, and specified form factor. Whereas the detector speed may be on the order of tens of nanoseconds, the event of interest may be in the picosecond range. It would not be possible to vary the spectral sensitivity of the detector in such a short time. Furthermore, spatially offsetting different wavelengths using diffractive elements would enhance the effects of shot noise, due to a reduced per-pixel light intensity. The key idea behind photonic time stretch is that by optically manipulating the time-scale properties of incident light, it is possible to fully preserve the signal information without ultrafast detectors such as APDs. The reader is directed to (Hau, 2011) and (Han and Jalali, 2003) for more fundamental discussions on the concept of *slow light* and time-stretch photonics.

There are four key steps associated with time stretching. (1) The optical signal is modulated onto a broadband chirped carrier wave. (2) The signal is stretched in time by use

5.7 Related Optical Imaging Techniques

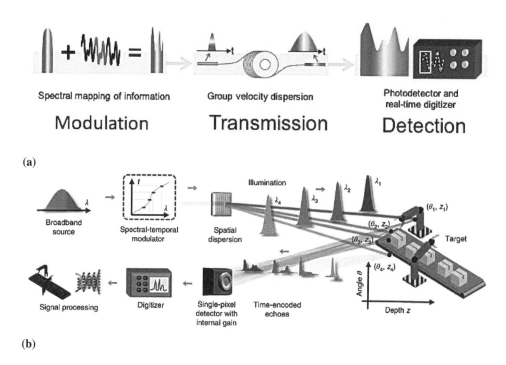

Figure 5.24 (a) Photonic time-stretch principle. (b) Imaging setup for time-stretch LiDAR. Reprinted from (Jiang et al., 2020).

of a dispersive element (a medium in which the index of refraction, and thus speed of light, vary as a function of frequency). (3) The signal is measured and digitized at the photodetector using coherent spectral interferometry. (4) Using digital signal processing and/or machine learning, the original signal is reconstructed using the temporal encoding of the frequency information (Mahjoubfar et al., 2017). This principle is illustrated in Figure 5.24a. Whereas such time stretching has been highly potent in spectroscopy and microscopy, it is also particularly consequential for LiDAR. Time-of-flight LiDAR using spectral scanning has not been feasible due to the lack of available tunable pulsed lasers. However, with use of a broadband source the light can be spatially dispersed by frequency into the scene, and the frequencies can be optically separated in time. This is shown schematically in Figure 5.24b. Other methods to effectively slow down light have used heterodyne (Kadambi and Raskar, 2017) and superheterodyne (Li et al., 2017b) interferometry to convert high-frequency modulated signals to measurable low frequencies for long-range and high-resolution time-of-flight imaging.

The phase stretch transform (PST) is inspired by the physics of time stretching and is based on the idea that phase primarily captures the information content in an image, more so than magnitude does. PST takes a 2D image and simulates it as propagating along an engineered diffractive medium. As the image is propagated along the medium, the different spatial frequencies of the image will propagate at different velocities. This effectively stretches the phase between different frequency components over the propagation distance. This results in a phase profile that reveals edges when applied with morphological and threshold operators, because edges are characterized by a high spatial frequency (Asghari and Jalali, 2015).

Chapter Appendix: Notations

Notation	Description
$m(x, y, t)$	Measurements at spatial coordinates (x, y) at time t
$p(t)$	Probing function
$\delta(t)$	Dirac delta distribution
$h(t, t')$	Scene response function (SRF)
$h_{SI}(t - t')$	Shift-invariant SRF
d_i	Distance from i^{th} to the sensor
t_i	Time-delay due to the object at distance d_i
Γ_i	Reflectivity of the object at distance d_i
$h_K(t, t')$	SRF accounting K light paths
$h^{Depth}(t, t')$	SRF due to depth in fluorescence lifetime imaging (FLI)
$h^{Decay}(t, t')$	SRF due to decay in FLI
μ	Emission strength of the FLI sample
λ	Lifetime of the FLI sample
$h_{Tr}(t, t')$	SRF in transient imaging
$h_D(t - t')$	SRF due to direct reflection
$h_I(t - t')$	SRF due to interreflections
$h_S(t - t')$	SRF due to subsurface scattering
$r(t)$	Reflected function
$\Psi(t, t')$	Instrument response function
$m(t)$	Continuous-time measurements
$y[n]$	Discrete-time measurements
S	Sampling operator
T	Sampling interval
$\otimes$	Cross-correlation operation
f	Frequency

5.7 Related Optical Imaging Techniques

ω	Angular frequency
θ_ω	Frequency dependent phase
$\overline{p}$	Time-reversed version of the probing function p
ϕ	Autocorrelation of the probing function p
T	Toeplitz/convolution matrix
m	Measurements vector
s	Sparse vector
$\{s_k, \lambda_k\}$	Scattering coefficients
$b_{\omega,\phi}$	Final measurements

Exercises

1. Depth estimation.
   Assume that we have the typical time-of-flight problem: a number of light pulses $p(t)$ are projected toward a scene and we need to measure the depth from the measurements of their reflections $y[k] = m(kT)$. Assume that the pixel grayscale intensities of an image I represent the depth values. Feel free to capture your own photo, or use the one shown in Figure E5.1.

(a) Grayscale image of Jupiter.

(b) Grayscale intensities converted to depth values.

Figure E5.1 Producing depth values from a grayscale image.

   a) Time-domain depth estimation. We define a set of probing functions given by
   $$p_{kl}(t) = e^{-t^2}, k = 1, \ldots, K, l = 1, \ldots, L,$$
   where K, L denote the dimensions of image I. Assume that we have a time-of-flight setting in which a scene is probed with functions $p_{kl}(t)$, which leads to reflection functions
   $$r_{kl}(t) = 2e^{-(t-I(k,l))^2}, k = 1, \ldots, K, l = 1, \ldots, L,$$
   where $I(k, l)$ denotes the grayscale intensity of pixel (k, l) in the image.
   We assume a lock-in sensor architecture that produces continuous measurements
   $$m_{kl}(t) = (r_{kl} \otimes p_{kl})(t) = \int_{\mathbb{R}} r_{kl}(t) p_{kl}(t - \tau) d\tau.$$

Let $y_{kl}[n] = m_{kl}(nT)$ denote the sampled measurements. Using an appropriate sampling time T, recover the depth values in the time domain by implementing

$$\widetilde{I}(k,l) = T\operatorname*{argmax}_{n}(y_{kl}[n]).$$

The depth recovery error is a matrix defined

$$I_{\text{error}}(k,l) = I(k,l) - \widetilde{I}(k,l).$$

Plot the error, similarly to the example shown in Figure E5.2.

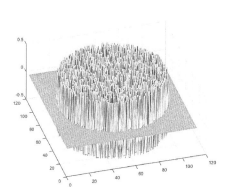

(a) Depth recovery error image.

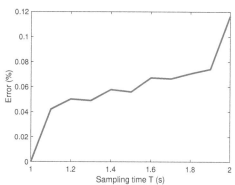

(b) Depth recovery error as a function of the sampling time T.

Figure E5.2 Example of depth recovery errors.

Now we can think of a real life situation where the samples are noisy and therefore the reflected measurements are corrupted by Gaussian white noise

$$r_{kl}(t) = 2e^{-(t-I(k,l))^2} + n(t), k = 1,\ldots,K, l = 1,\ldots,L,$$

where $n(t)$ is sampled from the normal distribution with mean 0 and standard deviation σ.

Plot the $\widetilde{I}(k,l)$ and $I_{\text{error}}(k,l)$ images for $\sigma = 0.4$. Given that the depth values of the image should be in the range 0–255, the result should be similar to the one shown in Figure E5.3.

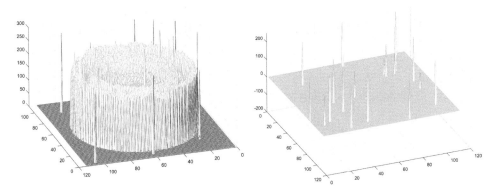

(a) Depth recovery error image.

(b) Depth recovery error as a function of the sampling time T.

Figure E5.3 Example of depth recovery errors for noisy measurements.

b) *Frequency-domain depth estimation.* As mentioned previously, in this case, the probing function is

$$p_{kl}(t) = 1 + p_0 \cos(\omega t), p_0 < 1, k = 1, \ldots, K, l = 1, \ldots, L,$$

and therefore the reflected function is

$$r_{kl}(t) = \Gamma_0 p_{kl}(t - t_{kl}), k = 1, \ldots, K, l = 1, \ldots, L,$$

Prove analytically that, for a lock-in sensor, the measurement function satisfies

$$m_{kl}(t) = \Gamma_0 \left(1 + \frac{p_0^2}{2} \cos(\omega t + \omega t_{kl})\right).$$

We now use the four bucket method presented previously to recover the depth values for $t_{kl} = I(k,l)$, using the same image used previously. Let $m_n(k,l) = m_{kl}(n\frac{\pi}{2\omega}), n = 0, \ldots, 3$. Show that

$$\begin{bmatrix} m_0(k,l) \\ m_1(k,l) \\ m_2(k,l) \\ m_3(k,l) \end{bmatrix} = \frac{\Gamma_0}{2} \begin{bmatrix} 2 + p_0^2 \cos(\omega I(k,l)) \\ 2 - p_0^2 \sin(\omega I(k,l)) \\ 2 - p_0^2 \cos(\omega I(k,l)) \\ 2 + p_0^2 \sin(\omega I(k,l)) \end{bmatrix}.$$

Exercises

Now define the complex number

$$z_\omega(k,l) = (m_0 - m_2) + J(m_3 - m_1), k = 1, \ldots, K, l = 1, \ldots, L.$$

Prove that the depth values can be recovered as

$$\widetilde{I}(k,l) = \frac{\angle z_\omega(k,l)}{\omega}.$$

Compute the depth values, and plot the error function $I_{\text{error}}(k,l) = I(k,l) - \widetilde{I}(k,l)$.

c) Depth estimation using Fourier series. Our probing function is not bandlimited, but it can be approximated arbitrarily closely with a bandlimited function

$$p_{kl}(t) \approx \widetilde{p}_{kl}(t) = \sum_{m=-K}^{K} \widehat{p}_{kl}^{(m)} e^{Jm\omega_0 t}.$$

Plot the probing function together with its Fourier series approximation using $K = 2, 5, 10$.

Write an analytical derivation showing that the depth can be estimated using the Fourier series approximation

$$I(k,l) = -\frac{1}{\omega_0} \angle \frac{\widetilde{m}_{kl}(t)}{\sum_{m=-K}^{K} |\widehat{p}_{kl}^{(m)}|^2 e^{Jm\omega_0 t}}.$$

d) Depth estimation for multipath interference.
Assume that due to imaging through a transparent object there are several reflections, which lead to a reflected signal of the form

$$r_{kl}(t) = \sum_{m=1}^{K-1} \Gamma_m \cos(\omega t - \omega t_{kl}^{(m)}).$$

Demonstrate that the measurement function takes the form

$$m_{kl}(t) = \frac{1}{2} e^{J\omega t} \widehat{h}_{kl}^*(\omega)$$

where $\widehat{h}_{kl}(\omega) = \sum_{m=0}^{K-1} \Gamma_m e^{-J\omega t_{kl}^{(m)}}$, and x^* denotes the complex conjugate of x. Here, $\widehat{h}_{kl}(\omega)$ is the Fourier transform of $h_{kl}(t) = \sum_{m=0}^{K-1} \Gamma_m \delta\left(t - t_{kl}^{(m)}\right)$, the scene response function of the time-of-flight imaging pipeline.

2. Fluorescence lifetime imaging with depth sensors. Given a fluorescent sample with lifetime parameter λ_0, the measurements resulting from the interaction of a continuous

wave and the samples can be modeled

$$m(t) = \delta(t) * \rho_0 e^{-\frac{t}{\lambda_0}} u(t),$$

where m is the measured signal, δ is the Dirac distribution, and u is the usual Heaviside function. In applications such as microscopy, such a sample is placed at a distance d_0 from the imaging plane, and consequently the scene response function may be written

$$h_{\text{FLI}}(t) = \underbrace{\Gamma_0 \delta(t - t_0)}_{\text{Direct reflection}} + \underbrace{\rho_0 e^{-\frac{t-t_0}{\lambda_0}} u(t - t_0)}_{\text{Fluorescent sample}}, \quad t_0 = 2\frac{d_0}{c}.$$

a) Fourier transform of scene response function
   What is the Fourier transform of the scene response function $h_{\text{FLI}}(t)$?

b) Continuous wave measurements. Consider the case of continuous-wave time-of-flight imaging where the probing signal is defined by $p(t) = 1 + p_0 \cos(\omega t)$ and a lock-in sensor is employed. Show that the measurements read

$$m(t) = \left|\widehat{h}_{\text{FLI}}(0)\right| + \left|\widehat{h}_{\text{FLI}}(\omega)\right| \frac{p_0^2}{2} \cos\left(\omega t - \angle \widehat{h}_{\text{FLI}}(\omega)\right),$$

where $\widehat{h}_{\text{FLI}}(\omega)$ is the Fourier transform of the SRF derived in this exercise.

c) Estimation of lifetimes. We have shown previously that for given pixel coordinates (k, l) we can estimate the Fourier transform of the scene response function by using the four bucket method. Hence, for a modulation frequency ω, we define measurements,

$$m_k = m\left(\frac{k\pi}{2\omega}\right), \quad k = 0, \ldots, 3,$$

and then we have

$$z_\omega = (m_0 - m_2) + j(m_3 - m_1) = p_0^2 \widehat{h}^*_{\text{FLI}}(\omega).$$

Supposing that we are give multiple frequency measurements, $y_n = z_{n\omega_0}/p_0^2$ and $\omega_n = n\omega_0$, then we have

$$y_n = \left(\frac{\Gamma_0 \lambda_0}{1 + j\omega_n \lambda_0}\right) e^{-jn\omega_0 t_0} \quad \text{and} \quad y_{n+1} = \left(\frac{\Gamma_0 \lambda_0}{1 + j\omega_{n+1}\lambda_0}\right) e^{-jn\omega_0 t_0} e^{-j\omega_0 t_0}.$$

The unknowns in this problem are lifetime λ_0 and time delay t_0. Starting with the ratio

$$\frac{y_{n+1}}{y_n}$$

show that this forms a linear system of equations in λ_0 and $e^{-j\omega_0 t_0}$ and can be solved with any four contiguous values such that $\ell \leqslant n < \ell + 4, \ell \in \mathbb{Z}$.

6 Light Field Imaging and Display

Capturing two-dimensional (2D) pictures has been the main interest since imaging began. However, humans, like most other animals, possess stereoscopic vision, which gives us a 3D-like perspective of reality. Light field imaging goes one step further, from a generalization of 3D stereo vision to the new realm of capturing four-dimensional (4D) photographs. In this chapter we present the principle of light fields, their numerous applications, and some of the devices used to capture them.

To grasp the basic idea of light fields, one must note that an ordinary, in-focus photograph adds up all the light rays that emit from a spatial location, regardless of their angle. In contrast, *light fields* enable us to capture brightness representations in both space and angle. Thus, the light field is characterized by all the light rays passing through space in all directions. The four-dimensional parametrization of light fields enables unique identification of rays—in contrast to the two-dimensional parametrization of ordinary photographs.

To simplify the representation in four dimensions, we will create an analogy of the light field in two dimensions, known as the Flatland analogy. This creates a projection of the 4D light field onto a hypothetical 2D plane, as depicted in figure 6.1, essentially dividing the dimensions in half. Most of the properties of the light field hold in the Flatland analogy. In Flatland, a light field camera is 2D, and an ordinary photograph is 1D.

We know from geometry that a line in a plane can be characterized either by its coordinates at two points in the plane, or by one point and an angle with one of the axes. The same principle is applied to parametrize the light field in Flatland. The light rays in Flatland can be parameterized using their intersections with two parallel lines, known as the *light slab parametrization*. Alternatively, we can use the point of intersection with one line and the angle between the ray and the line, which is called the *spatio-angular parametrization* (see figure 6.1).

In section 6.1 we discuss one of the first approaches for measuring the light field in the early twentieth century, the Lippmann light field camera. Next, section 6.2 presents the theoretical apparatus for modeling and processing light fields, and section 6.3 introduces

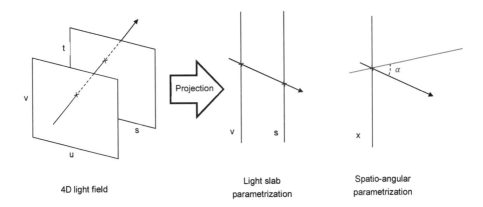

Figure 6.1 The 4D light field and two projections in Flatland. The 4D light field is quantified using the two-plane parameterization (*left*) and subsequently projected in Flatland to yield the light slab parametrization (*center*) and the spatio-angular parametrization (*right*).

some of the common setups for light field capture. Finally, section 6.4 gives an overview of the major steps in the development of the light field displays.

6.1 Historical Highlight: Lippmann Light Field Camera (1908)

The first attempt at capturing the light field was made by Gabriel Lippmann. In his research, Lippmann was inspired by the insect compound eye. He created an array of small cameras using a plastic sheet with spherical segments inserted. The array was placed in an opaque chamber. When exposed to light, each tiny camera captured a different perspective of the scene. The principle of capturing several perspectives at once applies to the compound eye, which is common in the animal kingdom, in particular for insects. A diagram of the compound eye and of the camera proposed by Lippmann in his original 1908 paper is shown in figure 6.2.

He designed the camera so that, when illuminated from behind, it would project the different perspectives that it recorded, so that a viewer would have the illusion of a 3D image. His attempts were not fully successful because he could not find materials with the suitable properties.

6.2 Light Field Processing

The light rays surrounding us carry an abundance of information about the 3D environment, and only a fraction is captured using traditional photography. As discussed previously, the

6.2 Light Field Processing

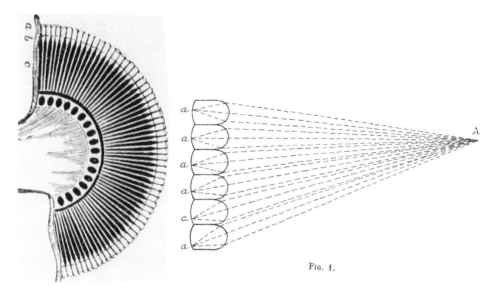

Figure 6.2 Diagram of the composite eye (*left*) and the light field camera proposed by Lippmann (*right*). Reprinted from (Carpenter, 1856) and (Lippmann, 1908).

pioneering work of Lipmann on lenslet arrays was experimental in nature, and it stagnated due to the lack of imaging hardware and materials to achieve his desired performance. Later, theoretical models of the propagation of light rays started to emerge in the work of (Gershun, 1939) and more recently in the work of Adelson and others (Adelson and Bergen, et al, 1991), which introduced the concept of the *light field*. The light field is described by the totality of all light rays passing through every point in a scene, along every possible direction. This gives a more comprehensive description of a scene than traditional photographs, which represent only a slice of the light field. Light fields therefore allow extracting more information from a scene, at macroscopic or at microscopic scales.

Light fields allow a multitude of fascinating capabilities. Even without access to a 3D model of the scene or knowledge of its texture, a light field can be used to compute novel views that could, for example, produce videos in which the camera moves freely through the static scene. The scene lighting can be adjusted, photographs can be refocused postcapture, and the depth of the scene points can be computed, which can be used to generate 3D models of the scene. To understand the principles behind these processing capabilities, we introduce the theoretical formulation of a light field.

6.2.1 Light Field Formulation

A very detailed model of the light field is given by the plenoptic function, which describes the light ray as a function of position, angle, wavelength, and time. The corresponding data

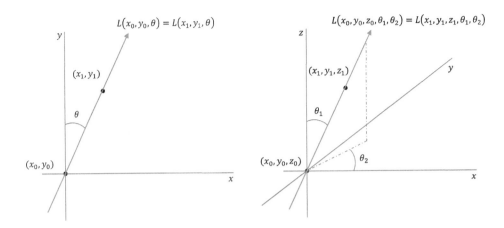

Figure 6.3 The plenoptic function in Flatland and 3D. The figure illustrates that the five-variable plenoptic function (*right*), or the equivalent three-variable function in Flatland (*left*) are both constant along light rays intersecting the origin. Hence the variables are not independent, and the plenoptic function can be expressed only as a function of four variables or in Flatland as a function of two variables.

would have seven dimensions, which is a significant challenge to acquisition. Therefore, a commercial light field capture device captures a maximum of five dimensions, three for position and two for angle, thus assuming the light to be monochromatic and time invariant. The plenoptic function in this case is $L(x, y, z, \theta_1, \theta_2)$. In a simplified Flatland scenario, the plenoptic function under the light slab parametrization is $L(x, y, \theta)$. The plenoptic functions, displayed in figure 6.3 for a given light ray, show how the dimension can be reduced by one in both cases to describe the light field.

This extended the applications from image-based rendering to a much larger range, including 3D reconstruction, segmentation, object recognition, and other applications. Big challenges remain for light field capture devices, such as the trade-off between dimensional resolution. Even so, the data sizes involved impose limits on the processing algorithms.

(Levoy and Hanrahan, 1996) and (Gortler et al., 1996) observed that one more dimension of the plenoptic function can be eliminated, for the following reason. A given light ray is captured by the plenoptic function for any 3D position located on that ray. This principle is depicted in figure 6.3. Therefore, the light field can be described by a function with only four coordinates $L(u, v, s, t)$, where u, v and s, t denote the intersection coordinates between the measured light ray and two predefined planes (Wu et al., 2017).

The plenoptic function $L(u, v, s, t)$ describes the radiance of light, and is measured in $W \cdot m^{-2} \cdot sr^{-1}$, that is, the number of watts per square meter per steradian, where the steradian is a measure of a solid angle, defining a field of view for the incident light.

6.2 Light Field Processing

Any light ray is uniquely described by two points on the two predefined planes, by defining coordinates u, v and s, t. This light field description can be interpreted as a series of cameras located in points s, t called *angular dimensions*, capturing rays coming from all points u, v called *spatial dimensions*. The slice determined by the rays detected with one of the cameras, denoted $I_{s^*, t^*}(u, v)$ is known as a *sub-aperture image*. Conversely, the slice selecting the rays captured by each of the cameras coming from a fixed point, denoted $I_{u^*, v^*}(s, t)$ is known as a *light field subview*. These functions are computed for two fixed spatial dimensions or angular dimensions. However, it is also possible to make an angular and spatial dimension constant, resulting in *epipolar planes* (EPIs). The two EPIs are $E_{v^*, t^*}(u, s)$ and $E_{u^*, s^*}(v, t)$. They are commonly used in fields such as multiview computer vision (Bolles et al., 1987).

An image is formed on the s, t plane by integrating the radiance $L(u, v, s, t)$. The image, characterized by its irradiance $I(s, t)$ satisfies

$$I(s, t) = \int_\Omega \int_\Omega L(u, v, s, t)\, du\, dv.$$

In this equation, Ω represents a subset of all angles u, v for which the light rays reach the point s, t, defined by the aperture of the camera. The irradiance is measured in $W \cdot m^{-2}$.

In the next subsections we go through some of the most important applications of light field processing.

6.2.2 Refocusing

The plenoptic function formulation of the light field can be used to generate new images focused on different points in the scene. To illustrate this, consider the simplified two-dimensional scenario, in which the s, t and u, v planes become the s line and v line, located one unit apart, as in figure 6.4. When the light is focused on the s line, the image is computed, as previously, with $I(s_0) = \int_\Omega L(v, s_0)\, dv$. The light focused on a line located at distance d no longer crosses the s line at a fixed point, but at a variable point $s_0 + dv$, as shown in figure 6.4. Therefore, the new refocused image $I_d(s_0)$ is computed

$$I_d(s_0) = \int_\Omega L(v, s_0 + dv)\, dv.$$

The irradiance is, mathematically speaking, a projection applied to the light field. To better understand how this transforms the light field, we interpret it in the Fourier domain using the *projection slice theorem*. This theorem states that the one-dimensional Fourier transform of $I_d(s)$ is equal to a slice of the two-dimensional Fourier transform of $L(v, s)$. In mathematical terms, this can be expressed

$$\widehat{I_d}(f_s) = \widehat{L}(f_s f_v - df_s).$$

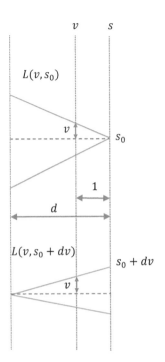

Figure 6.4 Refocusing via light field processing. Image focused on the s line (*top*) and refocusing on a new line located at distance d (*bottom*).

In this equation, $\widehat{I}_d$ and $\widehat{L}$ denote the Fourier transforms, and f_s, f_v are the variables in the Fourier domain. This gives us an additional way of computing a refocused image, by taking a slice of the Fourier transform of the light field. The principle behind the projection slice theorem is depicted in figure 6.5.

6.2.3 Generating Novel Views

After refocusing, the processed image has the same viewing perspective as the input image. However, the light field can be rendered to generate new views that were never captured by any of the physical cameras. These views correspond to theoretical devices called virtual cameras, explained as follows.

The plenoptic function variable pairs u, v and s, t are typically defined as coordinates in the two-plane representation of the light field. However, they can in some applications also be coordinates on spheres (Wu et al., 2017). These virtual views are generated by selecting the light rays passing through the new point of interest. A diagram of this process is shown in figure 6.6a, where the blue dot represents the view of a *virtual camera* rendered using the light rays captured by the cameras located in the red dots. However, in practice, it is

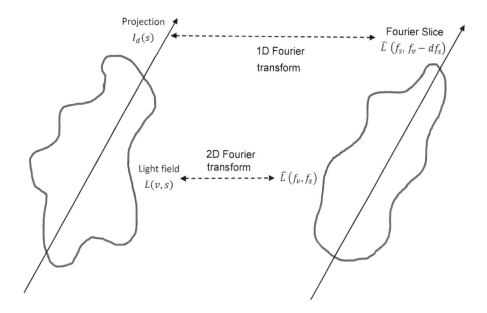

Figure 6.5 The projection slice theorem applied for image refocusing.

impossible to capture all the rays incident to one camera, and therefore the light field needs to be sampled at discrete locations. Therefore, a new ray is approximated by interpolating the sixteen closest rays, determined by any two blue dots located on different planes, as shown in figure 6.6b. An example of novel views generated with a camera array in (Levoy, 2006) is depicted in figure 6.7.

An important parameter for generating novel views is the number of acquired samples. If the number of samples is low, the novel interpolated views will be subject to distortion due to the ghosting effects. Too many samples, on the other hand, can lead to bandwidth and storage problems in transmitting the data. Due to the very large data quantity, the rendered light field was compressed using well-performing algorithms leading to compression rates of over 100:1 (Levoy and Hanrahan, 1996). It was shown that ghosting can be prevented if the neighboring views are closer than one pixel (Chai et al., 2000; Lin and Shum, 2004). However, when the geometry of the scene is known, the number of samples can be reduced significantly.

Sampling the light field can be translated into several subproblems of sampling the EPI functions describing it. Let us recall that the EPI functions $E_{v^*,t^*}(u,s)$ and $E_{u^*,s^*}(v,t)$ are derived from the plenoptic function, in which one spatial dimension and one angular dimension are kept constant.

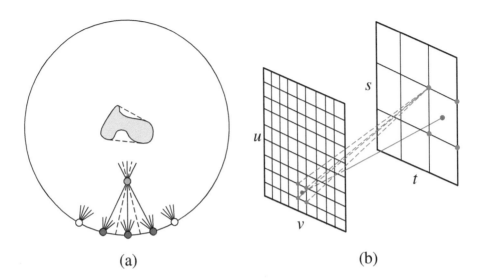

Figure 6.6 Rendering new views from the light field. (a) The blue dot represents the new view, computed using rays captured by the cameras in the red dots; and (b) a ray that is not captured by any camera (red ray) can be estimated by interpolation using the sixteen closest rays. Reprinted from (Wu et al., 2017).

To better gain insight into ghosting effects, we transition the sampling problem to the Fourier domain. We take the example of the EPI function $E_{v^*,t^*}(u,s)$ consisting of a single sloped line, as in figure 6.8a. When this function is sampled in the spatial domain, it is equivalent to copying the spectrum of the original function, shown in figure 6.8b, along the two frequency coordinates. The resulting periodic spectrum is then filtered to recover the original function, just as for one-dimensional signals. Using a rectangular filter, as in figure 6.8c, leads to aliasing, and hence ghosting effects. The reason is that parts of the shifted copies of the original spectrum are added up to the correct frequencies, causing distortion. This is prevented in this case by using shear reconstruction filters, such as shown in figure 6.8d, which do not cause any frequency interference with adjacent copies and lead to good reconstructions.

6.2.4 Depth Estimation

An important application of light field imaging is depth estimation. In order to estimate depth with an artificial camera, it is important to understand how depth is perceived by biological organisms. For example, the human visual system perceives scene depth via the disparity between perspectives on the left and right eyes. This disparity leads to horizontal *parallax*, which represents the different appearance of objects when viewed from the two

Figure 6.7 Generating novel views with a camera array. The view from one camera (*left*) and the synthetic aperture photograph generated with the views of the whole array (*right*). Reprinted from (Levoy, 2006).

perspectives. Essentially, the human eyes sample a portion of the light field to compute the depth.

By detecting the parallax as do biological organisms, one can measure depth using several cameras offering distinct viewing perspectives. Unlike human vision, the artificial camera setups offer an extra degree of generality, by including both horizontal and vertical parallax with the cameras placed in the corresponding positions relative to each other. In this way, depth can be estimated from one single exposure (Adelson and Wang, 1992).

There are situations in which the parallax is not enough for a good estimation, in such a case different cues can be used for additional accuracy. For example, when an object is located in front of a wall or screen, the position of its shadow can be used to work out the distance between the object and the screen. Taking shadow into account leads to significantly improved depth estimations (Tao et al., 2015). However, we could not use this method if the objects in the scene had textures that generate *specular* reflections. A specular reflection has a unique direction relative to the surface, and it is common with objects having a glossy or polished texture. To use the shadows in the estimation, the scene objects need to be *Lambertian*, which means that the light is diffused uniformly by the scene when illuminated, which is not always valid in real situations.

A possible scenario may involve occluding objects between the camera and the scene. If we tried to use any of the previously presented algorithms in this case, the results would be inaccurate. However, it is possible to identify the occluders and include them in the depth

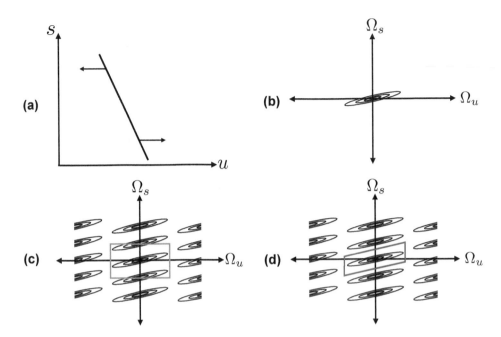

Figure 6.8 Sampling an EPI function in the Fourier domain. (a) EPI function consisting of a single line. (b) The Fourier spectrum of the continuous EPI function. (c) The Fourier spectrum of the sampled EPI function, processed with a rectangular filter (blue). (d) Spectrum processed with a shear filter. Reprinted from (Wu et al., 2017).

estimation process. This leads to a significant improvement in depth estimation, as shown in (Wang et al., 2016).

This leaves us with the problem of correctly identifying the occluding objects. A simple approach to this maps the points of the occluding objects to the pixels in the camera sensor on which they are projected. If a pixel corresponds to a point in the scene, its color will change significantly for different views. However, if the pixel maps to an occluding point, it is likely to have a similar color when viewed from different angles unless it is obstructed by an occluding spatial pixel.

It is possible, of course, to infer the depth by using the entire plenoptic function. This can be done using static images only, as shown in (Gortler et al., 1996), which can be interpolated to compute the *lumigraph*, another name for the plenoptic function. When specular surfaces are present in the scene, it is possible to formulate depth estimation as a constrained labeling problem on EPI 2D images, which also ensures consistent depth maps for all views (Wanner and Goldluecke, 2012).

6.2 Light Field Processing

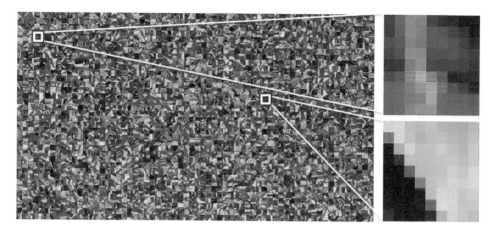

Figure 6.9 Overcomplete dictionary of light field atoms. Light fields can be recovered in a very noise-robust way, mostly as the linear combination of very few light field atoms.

6.2.5 Further Research

A widespread problem in light field photography is that it requires large data sets to reconstruct a high-resolution plenoptic function. This was addressed in (Marwah et al., 2013) by using *compressive sensing* techniques, which are methods to efficiently acquire and then reconstruct a signal using a small number of measurements. In (Marwah et al., 2013) a portion of the light field can be recovered from a single coded image, which is possible due to the interesting observation that a light field can be broken down into a large dictionary of atoms, called an overcomplete dictionary, depicted in figure 6.9.

This dictionary is retrieved from a large number of 4D spatio-angular light field patches of a predefined collection of light fields, called *training light fields* (Marwah et al., 2013). It is shown that using light field atoms allows recovering the light field with significantly higher resolution than other single shot methods.

The work in light field rendering has also attracted attention in the machine learning community. Rather than processing the light field on the basis of known mathematical models, machine learning extracts the required information from the data in a process known as *training*. For instance, the depth in a scene can be extracted automatically using a neural network. Whereas traditional machine learning methods would require training on measured depth values to predict new ones, the work of (Srinivasan et al., 2018) can train the network using images from a single perspective taken with a variable camera aperture.

Realistic scenes with complex geometry can be represented with high fidelity using triangle meshes, also called voxel grids. A distinct research direction attempts to use neural networks, also called *multi layer perceptrons* (MLPs) to directly map 3D coordinates to information about the objects in the scene (Park et al., 2019; Mescheder et al., 2019). The MLPs are part of a large class of generic *machine learning* (ML) models, which perform

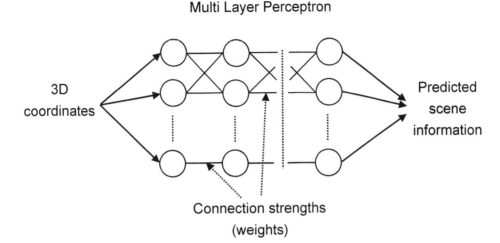

Figure 6.10 Multilayer perceptron. The connection weights are adjusted on a training data set such that it predicts the desired information from the scene on the basis of given 3D coordinates.

computations on data without being explicitly programmed to do so. This is accomplished through a process called training, in which the model parameters are adjusted to best fit the desired data. Figure 6.10 shows a diagram of an MLP. In section 3.4.2, we elaborate on neural networks and deep learning from a mathematical standpoint, as they are regarded an important aspect of computational imaging.

To map the scene with MLPs, it is necessary to have access to 3D geometry from the scene, which limits their application. The 3D geometry can be estimated directly from 2D images using neural networks, as shown in (Srinivasan et al., 2017). The authors of this work also introduced a very large publicly available database with light fields. Another method called neural volume (NV) predicts a voxel grid representation of a scene inside a bounded volume, with a clear background that has been prerecorded separately (Lombardi et al., 2019).

Another approach uses mesh-based representation of scenes using gradient descent (Waechter et al., 2014; Wood et al., 2000; Buehler et al., 2001; Debevec et al., 1996). In machine learning, *gradient descent* computes the derivative of the cost function, whose variables are the network weights. Subsequently, each weight is modified with a step along the direction in which the derivative of the cost function decreases fastest, hence comes the name *gradient descent*. The functioning principle of gradient descent is depicted in figure 6.11.

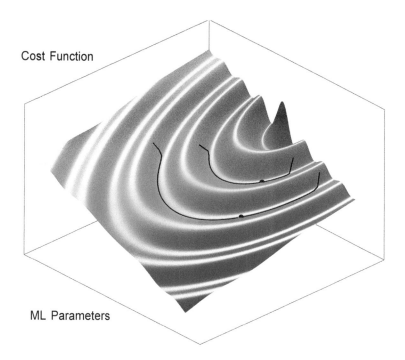

Figure 6.11 Optimization via gradient descent. The aim of the algorithm is to find the minimum value of the error function, plotted in yellow. At each iteration the parameters are changed in the direction in which the gradient descends fastest. Depending on the starting point, the algorithm might identify a local minimum instead of the global minimum.

A common problem in gradient descent is local minima. If the cost function is not convex, and has many local minima, it is very likely that the gradient descent algorithm would fall into such a minimum in which the gradient is very small. To address this, the networks are often initialized with random weights, and several training epochs are performed. However, even in this case there is no guarantee of fully avoiding local minima. An example is the case of mesh-based representations, for which the gradient descent fails often because these representations have a lot of local minima.

(Mildenhall et al., 2020) have trained a deep neural network (DNN) to generate novel views of a light field. Specifically, they propose a method to train the DNN to generate an RGB color corresponding to 5D coordinates of the light field $(x, y, z, \theta, \gamma)$. They call this a 5D neural radiance field representation.

The first step involves processing the light field coordinates $(x, y, z, \theta, \gamma)$ with eight fully connected layers into a density map σ and a feature vector with 256 dimensions. The

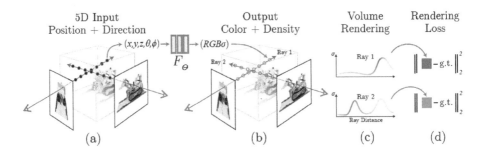

Figure 6.12 Novel view synthesis using neural radiance fields: (a) neural network input consisting of 5D light field coordinates; (b) predicted RGB value; (c) rendered volume; and (d) rendering loss function between the predicted volume and ground truth. Reprinted from (Mildenhall et al., 2020).

feature vector is then processed with a second network made up of four layers to output an RGB color that is view dependent. The work does not optimize only one network, but two at a time: a network to represent the course features of the scene, also called course network, and a fine network for representing the details.

The cost function for training is

$$\mathcal{E} = \sum_r \left[\left\| \widehat{C}_c(r) - C(r) \right\|_2^2 + \left\| \widehat{C}_f(r) - C(r) \right\|_2^2 \right],$$

where $C(r)$ is the true RGB color for a given light field sample r, and $\widehat{C}_c(r)$ and $\widehat{C}_f(r)$ are the predictions with the course and fine networks, respectively. The function approximated with each of the networks is naturally differentiable, and this makes it ideal for a training method such as gradient descent. A diagram of the processing pipeline proposed is shown in figure 6.12.

Deep neural networks can estimate incident illumination on objects in a 3D scene, therefore revealing content outside the observed field of view (Srinivasan et al., 2020).

Another interesting application of light field photography exploits the fact that the depth extracted from 2D images has generally poor depth resolution. However, depth can be estimated very precisely with an imaging paradigm called *time of flight*, which probes the scene with a stream of photons and measures the delay until they are reflected by the scene and captured by a camera. The method called *depth field imaging* measures the time of flight from different perspectives of the scene, enabling estimating the light field with improved depth accuracy (Jayasuriya et al., 2015).

Another application enabled by light fields is deblurring, which addresses the undesired blur effect, such as that caused by a 3D camera motion. In this case the deblurred light field is computed by estimating the camera motion curve (Srinivasan et al., 2017).

6.3 Light Field Capture

Thanks to technological advancements and the reduced dimensionality of the plenoptic function, discussed in the last section, the research on capturing light fields has progressed to the point at which plenoptic cameras are commercially available and are even produced in miniature sizes that can be integrated in cell phones. As the field has emerged, a number of applications have extended to areas such as microscopy and computer vision.

Even with reduced dimensionality, it is easy to see that the amount of data contained in light fields greatly exceeds the amount in conventional 2D images. Therefore, capturing the light field requires carefully selected imaging equipment, and also hardware capable of transmitting and storing the large amounts of data involved.

Just like the visual systems in the animal kingdom, the imaging devices available compute 2D projections of the 4D light field. Despite this restriction, the following techniques are available to acquire a complete light field:

1. **Using a camera array** In this case, the resolution in the s, t plane is determined by the number of cameras, and their distribution, and the resolution in the u, v plane is determined by the number of sensors per camera, which controls the number of angles of light rays detected by each unit.
2. **Capturing more images with one camera** Such cameras are mounted on computer controlled mechanical gantry devices. These systems require high-precision control and cannot capture dynamic scenes accurately. However, the amount of data transmitted is reduced.
3. **Multiplexed imaging** This technique uses a single camera that introduces a trade-off between the angle and spatial resolutions. For example, in spatial multiplexing, a lenslet array takes the place of a single lens, as shown previously. This effectively reduces the angular resolution while increasing the number of perspectives of the scene. This was also the idea of Lippmann, the pioneer of light field photography.

In the following, we go through the main techniques and challenges involved in capturing light fields.

6.3.1 Camera Arrays

To give a better insight into the scene properties, the images should come from different orientations and coordinates. This would be equivalent to acquiring several slices of the light field, which can allow its reconstruction. This leads to generating additional slices that show the scene from new perspectives. It is also possible to code information received by each camera. For example, (Inagaki et al., 2018) used a learning framework to capture a light field through a coded aperture.

In addition to providing more information, acquiring different images of lower resolution also proves more cost effective than using a high resolution camera. The same trend can be observed in other areas, such as computer processors, where using several average CPUs is more affordable than one single powerful CPU.

In imaging, this was first achieved using a single moving camera capturing images of a static scene. Later, Dayton Taylor updated it to a linear array of still cameras, which created the illusion of a camera moving through a static scene (Taylor, 1996), and Manex Entertainment introduced an adjustable trigger delay to simulate a high-speed camera moving around the scene (Wilburn et al., 2005). All of those applications were important steps forward, but they suffer from the limitation of being tied to a specific camera trajectory that can be produced by each setup.

A more general approach introduced later is based on an array of cameras and offers much more flexibility. The project that pioneered camera arrays in 1997 was called Virtualized Reality, and it aimed to generate new views of the scene by interpolating the captured videos. However, the downside of using camera arrays is very large data sets that increase the complexity for transmission, storage, and processing. There have been two main prototypes for this project. First, a videocassette recorder (VCR) enabled very large storage capabilities at the cost of low camera resolution. Second, an array of forty-nine higher resolution cameras was used, which generated a large amount of data and required a PC for every three cameras.

The projects mentioned so far required rather expensive equipment. To make the camera arrays more consumer friendly, Yang introduced distributed light field cameras (DLFCs), which used an 8×8 array of webcams (Yang et al., 2002). A reconfigurable array was made that, due to its mobility, boosts the quality of the interpolated view and therefore generates more realistic images (Zhang and Chen, 2004). The more accessible price meant that these arrays could not be synced as well as the more expensive ones, which led to artifacts in the interpolated views. Moreover, the large amounts of data restricted the size of the array compatible with these methods. To allow larger arrays, the data transmitted was decreased either by sending data corresponding to fewer views, or by using JPEG compression.

One of the greatest barriers in camera array imaging is the large amounts of data that require high bandwidth cables. This limits the quality of the footage captured in terms of number of simultaneously recorded images, image resolution, and number of frames per second. Therefore a research direction is focused on developing new and better compression algorithms that would ease the restrictions on the recorded data. It has been observed that there is a large amount of redundancy between the images recorded by each camera (Zhu et al., 2003). Therefore, even after the JPEG compression, the data transmitted by each camera can be greatly reduced with minimal effect on the accuracy.

Instead of transmitting the recorded bits, their system sends shorter sequences of bits computed using error-correcting codes. A centralized decoder receives these new bit sequences from each camera and can recover the data with very good performance. In

6.3 Light Field Capture

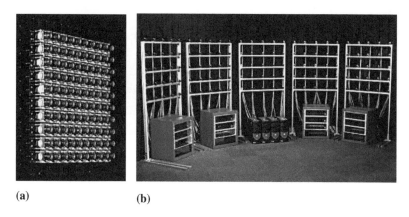

(a) (b)

Figure 6.13 Camera array recording system: (a) tightly packed configuration; and (b) widely spaced configuration. Reprinted from (Wilburn et al., 2005).

essence, this shifts the burden of high data bandwidth to a more complex centralized decoder, which in practice is easier to manage.

It has been possible to use MPEG2 compression for the video streams to allow transferring the complete camera recordings to PCs, which led to a 17.5:1 compression ratio (Wilburn et al., 2005). This was essential for the array used, which had one hundred video cameras with 640 × 480 resolution. Uncompressed video frames were also stored to verify the system performance offline.

The precise time synchronization between the cameras in a large array is essential, and dedicated cables were chosen for the clock signal (Wilburn et al., 2005). Using the system described here, the authors were able to capture video streams up to 2.5 minutes long, limited to 2 GB in size. Their prototype is depicted in figure 6.13 for two spatial configurations.

In order to get a good representation of the light field, some camera characteristics, such as geometry and color, require *calibration*. Full metric calibration is usually used in a generalized array. If the array is situated on a plane, then a simpler calibration technique can be performed that estimates the camera positions and achieves better results than full metric calibration (Vaish et al., 2004).

The camera arrays exploit an important physical phenomenon known as the *parallax*, which represents the apparent displacement of objects when viewed from different angles. This can be adjusted with complex camera setups, or it can be compensated for in software, which was the choice in (Wilburn et al., 2005).

Most camera arrays suffer from *variation in colour* across the array. Minimizing color variation is essential for maintaining the illusion of a single high-speed video camera. An automatic color calibration technique was proposed in (Joshi et al., 2005), which is based on placing a color chart in the field of view of all cameras. This was combined with the

(a)   (b)

Figure 6.14 Images captured with a camera array, for which (a) the exposure time is equal for all cameras, and (b) the exposure time is adjusted individually for each camera. Reprinted from (Wilburn et al., 2005).

geometrical calibration target. In the camera setting proposed, gains and offsets of the sensor response are adjusted to match the target response. Here, for each camera, the gain refers to the increase in sensor response for a given increase in light intensity, and the offset is the sensor response for each camera given the same constant light intensity.

Moreover, each sensor saturates at the high and low ends of the light intensity, meaning that the response is no longer a linear function of the incoming light intensity. This is compensated for by linearizing the responses of all cameras (Joshi et al., 2005). Another inconsistency is the *falloff*, which represents a variation in intensity between the center and extremities of an image. This is corrected using color checking cards placed in the field of view of the cameras. Because the exact colors on these cards are known, they can be used to quantify the falloff. Then a global error correction is performed by computing transforms between the color patch values for each camera and the average patch values for all cameras and then applying the transform to the data filmed by the camera.

The camera calibration can give more control over the image appearance, which produces better results than using a single high-resolution camera. It has been demonstrated how an image mosaic, which refers to joining small resolution images into a single high-resolution picture, can benefit greatly from exposure calibration (Wilburn et al., 2005). By overlapping the fields of views of all cameras one can produce a high dynamic range image, in which each camera can increase or dim the brightness in different locations. An example is shown in figure 6.14, in which the features in the bright and dark areas of the image are much more visible when the exposure is calibrated.

Rather than generating new views of the captured scene, there has been research into computing image collages, which give a more comprehensive view of the scene. The collages can be viewed as a form of geometric calibration, in which the 2D points in the camera's image plane are matched via the features in the scene. An automatic method was created for rotating, translating, and scaling images by applying the scale-invariant feature transform to match common features between images (Nomura et al., 2007). An example

6.3 Light Field Capture

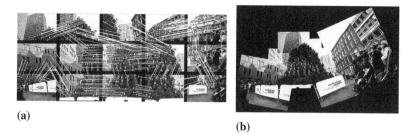

(a)

(b)

Figure 6.15 Collage computed from images generated with a flexible camera array: (a) the original images and the matching features, and (b) the collage generated through rotations, translations, and scalings. Reprinted from the presentation slides in (Nomura et al., 2007).

of such a collage is shown in figure 6.15. The authors also introduced a flexible camera array that can be assembled in a new configuration in a matter of minutes to allow new scene perspectives.

6.3.2 Dappled Photography

The capturing methods described in the preceding section are based on *spatial multiplexing*, so that each image represents a spatial slice of the plenoptic function. A different approach is *Fourier multiplexing*, which encodes slices of the plenoptic function in different frequency bands. The advantage here is that a single capture includes several dimensions of the plenoptic function, and therefore it leads to better light transmission (Wetzstein et al., 2013).

A popular Fourier multiplexing method called *dappled photography* is based on a mask that does not bend light as lenses do but rather attenuates it in a shadowlike pattern (Veeraraghavan et al., 2007). This allows acquiring a high-resolution 2D image by accounting for the shadow patterns, and also a low-resolution 4D image that can be recovered by Fourier domain decoding.

The functioning principle of dappled photography is based on the modulation theorem, which is routinely used in telecommunications. Given a signal $u(t)$, the theorem states that the modulated signal $y(t) = u(t)\cos(2\pi f_0 t)$ has a Fourier transform

$$\mathcal{F}[y](f_t) = \mathcal{F}[u](f_t) * \mathcal{F}[\cos(2\pi f_0 t)](f_t)$$
$$= \frac{1}{2}[\mathcal{F}[u](f_t - f_0) + \mathcal{F}[u](f_t + f_0)].$$

In other words, the theorem states that the spectrum of the modulated signal is made up of copies of the original spectrum shifted at the frequency of the cosine, also known as the carrier function.

In the case of dappled photography, the incoming light field passes through an aperture, is filtered by a mask, and is then recorded by the camera sensor (see figure 6.16). Let

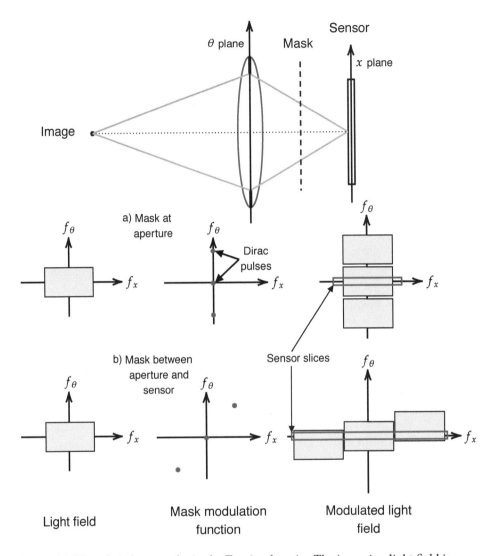

Figure 6.16 Dappled photography in the Fourier domain. The incoming light field is parametrized on the aperture plane and sensor plane (*top*). Illustration of the modulation theorem when the mask is on the aperture and is between the aperture and sensor (*bottom*). The chosen parametrization defines the sensor measurements as a horizontal slice from the modulated light field spectrum.

$L(x, \theta)$ denote the incoming light field and $M(x, \theta)$ denote the mask modulation function. According to the modulation theorem, the Fourier transform of the light field after the

6.3 Light Field Capture

aperture and mask is

$$\widehat{L}_M(f_x, f_\theta) = \left(\widehat{L} * \widehat{M}\right)(f_x, f_\theta),$$

where $\widehat{L}(f_x, f_\theta)$ and $\widehat{M}(f_x, f_\theta)$ are the Fourier transforms of the light field and mask modulation function, respectively.

Due to the chosen parametrization of the light field, the measurements at the sensor $S(x)$ do not change with θ, and therefore

$$\widehat{S}(f_s) = \mathcal{F}[S](f_s) = \widehat{L}_M(f_x, f_\theta)\Big|_{f_x = f_s, f_\theta = 0}.$$

Therefore the spectrum of the sensor measurements represents a horizontal slice from the modulated light field spectrum $\widehat{L}_M(f_x, f_\theta)$. The position of the mask relative to the sensor is an important parameter. When the mask is on the θ plane, then the content of $\widehat{M}(f_x, f_\theta)$ is only on the f_θ axis. When the mask is placed between the planes θ and x, then $\widehat{M}(f_x, f_\theta)$ is zero apart from a line crossing the origin (see figure 6.16). Typically, $\widehat{M}(f_x, f_\theta)$ is chosen as a train of Dirac pulses on the corresponding line, which according to the modulation theorem leads to copying the light field spectrum in the center of each Dirac, as depicted in figure 6.16.

With no modulation, the sensor measurements satisfy $\widehat{S}(f_s) = \widehat{L}(f_s, 0)$, so that a whole dimension of the incoming light field is lost. With a carefully chosen modulation, the sensor slices from the modulated light field contain information about the slices of the light field for different f_θ, which can then be used to approximate $\widehat{L}$ with much higher accuracy.

The prototypes used for dappled photography in (Veeraraghavan et al., 2007) are depicted in figure 6.17.

A different way to recover the light field in a single image capture is given by interposing a phase plate between the lens and the camera sensor (Antipa et al., 2016). The phase plate, also called a light shaping diffuser, consists of a plate of polymer on which the input side is flat, and the output side is a random Gaussian surface, therefore refracting the light in a diffusive pattern. The output surface is modeled as random Gaussian noise filtered with a smoothing kernel.

In this way, the light field is encoded into the phase of the incoming light, which is then decoded from the sensor data using an algorithm called *phase retrieval*. The advantage of such a setup is a much higher light throughput than given by dappled photography (Antipa et al., 2016).

6.3.3 Microscopic Light Field Imaging

Capturing multiple perspectives attracted a lot of interest not only for macroscopic objects but also for microscopic ones (Levoy, 2006; Levoy et al., 2006). This can address some of the known drawbacks of microscopic imaging. One of those drawbacks is that microscopes are orthographic projection devices, so that they display images only from perspectives

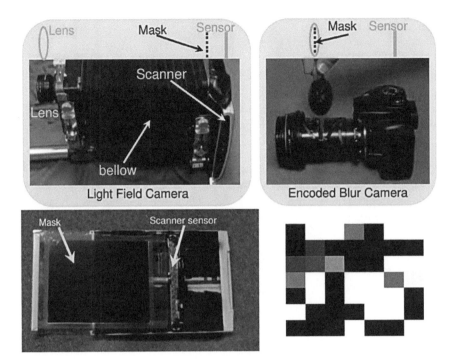

Figure 6.17 Dappled photography setup using two camera designs: the proposed cameras (*top*) and the associated masks (*bottom*).

directly above the specimen. They also suffer from a limited depth of field, which allows them to view a thin section of the specimen. The section can be adjusted, but this is time consuming and assumes a static scene.

The development of light field microscopy started with the early experiments of Gabor, and was later improved with the development of lasers. However, it was only after the work of Lippmann, discussed in section 6.1, that this technique started making use of microlens arrays. As in macroscopic photography, a typical application of lens arrays in microscopy is to increase the field of view.

In light field microscopy imaging, the lens array is used to generate several perspectives, at the cost of a lower spatial resolution. This has two important implications. First, it means that the perspective of the object can be changed after the image has been captured. Second, it allows generating tomographic images with one capture, as it is explained in the following. Unlike macroscopic photography in which the objects are mostly opaque, in microscopy the specimens are more translucent. This allows capturing images with changing focus, generating 3D slices of the object, also known as *focal stacks*, which can subsequently be processed into 3D tomographic images of the specimen. One can imagine

6.3 Light Field Capture

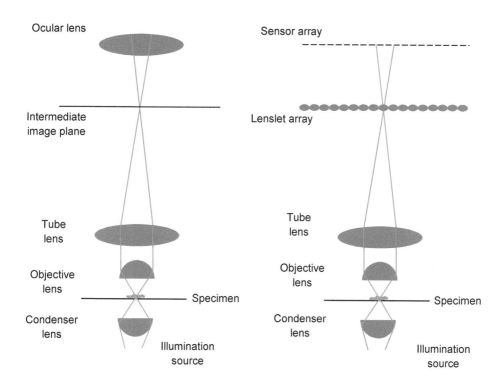

Figure 6.18 Comparative diagram of the traditional and light field microscopes.

that each slice in a focal stack originally contains the contribution from the out-of-focus parts of the specimen. This is addressed with an algorithm known as *3D deconvolution*, which subtracts these contributions using inverse filtering (Agard, 1984). This can also be done with traditional microscopes, but with light field microscopy the focal stack can be computed from a single capture.

The functioning principle of a conventional microscope in comparison with a light field microscope is depicted in figure 6.18. In a conventional microscope, the specimen is illuminated through a *condenser lens*. The light is then bent by an *objective lens* into parallel rays that are further bent by a *tube lens* such that the specimen is in focus on the intermediate *image plane* located inside the microscope. Early microscopes did not have tube lenses, which were later introduced to allow a variable tube length. A viewer can see the specimen through an additional *ocular lens*, which magnifies the view created on the image plane.

The light field microscope relies on the same basic principles as the classical one. The condenser and objective lens are not changed. However, a *lenslet array* is placed in the image plane, and a *sensor array* replaces the ocular lens. This is a slightly different setup from the typical camera array. To explain this, we can use the two plane interpretation of light fields given in section 6.3.1. Cameras are located in all points with coordinates s,t from one plane, oriented toward all points with coordinates u, v, located on the second plane. Thus the variables s, t dictate the spatial resolution of the light field and u, v the angular resolution. Typically, in a camera array, the spatial resolution is given by the number of pixels, and the angular resolution depends on the number of lenslets. However, the setup shown in figure 6.18 has the additional objective and tube lenses positioned such that the specimen is focused on the lenslet array rather than the sensors. This swaps the role played by each array, such that the lenslet array dictates the spatial resolution and the sensor array controls the angular resolution. This modified setup has the disadvantage of a slightly larger size. The advantage, however, is given by the objective lens, which is already present in most microscopes. Making a lens array of similar quality would be complex and costly (Levoy, 2006).

As in macroscopic light field photography, in microscopy there is a trade-off between the angle (or axial) and spatial resolution. The work of (Prevedel et al., 2014) accommodates this by oversampling the data and then eliminating partly the aliasing effect introduced by using the algorithm 3D deconvolution.

6.3.4 Further Research and Applications

The level of flexibility given by a plenoptic camera can be observed by analyzing an image generated with such a camera, such as the image shown in figure 6.19. The image contains several perspectives of each scene point captured, all grouped in small circular pixel patches.

Because each pixel location in a patch corresponds to a viewing perspective, this image can be used to access a desired view of the scene by picking the same pixel location in all patches (see figure 6.20). Similarly, by summing up several pixels in each patch, one can simulate a larger aperture size. The depth of field can be adjusted by shifting the selected pixels relative to each other, as depicted in figure 6.20.

Figure 6.21 illustrates a few examples of the light field cameras (macroscopic and microscopic) that were described in this section, and a few example images. Figure 6.21a depicts a motorized gantry with a single camera (*top*) and two frames from the captured light field (*center* and *bottom*). Figure 6.21b shows an array of 128 cameras (*top*), a view from the array (*center*), and a synthetic aperture photograph computed from the light field, setting the virtual viewpoint behind the foliage (*bottom*). Figure 6.21c displays a plenoptic camera (*top*) and two images computed by refocusing the light field postcapture (*center* and *bottom*). A light field microscope is depicted in figure 6.21d (*top*), followed by an embryo mouse lung viewed from two perspectives using a single snapshot (*center* and *bottom*).

Light field cameras are not yet as prevalent as traditional 2D cameras. However, as new phones already have multiple cameras, using light field cameras becomes technologically possible. There are also specialized devices such as the Light L16 camera that are already

6.3 Light Field Capture

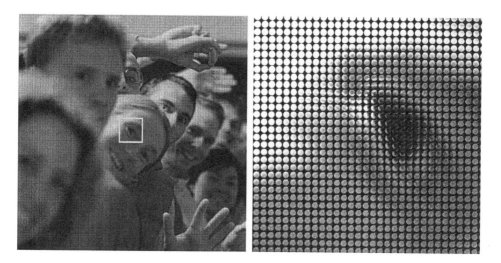

Figure 6.19 Image captured with a plenoptic camera. The image consists of small circular patches, each containing pixels with different perspectives of a point in the scene (Gkioulekas, 2018).

Figure 6.20 Computing 2D images from the plenoptic image. By picking only the pixels marked with red in each circular patch, one can simulate a desired viewing angle (*left*), aperture size (*center*), or focus depth (*right*) (Gkioulekas, 2018).

commercially available. The device has sixteen imaging modules with thirteen-megapixel resolution and three focal lengths of 28 mm, 70 mm, and 150 mm, respectively, which allows zooming optically by interpolating the light field after an image is captured. The sixteen cameras also have three different fields of view of 75°, 35°, and 17°. After processing, the images captured with the camera can be combined into a single fifty-two-megapixel high dynamic range image, as depicted in figure 6.22.

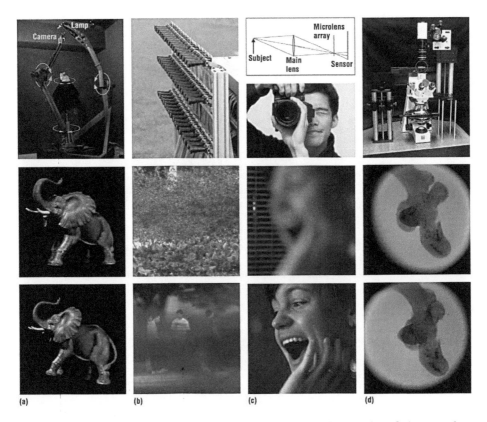

Figure 6.21 (a)–(d) Four of the typical light field cameras and examples of photographs sliced from each light field. Reprinted from (Levoy, 2006).

The multiplexed light field capture principle was used to design a handheld plenoptic camera, which can be operated as a traditional camera (Ng et al., 2005). Today plenoptic cameras come in rather compact forms. A modern commercial and an industrial camera are depicted in figure 6.23.

Because the concept of a plenoptic camera has existed since 1908, one may wonder why it has taken one hundred years to produce commercially viable products. An answer is that plenoptic cameras record many images from many perspectives simultaneously, and therefore they require high resolution to produce good results. Such sensors have been made available in the last decades. Moreover, recent results have shown that a plenoptic camera can be used for a lot more than just depth detection, which was not common knowledge in the early twentieth century. Finally, the optical part consisting of large arrays of lenslets could not be easily manufactured until recently.

6.3 Light Field Capture

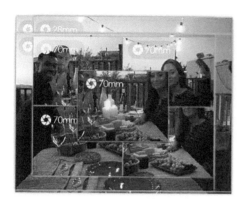 

Figure 6.22 Light field sample interpolation using the commercially available Light L16 camera. The thirteen-megapixel images captured with its sixteen camera modules (*left*). Final fifty-two-megapixel image (*right*). Reprinted from (Sahin and Laroia, 2017).

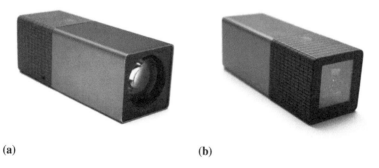

(a)   (b)

Figure 6.23 Modern commercial light field camera: (a) front, and (b) back. *Source:* Lytro (https://en.wikipedia.org/wiki/Lytro).

A method requiring a single 2D image to capture the light field was proposed in (Antipa et al., 2016). However, multiple images are needed during calibration. The 3D shape of an object can also be captured via a digital holographic recording, which is based on a sensor array comparing a reference wave to a wave reflected from the object (Schnars et al., 2015). It was shown that using holography can provide a better spatial resolution for refocusing than capturing the 4D light field (Cossairt et al., 2014). More comprehensive reviews of light field capture can be found in (Levoy, 2006; Ng, 2006).

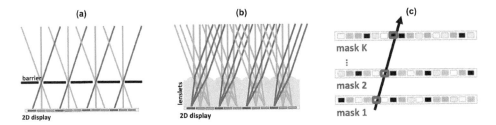

Figure 6.24 Traditional and multilayer light field displays: (a) a traditional display based on a slitted barrier, (b) a traditional display using a front layer based on lenses, and (c) a multilayer display. Reprinted from (Wetzstein et al., 2012).

6.4 Light Field Displays

To exploit the full benefit of a captured light field, one needs to display it in a way that presents its true potential. The conventional 2D displays can show 2D images only, which as stated previously, represent only one slice of the light field at a time.

The idea of a 3D display spans all the way to the seventeenth century, when the French painter Gaspar Antoine de Bois-Clair created a portrait that would show two different people depending on the viewing angle, using *parallax barriers*, which are a series of occluding bars. When viewed from an angle, the bars occluded a part of the painting that is designated for the second viewing angle. This concept was further developed in 1903 by Frederic Ives, who created a patent called the *parallax stereogram*, which enabled viewing several images using parallax barriers. In 1908 Gabriel Lippmann proposed the concept of 3D display using a lenslet array as part of his work called *integral photography* (Masia et al., 2013).

Since then, the concept of parallax barriers has been developed and extended to produce improved 3D displays. However, there are several challenges in designing a 3D display, such as low light throughput, reduced angular resolution, and crosstalk between distinct views. With the advent of digital computers and liquid crystal displays (LCDs), many of these problems can be alleviated, as subsequently discussed in this section.

We first introduce traditional 3D displays and their limitations, which motivate their replacement with more complex displays such as multilayer, multiframe, and tensor displays.

6.4.1 Traditional 3D Displays

The more advanced 3D stereoscopic displays show two light field slices at once, one for each eye. The challenge for a light field display is to present many slices of the light field at the same time, without the need to wear any viewing equipment.

Traditional light field displays rely on very simple preprocessing. They are based on a two-layer system: the back layer displays a number of views, or light field slices, and the front layer acts as a barrier, allowing the viewer to see only one slice at a time, depending on the viewer's position relative to the display. The front layer can be based simply on a slitted opaque screen or on a more sophisticated layer of lenses. This architecture is called a parallax barrier display, due to the parallax effect created by the barrier when the observer changes the viewing perspective. The schematic of these displays is depicted in figure 6.24a and 6.24b. The barriers can consist of bars, supporting one-dimensional parallax, or of a mask with pinholes in a lattice pattern, supporting full parallax (left-right and up-down).

There are a few drawbacks of these architectures. First, the viewer will see only a fraction of the emitted light at a time, that is, the light throughput is very low. Second, the spatial resolution is limited, as the pixels from different views need to be adjacent. Third, the field of view is limited, with only a reduced range of visible angles. Fourth, despite using the front layer to separate light field slices, there is crosstalk between views, which decreases the view quality.

Subsequently, the displays have a limited angular resolution. Similarly to sampling signals in time, if there are not enough angle samples of the light field, the views will be affected by aliasing. This means that the high frequencies, which are not captured by the samples, lead to distortions in the displayed image. Aliasing can be addressed by prefiltering the light field before sampling. The angular resolution is in direct connection with the maximum depth of the display. Therefore, the scene depth is regularly adapted to the display depth with methods called depth retargeting techniques.

6.4.2 Multilayer and Multiframe Displays

The problems of traditional displays, such as parallax barriers and microlens arrays, were addressed by generating compressed representations of the light fields. Even though they are not identical replicas of the light field, these representations are decoded by integration in the human eye, and therefore are perceived as good estimations. The displays generating such representations are known as *compressive displays* (Banks et al., 2016; Wetzstein et al., 2012).

Traditional displays suffer from low spatial resolution and low light throughput. To address these problems, pinhole masks can be replaced by LCD screens. Unlike the masks whose patterns are static, the LCDs can adapt the pattern dynamically. Therefore, the display proposed is made up of a rear and front LCD, illuminated by backlight, as illustrated in figure 6.25a. From each viewing perspective, an observer can see the pixels from the rear LCD that are not blocked by the pixels of the front LCD, acting as a parallax barrier. So far, the setting still suffers from the problems mentioned previously, as many of the pixels in the image on the rear LCD will be blocked.

This can be fixed in the case of LCDs using a well known phenomenon in the human visual system called *flicker fusion*. This phenomenon manifests through the perception of a rapid sequence of images as their temporal average. Essentially, any sequence with a faster

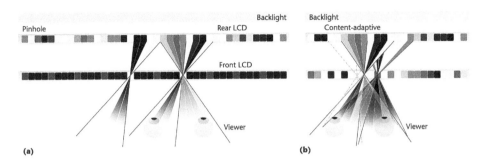

Figure 6.25 Content-adaptive light field displays. (a) A parallax barrier implemented with a dual-stacked LCD. The viewer sees only the light crossing the front LCD, and (b) a content-adaptive dual-stacked LCD, displaying several time-multiplexed frames corresponding to the viewing perspective. Reprinted from (Wetzstein et al., 2012).

frequency than 60 Hz is no longer perceived by the eye as separate events but merged into a continuous event.

The front LCD uses this phenomenon by displaying high speed content-adaptive patterns, which allows the observer to see more pixels for every viewing angle, which leads to higher resolution and brightness, as depicted in figure 6.25b. The shifting masks displayed by the front LCD thus allow the viewer to see an increased resolution for each viewing angle. At each time, the mask covers a number of pixels displayed by the rear LCD to show only one viewing angle. However, these pixels are uncovered when the mask shifts to show additional pixels from the same angle.

The display method based on time multiplexing presented here is called *high rank 3D (HR3D)* (Wetzstein et al., 2012). This name is derived from the mathematical interpretation of the process proposed. In the simplified scenario of a 2D light field, we can construct a matrix in which the row and column indices represent the points of intersection of a light ray with each of the two LCDs. It turns out that such a matrix has rank 1, which corresponds to a poor approximation of the light field. HR3D can achieve a much more accurate reconstruction in which the light field matrix has rank 3.

The next advancement consisted of a 3D display method called *layered 3D*, which involves stacks of multiple LCDs (Wetzstein et al., 2011). In this case, rather than using one layer to mask an image, a whole stack of LCDs is used to attenuate light generated by a backlight-emitting layer. This approach increases the computational complexity, but alleviates some of the drawbacks of traditional displays. The functioning principle of such a display is depicted in figure 6.26.

In layered 3D, the light rays that pass through the whole stack are attenuated cumulatively by the corresponding pixels in each layer. In designing such a display, the attenuation of each light ray in each direction should match the light field that is being displayed. The

6.4 Light Field Displays

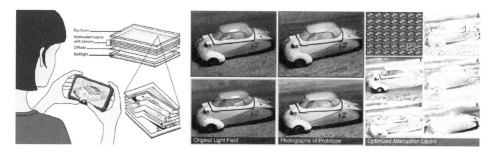

Figure 6.26 Layered 3D display. The display design is based on five attenuation layers (*left*), the scene (*center*), and the corresponding light field and five optimal layers (*right*). Reprinted from (Wetzstein et al., 2011).

principle is very similar to that of computed tomography (CT) scanning. The body of a patient is illuminated, and a detector measures the accumulated light attenuation through the patient's body. Similarly, the LCD stack is illuminated from below, and the pixels attenuate each light ray such that, when viewed from different perspectives, the intensity of the light observed reflects the 3D structure of the object. The layered 3D display computed for a real scene is depicted in figure 6.26. The display has five layers that were optimized to best approximate the light field. The tomographic light field synthesis in layered 3D gives additional display capabilities such as increased depth of field and better images in terms of brightness and resolution (Wetzstein et al., 2011). A new method based on adaptive sampling was proposed to decrease the computational resources required (Heide et al., 2013).

Let $L(s, v)$ be the target light field. Then the layered 3D display illuminated by backlight generates an approximation given by

$$\widetilde{L}_N(s, v) = \prod_{k=1}^{N} f^{(k)}\left(s + \frac{d_k}{d_r}v\right),$$

where s, v denote the variables of the light field, measured as coordinates on two lines placed along the backlight emitting layer and the layer near it, respectively; $f^{(k)}(s) \in [0, 1]$ is the transmittance at point s of layer k; d_k is the distance of layer k to the s-axis; and d_r is the distance between the s-axis and v-axis.

Therefore, the layered 3D display generates an Nth order approximation of the light field. For $N = 3$, the points $\{s + v \cdot d_1/d_r, s + v \cdot d_2/d_r, s + v \cdot d_3/d_r\}$ lie on a plane with equation

$$\mathsf{P} = \{(d_3 - d_2)x_1 + (d_1 - d_3)x_2 + (d_2 - d_1)x_3; \ x_1, x_2, x_3 \in \mathbb{R}\}.$$

Using this observation, it is convenient mathematically to parametrize the estimated light field in a three-dimensional space as function $\widetilde{L}(x_1, x_2, x_3)$, such that

$$\widetilde{L}\left(s + v \cdot \frac{d_1}{d_r}, s + v \cdot \frac{d_2}{d_r}, s + v \cdot \frac{d_3}{d_r}\right) = \widetilde{L}_N(s, v),$$

$$\widetilde{L}(x_1, x_2, x_3) = 0, \quad \{x_1, x_2, x_3 \notin \mathsf{P}\}.$$

To make the expression more compact, we can use the concept of tensor, which is the generalization of a vector.

We define the estimated light field tensor as $\mathbf{L}$, such that $[\mathbf{L}]_{x_1, x_2, x_3} = \widetilde{L}(x_1, x_2, x_3)$. In the tensor formulation, we assume that $x_1, x_2, x_3 \in \mathbb{Z}$ are the coordinates of the pixels on each layer. We also define the tensor $\widetilde{\mathbf{T}}$

$$\widetilde{\mathbf{T}} = \frac{1}{M} \sum_{p=1}^{M} \mathbf{f}_p^{(1)} \circ \cdots \circ \mathbf{f}_p^{(N)},$$

where $\circ$ denotes the vector outer product, M denotes the total number of frames, and (Wetzstein et al., 2012)

$$\left[\mathbf{f}_p^{(k)}\right]_{x_k} = f_p^{(k)}(x_k).$$

Essentially tensor $\widetilde{\mathbf{T}}$ represents the set of all possible combinations of N pixel values, taken from N different layers. We can then state that

$$\widetilde{\mathbf{L}} = \mathbf{W} \odot \widetilde{\mathbf{T}},$$

where $\odot$ is the elementwise product and $\mathbf{W}$ is a binary mask such that $[\mathbf{W}]_{x_1, x_2, x_3} = 1$ if $(x_1, x_2, x_3) \in \mathsf{P}$ and 0 otherwise.

In other words, the estimated N-dimensional light field is modeled by the successive N attenuations of any possible N pixels from different layers, which are then limited by $\mathbf{W}$ to only the groups of pixels that have light rays passing through them.

6.4.3 Tensor Displays

Although they are crucial steps for the advancement of 3D displays, these techniques cannot display a light field for multiple observers located in a wide viewing zone. The coupling of time-multiplexed (HR3D) and light attenuating layers (layered 3D) led to the introduction of the *tensor display* (Wetzstein et al., 2012). The tensor display exhibits both multiframe and multilayer capabilities, which reduces artifacts significantly in comparison with layered 3D and HR3D. Due to additional degrees of freedom, tensor displays achieve a wider field of view and wider depths of field (Wetzstein et al., 2012).

6.4 Light Field Displays

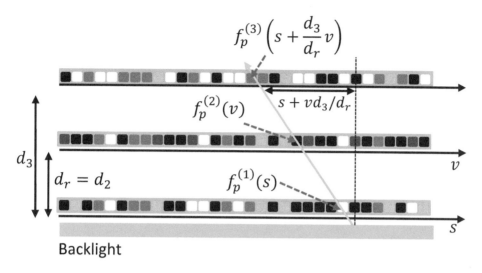

Figure 6.27 Tensor display with three layers. The light illuminating the rear LCD is attenuated cumulatively by each layer.

In mathematical terms, the tensor display decomposes the light field $L(s, v)$ into a low-rank approximation

$$\widetilde{L}_{N,M}(s, v) = \frac{1}{M} \sum_{p=1}^{M} \prod_{k=1}^{N} f_p^{(k)}\left(s + \frac{d_k}{d_r} v\right),$$

where $f_p^{(k)}(x) \in [0, 1]$ is the transmittance at point s of layer k during frame p; M denotes the total number of frames; and N is the order of approximation for the light field. A diagram of a tensor display is shown in figure 6.27. The estimated light field is computed by solving

$$\left\{f_p^{(k)}(x)\right\}_{\substack{p=1,\cdots,M \\ k=1,\cdots,N}} = \arg \min_{f_p^{(k)}(x) \in [0,1]} \int_{\mathbb{R}} \int_{\mathbb{R}} \left(L(s, v) - \widetilde{L}_{N,M}(s, v)\right)^2 ds\, dv.$$

In other words, the estimation computes the transmittance values $f_p^{(k)}(x) \in [0, 1]$ that minimize the error between the original and estimated light fields.

As previously, we can represent the light field as a tensor, assuming that the coordinates on each layer are integer values denoting the pixel coordinate. In the case of tensor displays, however, the tensor has order N and rank M, therefore allowing a much better approximation of the light field

$$\widetilde{\mathbf{T}} = \frac{1}{M} \sum_{p=1}^{M} \mathbf{f}_p^{(1)} \circ \cdots \circ \mathbf{f}_p^{(N)}.$$

The estimation of the light field is then written in a more compact form as

$$\widetilde{\mathbf{T}} = \arg \min_{\widetilde{\mathbf{T}}} \left\| \mathbf{L} - \mathbf{W} \odot \widetilde{\mathbf{T}} \right\|^2,$$

where $\mathbf{L}$ is the tensor representation of the true light field and $\|\cdot\|^2$ is the squared tensor norm defined for a tensor $\mathbf{Q}$

$$\|\mathbf{Q}\|^2 = \sum_{x_1 \in \mathbb{Z}} \sum_{x_2 \in \mathbb{Z}} \sum_{x_3 \in \mathbb{Z}} [\mathbf{T}]^2_{x_1, x_2, x_3}.$$

The search space is much larger here than for layer 3D, because the values of the attenuations $f_p^{(k)}(x)$ depend on the frame number p in addition to layer number k. Therefore the problem is more complex, but leads to better estimations of the light field.

An additional degree of freedom is given by controlling the direction of the LCD illumination source, known as *directional backlighting*. This allows sweeping through several light field views sequentially, adding an additional boost to the field of view and depth of field (Wetzstein et al., 2012).

The estimated light field in the case of directional backlighting is

$$\widetilde{L}_{N,M}(s, v) = \frac{1}{M} \sum_{p=1}^{M} b_p(s, v) \prod_{k=1}^{N} f_p^{(k)} \left(s + \frac{d_k}{d_r} v \right),$$

where $b_p(s, v)$ models the backlight field parameterized by s, v emitted at time frame p. The tensor formulation is

$$\widetilde{\mathbf{T}} = \frac{1}{M} \sum_{p=1}^{M} \mathbf{b}_p \circ \mathbf{f}_p^{(1)} \circ \cdots \circ \mathbf{f}_p^{(N)},$$

where $\mathbf{b}_p$ is the vectorized form of the backlight field. The estimation of $\widetilde{\mathbf{T}}$ is done as before to estimate the desired light field. However, in this case, the additional degrees of freedom given by $\mathbf{b}_p$ lead to a higher performance.

6.4.4 Open Problems with Light Field Displays

Some techniques have been put in place to address the most common problems with light field displays. The crosstalk between views, which is a distortion of a viewing angle by adjacent views, has been incorporated in the model to alleviate the negative effect. This is done by adjusting the luminance of the images displayed so that the final perceived image is as close to the target one as possible.

Unlike traditional displays, the compressive displays are based on a stack of LCDs and optical elements such as microlens arrays (Banks et al., 2016; Wetzstein et al., 2012). This causes a *multiplicative effect* on the incident light which, in general, allows displaying more viewing angles than additive displays such as multiplane or volumetric.

6.4 Light Field Displays

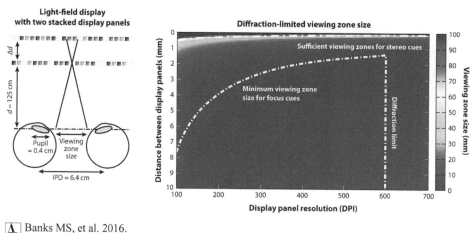

Banks MS, et al. 2016.
Annu. Rev. Vis. Sci. 2:397–435

Figure 6.28 Viewing zone size for two-stacked displays. The display is a fixed distance ($d = 125$ cm) from the viewer, who has an interpupillary distance (IPD) set to 6.4 cm (*left*). The viewing zone can be computed as a function of the display resolution and interdisplay distance (*right*). A resolution beyond 600 dpi leads to significant blur. For a resolution below that, the interdisplay distance should be large enough that two views enter the same pupil (so that focus cues can be achieved). Reprinted from (Banks et al., 2016).

An important problem with multiplicative displays is their limited resolution. This occurs because a common approach in designing these displays is to use two LCD panels between which the frontal one acts as a parallax barrier. When the pixel size of the frontal LCD is close to the wavelength of the light, this causes diffraction, which leads to a significant blur.

There is evidence that focus cues, such as blur and accommodation, affect both 3D shape perception and the apparent scale of the scene. Figure 6.28 depicts the conditions under which focus cues can be achieved by a two-layer display. Specifically, this can happen if the viewing zone is smaller than the size of the pupil, allowing two different views to enter the same pupil.

Even so, in attempting to display 3D data, such as tomography scans, there are different methods used to view the data. These methods are based on extracting continuous objects using boundary detection, performing 2D boundary detection on every slice, or using voxels to represent the data (Herman and Udupa, 1983). The latter is less complex because it avoids the preprocessing steps of boundary detection (Frieder et al., 1985). 3D displays can also be designed as holographic displays, which exploit the diffraction phenomenon using appropriately chosen optical elements (Peng et al., 2017).

  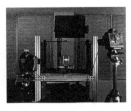 

Figure 6.29 3D autostereoscopic light field display. Two perspectives of the display, each a pair for stereo vision (*left* and *right*). The object shown is photographed by a stereo camera system (*center*). Reprinted from (Jones et al., 2007).

An example of a light field display is given in (Jones et al., 2007). This is an autostereoscopic display, which adds binocular perception of 3D depth. It includes a user tracking mechanism to adjust the display to the viewer's height and distance. The system is compatible with multiple viewers placed around the display at the same time. An object displayed on this device is depicted in figure 6.29.

Applications such as virtual reality or augmented reality have pushed the advancement of near-eye displays, which are based on a headset providing two views for each eye. Most near-eye displays are based on two separate microdisplays, that is, a screen split optically with two lenses. In 2015, a new near-eye display based on two LCDs was proposed, allowing each eye to see freely a 4D light field (Huang et al., 2015).

Chapter Appendix: Notation

Notation	Description
(x, y, z)	Spatial position
(θ_1, θ_2)	Angles
$L(x, y, z, \theta_1, \theta_2)$	Plenoptic function (that a light field device captures)
(s, t)	Angular dimensions
(u, v)	Spatial dimensions
$L(u, v, s, t)$	Plenoptic function (reparametrized)
$I_{s^*, t^*}(u, v)$	Subaperture image
$I_{u^*, v^*}(s, t)$	Light field subview
$E_{v^*, t^*}(u, s)$,	Epipolar planes (EPIs) in (u, s) co-ordinates
$E_{u^*, s^*}(v, t)$	Epipolar planes (EPIs) in (v, t) co-ordinates
$I(s, t)$	Irradiance
Ω	Subset of all angles (u, v) for which the the light rays the point (s, t)

6.4 Light Field Displays

I_d	Refocused image
$\widehat{I_d}$	Fourier transform of refocused image I_d
(f_s, f_v)	Frequency variable corresponding to (s, v)
$\mathcal{F}[\cdot]$	Fourier transform operator
$M(x, \theta)$	Mask modulation function
$S(x)$	Light field sensor measurements
$f^{(k)}(s)$	Transmittance at point s of layer k
$\circ$	Vector outer product
$\mathbf{L}$	Light field tensor
$\odot$	Elementwise product
$\mathbf{W}$	Binary mask
P	Set of points in a plane

Exercises

1. Motivation

    At the top level, this problem set enables you to turn your cell phone into a 4D light field camera.

    Shallow depth of field is a desirable aesthetic quality in a photograph. Unfortunately, this effect requires a large aperture, that is, the lens is going to be big and bulky! But what if it were possible to turn your cell phone into a camera with a large aperture? What if we could selectively focus on objects in postprocessing?

    The goal of this exercise is to synthesize images with smaller depths of field, thus making it appear to have been taken from an expensive camera with a larger aperture (Lumsdaine and Georgiev, 2009; Levoy et al., 2004). Figures E6.1a and E6.1b show a scene image with the corresponding synthetic aperture image with lower depth of field.

    (a) (b)

    Figure E6.1 Turning a cell phone into a light field camera. (a) An all-in-focus image taken with a cell phone camera. (b) A light field stack is postprocessed to blur out the background. Notice how the helmet stands out from the background.

2. Experimental component

    We will capture a video by moving the camera along a zigzag path, as shown in Figure E6.2 in front of the static scene. We recommend using MATLAB for all codes.

    Please note the following:

    - The algorithm being implemented does not take camera tilt into account. As much as possible, avoid tilting and rotating the camera.

Exercises

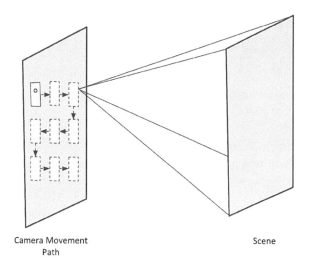

Figure E6.2 Zigzag planar motion of the camera in front of the static scene to capture a video.

- The instruction set use a planar zigzag path for camera motion as shown in Figure E6.2. However, you may wish to try different paths such as circular or polyline.
- The number of frames in the video captured will determine the time required to compute the output. Make sure the video is not too long.

Solve all the following problems below:

a) Setting up a static scene
   Set up a static scene similar to the one shown in Figure E6.1a. Try to have objects at different depths.

b) Capturing a 4D light field
   Take a video by waving your camera in front of the scene by following a specific planar motion. The more you cover the plane, the better will be your results. Ensure that all objects are in focus in your video. Generate three frames of the video. These frames differ in their *parallax*, which is the effect in which object positions change in response to view.

c) Acquiring the data
   Write a function to read your video file and convert the video into a sequence of

frames. Because this was captured from a cell phone, each frame image is in RGB color. Write a script to convert each frame to grayscale.

d) Registering the frames

   i. Template and window
      From the first frame of your video, select an object as a template. We register all other frames of the video with respect to this template. Once a template has been selected in the first frame, we search for it in the subsequent frames. The location of the template in a target frame image gives us the shift (in pixels) of the camera. Because we do not have to search for the template in the entire target frame image, we select a window to perform this operation. Note, however, that selecting a window is optional and is done only to reduce the computation time.

   ii. Normalized cross-correlation
       Perform a normalized cross correlation of the template with the extracted search window.

       Let $A[i,j]$ be the normalized cross-correlation coefficient. If $t[n,m]$ is our template image and $w[n,m]$ is our window, then from (Lewis, 1995) we have:

       $$A[i,j] = \frac{\sum_{m=1}^{T}\sum_{n=1}^{T}(w(n,m) - \overline{w}(i,j))(t(n-i, m-j) - \overline{t})}{\sqrt{\sum_{m=1}^{T}\sum_{n=1}^{T}(w(n,m) - \overline{w}(i,j))^2(t(n-i, m-j) - \overline{t})^2}} \quad (6.1)$$

       where, $\overline{t}$ is the mean of the template, and $\overline{w}_{i,j}$ is the mean of the window $w[n,m]$ in the region under the template. Plot the cross-correlation coefficient matrix $A[i,j]$ for one of the frames.

   iii. Retrieving the pixel shifts
        The location that yields the maximum value of the coefficient $A[i,j]$ is used to compute the shift (Georgeiv and Intwala, 2006). The shift in pixels for each frame can be found by

        $$[s_x, s_y] = max_{i,j}\{A[i,j]\}. \quad (6.2)$$

        Then, generate the plot of s_x v/s s_y

e) Synthesizing an image with synthetic aperture
   Once you have the pixel shifts for each frame, you can synthesize a refocused image by shifting each frame in the opposite direction and then summing up all the frames.

Suppose the pixel shift vector for frame image $I_i[n,m]$ is $[s_{x_i}, s_{y_i}]$. Then, the image output $P[n,m]$ with synthetic aperture is

$$P[n,m] = \sum_i I_i[n - s_{x_i}, m - s_{y_i}]. \tag{6.3}$$

f) Repeating the experiment for different templates

Now we exploit the fact that we can synthetically focus on different depths. To do this, select a new object as your template and repeat all the steps to generate an image that is focused on this new object. Here, we have selected the cup as our new object. Generate a defocused image with a different template object in focus.

3. Assessment

a) Deriving the blur kernel width

The goal is to understand how much blur is synthetically added by using a model of pinhole cameras. Consider the coordinate diagram shown in Figure E6.3. Here, $[X_1, Z_1]$ is a scene point of an object in the template; $[X_2, Z_2]$ is a scene point of an object in the background; and $C^{(i)}$ for $i = 1, \ldots, k$ are positions of the apertures of cameras at which the scene is captured. The maximum camera translation is Δ, and f is the focal length of the cameras (all are assumed to be the same).

We use the shift-and-add method for light field imaging such that X_1 is the point in focus, that is, the template that we shift and add. Derive a mathematical expression for the full-width half maximum (FWHM) of the blur kernel (W) applied to X_2. You should not need figures but may include them. (Hint: Our solution to derive W was about a half page. To check your solution, if $Z_1 = Z_2$ the width of the blur kernel should be zero.)

b) Blur kernel shape

Now that you have derived the FWHM of the blur kernel, please write the functional expression for the blur kernel. For example, is it a Gaussian blur?

c) Blur and scene depth

Plot the width of the blur kernel, W, as a function of the difference in depth planes, $|Z_2 - Z_1|$. Comment on the relationship between these variables.

d) Blur and focal length

Plot the width of the blur kernel, W, as a function of the focal length of the camera, f. Comment on the relationship between these variables.

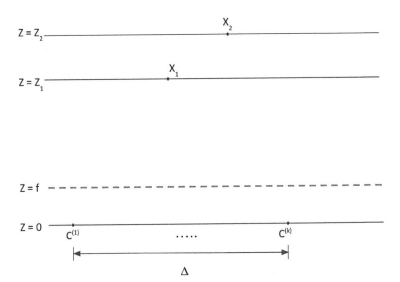

Figure E6.3 Example of coordinate system and notation. The dashed plane is the virtual film plane, placed one focal length *above* the apertures located at $C^{(1)}, \ldots, C^{(k)}$. This is a common shorthand convention so that we do not have to flip the camera images. In reality, the actual film plane would be one focal length below the aperture location. This coordinate system is used as a guide, and you are welcome to modify it as needed.

7 Polarimetric Imaging

In this chapter, we discuss coding strategies in *optical polarization* (hereafter *polarization*). Computational coded imaging using polarization has spawned a variety of imaging systems with never-before-seen capabilities. Polarization images of synchrotron radiation around active galactic nuclei provide valuable insight into the physics of supermassive black holes (Chael et al., 2016). Imaging systems have also been inspired by the vision systems of marine creatures, whose polarization-based vision enhances the contrast of their predators' appearance and helps them navigate through the seas (Powell et al., 2018). In this chapter we discuss systems that can acquire 3D shape with unprecedented quality, analyze the stress and strain of a material, or enable lost travelers to navigate the seas.

7.1 Principles of Polarization

We define polarization in 7.1.1 and then describe how it is leveraged in the contexts of coding in 7.1.2 and of information in 7.1.3. The reader is also welcome to study (Andreou and Kalayjian, 2002; Walraven, 1977; Hecht, 2012) for additional introductions to polarization.

7.1.1 Formal Definition of Polarization

Recall that light is an electromagnetic wave, which means that it has an electric field and magnetic field component (see Figure 7.1a). Polarization refers to the orientation of the electric field of light. When describing polarization, by convention we ignore the magnetic field orientation. The polarization can therefore be described by the plane in which the E-field oscillates. Consider a light wave propagating in the $+z$ direction. The electric field will be confined in the x–y plane, and have vector components E_x and E_y such that $\mathbf{E}(z) = \hat{\mathbf{x}} E_x + \hat{\mathbf{y}} E_y$. The relative phase and magnitudes of E_x and E_y as a function of time determine the polarization state.

Perhaps the simplest example of polarization is *linear polarization*, shown in Figure 7.1b. The electric field oscillates in a plane, and the orientation of that plane describes the linear polarization state. Another type of polarization is known as *circular polarization*. Intuitively, circular polarization occurs when the electric field vector changes its orientation

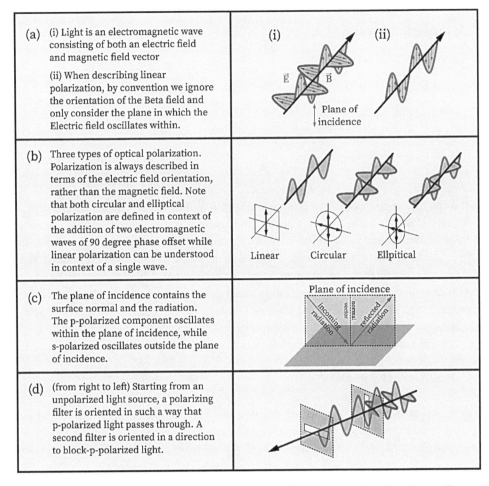

Figure 7.1 Electromagnetic waves and polarization. Polarization describes the oscillation of the electric field of an EM wave over time as it propagates through space.

as the light propagates through space, as shown in Figure 7.1b. In contrast to linear polarization, circular polarization is described when two electromagnetic plane waves are added together. These two waves are identical except for a 90° difference in phase. The resultant electromagnetic wave has an electric field whose direction changes along the axis of propagation. We can plot the tip of the E-field as a circle when viewed axially.

The third foundational class of polarization is *elliptical polarization*. This type of polarization is merely a generalization of circular polarization in which two electromagnetic plane waves have unequal amplitudes or a phase offset that is not 0° or 180°. If the polar-

7.1 Principles of Polarization

ization is not circular or linear, then it is elliptical. Figure 7.1b shows a summary of the types of polarization states discussed.

Polarization can be decomposed into two orthogonal axes with respect to a *plane of incidence* (i.e., when striking a material), shown in Figure 7.1c. The *p-polarized* axis is parallel to the plane of incidence, and the *s-polarized* axis is perpendicular to the plane of incidence. This nomenclature stems from the German word for parallel (*parallel*) and the German word for perpendicular, (*senkrecht*). It is common even for experts in the field to confuse the orientations of s-polarization and p-polarization. One helpful mnemonic is that *p-polarization* is *in-plane*. The s- and p-polarized light can also be referred to as perpendicular and parallel polarized light, respectively.

We use this decomposition with respect to the plane of incidence to find a set of transmission and reflection coefficients known as the *Fresnel coefficients*. The s-polarized and p-polarized light behave differently at the interface of two media. In other words, the amount of light reflected and transmitted at the interface is dependent on the polarization states of the light. These coefficients are

$$\Gamma_\perp = \frac{\eta_2 \cos\theta_i - \eta_1 \cos\theta_t}{\eta_2 \cos\theta_i + \eta_1 \cos\theta_t}, \quad \tau_\perp = 1 + \Gamma_\perp,$$

$$\Gamma_\| = \frac{\eta_2 \cos\theta_t - \eta_1 \cos\theta_i}{\eta_2 \cos\theta_t + \eta_1 \cos\theta_i}, \quad \tau_\| = \left(1 + \Gamma_\|\right) \frac{\cos\theta_i}{\cos\theta_t},$$

where Γ and τ are the reflection and transmission coefficients, corresponding to the proportion of incident light that is reflected and transmitted. η_1 and η_2 are impedances of the media, and θ_i and θ_t are the incident and transmitted angles of the wave, related together by Snell's law. The incident angle θ_i at which all parallel polarized light is transmitted (i.e., $\Gamma_\| = 0, \tau_\| = 1$) is known as *Brewster's angle*.

7.1.2 Coding with Polarization

The polarization state of light can be manipulated using polarizing filters. Suppose we have light that is initially *unpolarized*, meaning that it contains a mixture of both s-polarized and p-polarized light. Perhaps we want to filter out the s-polarization. A *polarizing filter* is a special material that can preferentially transmit light of a specific polarization state. As shown in Figure 7.1d, such a filter has a vertical spacing that allows p-polarized waves to pass. However, the s-polarized light cannot fit through the vertical spacing and is thus blocked. The resultant transmitted light is only p-polarized.

Now, let us consider how circularly polarized light could be obtained. Recall from the previous section that in circularly polarized light, the s- and p-polarization states are of identical amplitude but phase shifted by 90°. A device known as a *waveplate* has a slightly different refractive index for light at s-polarization versus p-polarization. Therefore, it is possible to introduce a controllable phase shift to light at different polarization states. Circular polarization is commonly produced by first linearly polarizing an unpolarized

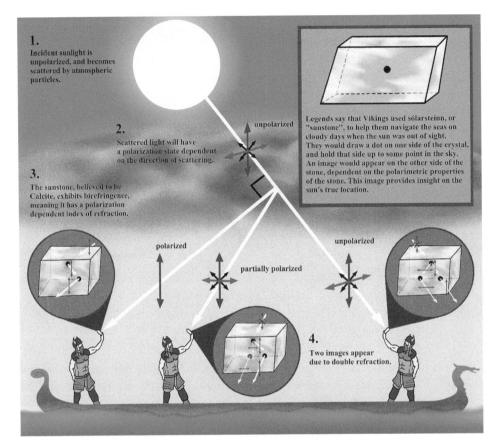

Figure 7.2 Vikings used what they called the *sunstone* for navigating the seas on cloudy days when the sun was out of sight. Historians believe that this navigation was enabled by the polarimetric properties of the stone, believed to be calcite.

beam (using a filter such as that shown in Figure 7.1d) and subsequently passing it through a *quarter-wave plate*, which introduces a phase shift of 90° between the two polarization components.

Today, polarizing filters are among the simplest, most reliable, and most inexpensive way to obtain complex control of light. There are several categories of materials that have a polarization-dependent behavior. The first approach is to use a *wiregrid polarizer* made of finely spaced parallel wires. Light polarized parallel to the wires will induce charge movement along the wires, causing energy dissipation. The result is the annihilation of the electric field component parallel to the wires and transmission of polarization states perpendicular to the wires, as shown in Figure 7.3. A second approach is to use a thin film coating applied to glass. This thin film is usually a fine layer of metal with anisotropic properties,

7.1 Principles of Polarization

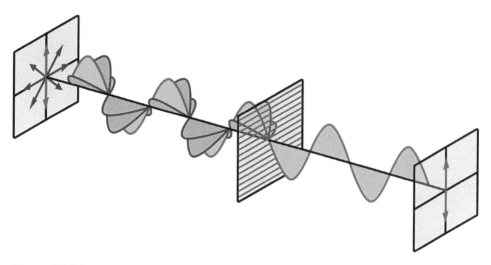

Figure 7.3 Wire grid polarizer, by which light polarized perpendicular to the wires is transmitted. In other words, the transmission axis of the polarizer is perpendicular to the wires.

fabricated carefully using principles from Fresnel coefficients for selective reflectance and transmittance. For further information on polarization-based optical coatings, readers are directed to (Macleod, 2005). Another type of polarizing filter is a piece of crystal with anisotropic properties. Such crystals can be found in nature, and the light-material interactions fall under the field of *crystal optics*. For a more detailed reference on crystal optics, readers are directed to (Yariv and Yeh, 2006).

We use polarization filters in our everyday lives. An ordinary liquid crystal display (LCD) uses polarization to adjust the brightness of each pixel. At the heart of the LCD is a crystal that acts as a polarization filter with controllable orientation (through applied voltage or another control input). The amount of linearly polarized light that passes through the crystal is a function of the orientation of the crystal and therefore a function of the control input (e.g., applied voltage). This enables the display to have a different brightness at each pixel. Not only 2D televisions use polarization but also 3D televisions at the movie theater. To obtain the sense of 3D, a display needs to transmit slightly different images to a human's left eye versus the right eye. A special television emits two different images, of which one is made up of s-polarized light and the other of p-polarized light. The use of 3D glasses with s- and p-polarization filters enables the human visual system to receive different light. These concepts are shown in Figure 7.4.

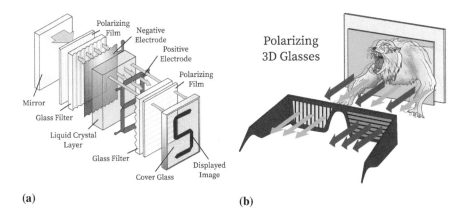

Figure 7.4 (a) Working principles of liquid crystal displays (LCD). (b) How 3D movies are projected.

7.1.3 Information in Polarization

We have thus far described the elementary principles of polarization and the modulation of polarization state. Now we consider the information within polarization. Polarization is a rich source of information. It is helpful to separate polarimetric information into two types: *engineered information* and *natural information*.

Engineered information occurs when human-made systems forcibly transmit data across polarization channels. Figure 7.4b shows a simple example of such information multiplexing. At the movie theater, we see a display encoding different streams of video information at different polarization states. Such multiplexing of information across separable polarization states is routinely exploited in fiber-optic and other optical telecommunication systems, transmitting information by using two waves of separable polarization states. This could be s and p components, or left circular and right circular beams transmitted through an optical fiber. For further details, we direct readers to the term *polarization-division multiplexing* (PDM) and the text material, "Polarization Optics in Telecommunications" (Damask, 2004).

Natural information is the information that the physical world encodes into the polarization state. For example, sunlight is partially polarized and changes based on the time of day and bearing and concentration of the atmosphere. Measuring the polarization state can enable one to estimate one's bearing, as the ancient Vikings did in the seventh century. Figure 7.2 shows how Vikings used a calcite crystal for navigation in a time preceding the magnetic compass. Photographs can use a similar polarization effect to increase the contrast of clouds in the sky. Deep sea animals such as the mantis shrimp have a polarization-sensitive vision system, which perhaps enhances the contrast of their prey underwater. Now, let us journey above water. On a sunny day, a sailor might be bothered

7.1 Principles of Polarization

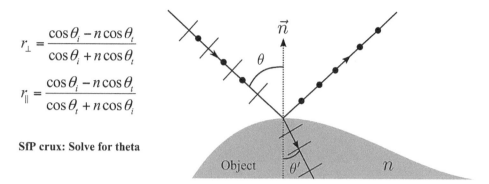

$$r_\perp = \frac{\cos\theta_i - n\cos\theta_t}{\cos\theta_i + n\cos\theta_t}$$

$$r_\parallel = \frac{\cos\theta_i - n\cos\theta_t}{\cos\theta_t + n\cos\theta_i}$$

SfP crux: Solve for theta

Figure 7.5 Shape from polarization problem. We can determine the surface normal if we have information about the reflected polarization and the material's index of refraction.

by the glare from the surface of a placid lake. The glare is a polarized reflection that can be removed through a polarization filter, revealing the background underneath the water. The stress and strain of a material can also be imaged through polarization cues.

Although visually appealing, the interaction between polarization and the scene can be quite complex. Even the seemingly simple reflection of light involves an intricate geometry, material, and polarization, as shown in Figure 7.5. Here, the object is in blue and has two properties of interest. First it has a refractive index n and a local geometry, described by a surface normal $\vec{N}$. When light strikes the surface at an angle θ with respect to the surface normal, there is a change in the polarization state. For instance, as shown in the figure, the incident light has both s- and p-polarization, whereas the reflected light has s-polarization and the transmitted light has p-polarization (the dots indicate that the oscillation is out of the incident plane and the hashes indicate that it is within the incident plane. A famous problem known as *shape from polarization* in computer vision uses measurements of the polarization state of the incident and reflected light to estimate $\vec{N}$, a proxy for shape.

If polarized light is shined upon a surface, for almost any angle a specular or mirror-like reflection will be linearly polarized at some angle θ. This angle can be filtered out to remove glare. A technique known as *cross-polarization* leverages this principle. As shown in Figure 7.6, two crossed filters are used to remove glare. The first filter vertically polarizes the light, and the specular reflection will therefore be vertically polarized. The light of interest (e.g., from the diffuse reflections), will have a varied polarization state, which is discussed in detail in section 7.5.1. Simply filtering out the specular reflection by using a horizontal polarizer will eliminate glare. Cross-polarization generates compelling results and is used commercially by art photographers.

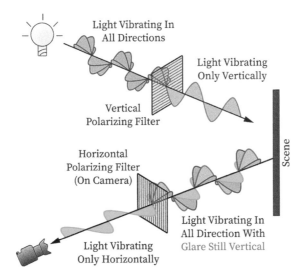

Figure 7.6 Glare removal using cross-polarization.

7.2 Full Stokes Imaging

7.2.1 Parametrization of Polarization

The polarization state of any light wave (polarized or not) can be compactly expressed as a single vector quantity known as the *Stokes vector*. The Stokes vector contains four components (S_0, S_1, S_2, S_3). The vector (S_0, S_1, S_2) describes the linear polarization of the wave, and S_3 describes the circular polarization of the wave. Each parameter can be extracted as intensity measurements (I_0, I_1, I_2, I_3) of the light passing through one of four different polarization filters. I_0 is measured by feeding the light through a filter that indiscriminately absorbs half the energy of all polarization states. I_1 is measured by a linear polarization filter at $0°$; I_2 is measured by a linear polarization filter at $45°$; and I_3 is measured by a right-hand circular polarized filter. Each of the four filters attenuates approximately half the incident intensity. Without considering the transmittance of the polarization filter for simplicity, the Stokes parameters are

$$S_0 = 2I_0,$$
$$S_1 = 2I_1 - 2I_0,$$
$$S_2 = 2I_2 - 2I_0,$$
$$S_3 = 2I_3 - 2I_0.$$

These four Stokes parameters are rather informative on how light will interact with a certain medium. Suppose light is propagating in medium 1 with electric field $E_i(z,t)$. This light is then incident on medium 2, and the transmitted electric field is given by $E_t(z,t)$. $E_i(z,t)$ and $E_t(z,t)$ have corresponding Stokes vectors $\mathbf{S}_i$ and $\mathbf{S}_t$. We can relate these two Stokes vectors together by a *Mueller Matrix*. Every optical medium has a Mueller matrix, which describes how incident light will be transformed by the media with respect to its polarization states. Matrix multiplication of the Mueller matrix with $\mathbf{S}_i$ yields $\mathbf{S}_t$, providing a full understanding of how the polarization state between the two media is affected. The reader is directed to (Hecht, 2012) for examples of Mueller matrices in different optical media.

If we know the Stokes vector for every pixel in an image, we are subsequently able to characterize the geometrical, chemical, and physical properties of scene surfaces. The first three parameters are often useful for improving visibility in scattering media, and the last parameter is often used for improving the contrast of images. These parameters also enable understanding of the surface smoothness, shape, size, color, and orientation.

Another useful way to represent light waves is through the *Poincaré sphere*. If we neglect S_0, we can plot the latter three terms of Stokes vector as a point contained within a 3D sphere. By dividing (S_1, S_2, S_3) by S_0, we obtain the normalized Stokes components (S'_1, S'_2, S'_3), which we can plot along in 3D Cartesian space, centered at the origin. (S'_1, S'_2, S'_3) can be plotted as (x, y, z), respectively. The $+z$-axis refers to right circularly polarized light, and the $-z$-axis refers to left circularly polarized light. All points along a circle for a given z value have the same ellipticity, but at a different orientation angle. Therefore, at $z = 0$, the polarization is linear.

7.2.2 Measuring Stokes Parameters

Imaging systems typically capture only a subset of the Stokes parameters, depending on the application. Imagers that do capture the full Stokes vector utilize a combination of sensors, controlled linear polarizers, retarders, and computing hardware. The most primitive of setups measures a scene using four different polarization filters, with a retarder appended to the fourth filter to measure circularly polarized light. The problem with these setups is that they often require capture of multiple images, which becomes challenging for dynamic scenes. We explore some alternative methods to obtain a full Stokes image.

One such way to efficiently obtain the Stokes parameters is to place a *micropolarimeter array* in front of the focal plane. A micropolarimeter array has a micropolarizer element at a given orientation in front of each pixel, filtering only light in that polarization state, as shown in Figure 7.7. Thus, each pixel senses exactly one polarization state. The other polarization states for that pixel are interpolated using the intensities from neighboring pixels, which measure intensities from different polarization states. This process is analogous to demosaicking the intensity output from a Bayer filter into an RGB image. Such micropolarimeter arrays are also fabricated using industry standard complementary metal-

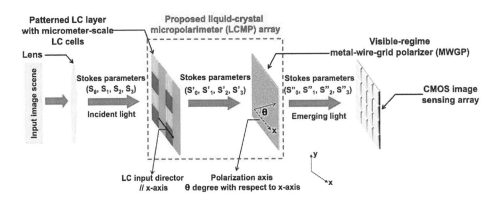

Figure 7.7 Full Stokes imaging with a micropolarimeter array. Reprinted from (Zhao et al., 2010).

oxide semiconductor (CMOS) fabrication processes, making them easy to manufacture on a single chip.

In designing such an integrated polarization imaging system, there are three key components: optical components, polarizing elements, and photodetectors. Figure 7.7 shows the image capture schematic for such a setup. Light is first directed from a scene into a patterned liquid crystal layer, which acts as either a polarization rotator, a retarder, or a neutral density filter. A metal wire grid polarizer then selectively allows certain polarization states to pass through to the focal plane array. As shown in Figure 7.7, the Stokes vector changes as it passes through each optical element. We can determine the Mueller matrix for each element, and properly invert the matrices to obtain the original pixelwise Stokes vector as it enters the micropolarimeter (Zhao et al., 2010). For information on how intensity measurements with polarization filters are calibrated to yield Stokes parameters, the reader is directed to (Vedel et al., 2011) and (Zhang et al., 2013).

When we want to capture the full Stokes vector, we typically need a minimum of four measurements. Provided that these four measurements are not coplanar in the Poincaré sphere, these measurements are sufficient to fully characterize the polarization of the light. Figure 7.8 shows a representation of two different ways to capture the Stokes parameters. Figure 7.8a depicts measurements consistent with the way we have defined the Stokes parameters.

When we consider the design of the micropolarimeter array, it is important to consider the implications of the demosaicking (interpolation) process in which we extract the Stokes parameters by interpolating neighboring pixel values. The space occupied by the four measurements is only a quarter of the sphere's volume. This is problematic because the SNR is proportional to the volume occupied by the four points on the Poincaré sphere. Therefore, measurements at points shown in Figure 7.8b would yield a better SNR for the

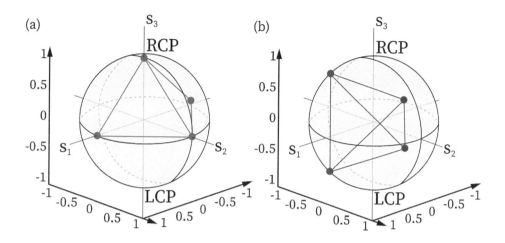

Figure 7.8 Poincaré representation of polarization. (a) Polarization measurements consistent with the definition of Stokes parameters vs. (b), a more robust measurement scheme for determining Stokes parameters with high SNR.

Stokes vector reconstruction. These points form a tetrahedron inside the Poincaré sphere, and correspond to four elliptical polarizers.

Points that are offset along the S_3 axis are measurements made using a combination of a microretarder and a micropolarizer. The more points inside the Poincaré sphere, the more volume occupied and the better the SNR is. However, for practical considerations, we are restricted to fewer measurements (Hsu et al., 2014).

7.3 3D Shape Reconstruction

As previously alluded, the polarization state of light often changes as it reflects off surfaces, depending on the geometry and index of refraction of the surface. The polarization state can be measured to determine the surface normal $\vec{N}$, which is then used as a proxy for the local shape of the object. Consumer cameras like Microsoft Kinect measure depth maps of their surroundings, but are heavily impacted by noise. To an extent, this noise can be computationally reduced, but filtering noise also filters away detail. This is where natural information from polarization can come into play. Polarization can also be particularly useful for analyzing the shape of glossy surfaces, in which the specular reflection is significant. As we see in section 7.5.1, specular reflections have certain exploitable polarimetric properties (Koshikawa, 1979).

There is, however, a key issue with relying exclusively on polarization measurements for shape information, known as *azimuthal ambiguity*. At some polarizer orientation φ_{pol}, the

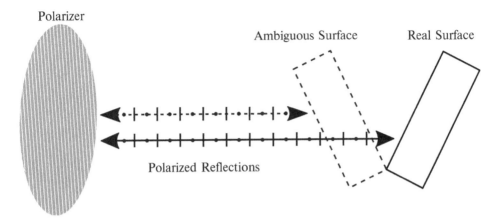

Figure 7.9 Azimuthal model mismatch in shape from polarization (Ba et al., 2020).

measured intensity, assuming unpolarized light, can be modeled as

$$I(\varphi_{\text{pol}}) = \frac{I_{\max} + I_{\min}}{2} + \frac{I_{\max} - I_{\min}}{2} \cos(2(\varphi_{\text{pol}} - \phi)),$$

where ϕ is the angle of polarization. Both ϕ and $\phi + \pi$ can satisfy this equation, leading to ambiguity that can cause reconstruction errors. Once ϕ is recovered, the azimuth angle can be recovered. The uncertainty in measuring the azimuth angle φ is related to the surface reflectance, in what we term *azimuthal model mismatch*. A measured azimuthal component can result in two different surface normals, offset by $\pi/2$, as shown in Figure 7.9. A common assumption is that if the intensity is dominated by diffuse reflections, then $\varphi = \phi$. If the reflection is dominated by specularity, $\varphi = \phi - \pi/2$.

Meanwhile, the zenith component θ of the reflection can be calculated from the degree of polarization ρ using Fresnel equations, assuming the index of refraction of the material is known,

$$\rho = \frac{\left(n - \frac{1}{n}\right)^2 \sin^2 \theta}{2 + 2n^2 - \left(n + \frac{1}{n}\right)^2 \sin^2 \theta + 4 \cos \theta \sqrt{n^2 - \sin^2 \theta}}.$$

However, the refractive index is often unknown, and approximations are often used, a common one being $n \approx 1.5$, resulting in *refractive distortion*. Up to this point, the described method of obtaining surface normals falls under the category of *shape from polarization (SfP)*. A reconstructed surface normal image using SfP cues is shown in Figure 7.10c. Although the shape does resemble the approximate shape of the cup, we can clearly see a distinction from the ground truth. This discrepancy is caused by refractive distortion and azimuthal ambiguity.

7.3 3D Shape Reconstruction

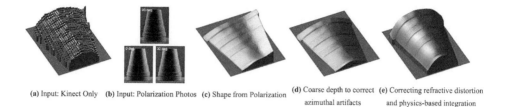

(a) Input: Kinect Only (b) Input: Polarization Photos (c) Shape from Polarization (d) Coarse depth to correct azimuthal artifacts (e) Correcting refractive distortion and physics-based integration

Figure 7.10 Shape reconstruction using polarization cues. Reprinted from (Kadambi et al., 2015).

The depth maps are far too noisy, and polarization information still has some shape ambiguity. However, both these information channels can be fused strategically to yield high quality surface normals. We will now consider a depth-normal fusion, a method known as *Polarized 3D*, to deal with the azimuthal ambiguity (Kadambi et al., 2015). We denote our depth map as $\mathbf{N}^{\text{depth}} \in \mathbb{R}^{M \times N}$ and our surface normals from polarization as $\mathbf{N}^{\text{polar}} \in \mathbb{R}^{M \times N \times 3}$, assuming there are three measured polarization images. To deal with azimuthal ambiguities as shown in Figure 7.10c, we search for a binary operator $\mathcal{A}$ to solve for the optimization problem

$$\widehat{\mathcal{A}} = \arg\min_{\mathcal{A}} \left\| \mathbf{N}^{\text{depth}} - \mathcal{A}\left[\mathbf{N}^{\text{polar}}\right] \right\|_2^2.$$

The binary operation corresponds to either rotating the azimuth angle by π or not rotating. Solving such an optimization deals with the azimuthal ambiguity and yields a corrected normal image $\mathbf{N}^{\text{corr}} = \widehat{\mathcal{A}}\left[\mathbf{N}^{\text{polar}}\right]$. Because the coarse depth map consists of low-frequency information, it cannot correct higher frequency ambiguities. To deal with high-frequency azimuthal ambiguity, the shape inside the high-frequency region is often assumed to be convex. The output after dealing with azimuthal ambiguity is shown in Figure 7.10d.

Finally, we must deal with refractive distortion, caused by the earlier $n = 1.5$ assumption. Although this assumption works relatively well for dielectric surfaces, the zenith angle becomes noticeably distorted in the presence of nondielectrics. We selectively make modifications to regions of the image using $\mathbf{N}^{\text{depth}}$ and $\mathbf{N}^{\text{corr}}$ as criteria. For a given point, if the depth map and polarization data both have low divergence, then we can use the surface normal predicted by $\mathbf{N}^{\text{depth}}$. This suggests that there is no high-frequency detail in that region. However, if one of the maps has high divergence, it suggests that there is fine detail in that region.

For a small patch P in the image, we search for a rotation operator $\widehat{\mathcal{R}}_P$ that solves the optimization problem

$$\widehat{\mathcal{R}}_P = \arg\min_{\mathcal{R}} \sum_{i=1}^{P} \mathbf{M}_{x_i,y_i} \left| \theta^{\text{depth}}_{x_i,y_i} - \mathcal{R}\left[\theta^{\text{corr}}_{x_i,y_i}\right] \right|^2,$$

where $\mathbf{M}_{x_i,y_i}$ is a binary mask corresponding to the divergence criteria described previously, and $\theta_{x_i,y_i}^{\text{depth}}$ and $\theta_{x_i,y_i}^{\text{corr}}$ are the zenith angles obtained from $\mathbf{N}^{\text{depth}}$ and $\mathbf{N}^{\text{corr}}$, respectively. Note that we work with patches here because the problem is spatially varying. Applying the $\widehat{\mathcal{R}}_P$ operator to every patch yields a surface normal corrected for refractive distortion, which we can use to extract the shape of the object. The 3D shape can be obtained by integrating the depth maps with the surface normals, as shown in Figure 7.10d (Kadambi and Raskar, 2017). The work in (Ba et al., 2020) expands on these physics-based polarization models, introducing a polarization-learning fusion to solve the azimuthal ambiguity, dubbed *Deep SfP*.

Structured light, made possible by active illumination, is also another successful way that has enabled 3D image reconstruction. The temporal and spatial coding of the illumination source, as discussed in previous chapters, provide another useful dimension to approximating the plenoptic function. However, using structured light is still vulnerable to complex, difficult-to-model ambient noise. It is useful to consider that light behaves differently at different surfaces depending on the polarization. However, an active polarization-based approach would be even more robust, because ambient polarization cues are often approximated and can be weak.

Many intensity-based structured light systems incorporate *gray code (GC) patterns*, in which the light pattern is encoded as either 0 for dark or 1 for bright. Whether spatially or temporally coded, the forward image model is inverted, and signals that have a weak response to the GC intensity modulation are filtered out. Similarly, we can encode the horizontal polarization state as 1 and vertical polarization state as 0. The use of polarization adds another layer of information to normal intensity-based coding, by providing information about the surface of the objects. For spatial modulation, we can use a micropolarimeter, and for temporal modulation we can use liquid crystal polarizers (Huang et al., 2017).

Another way to extract the shape for objects is by using the Stokes parametrization. Many setups discard the circularly polarized parameter to keep the imaging system cheap and/or compact. However, for many dielectric materials there is a nonnegligible circular polarization component that contains valuable surface normal information. One possible source of this circularly polarized light can be from subsurface scattering. Generally subsurface scattering is neglected, but accounting for it can improve model accuracy. The *Stokes reflectance field* is defined as the Stokes vector that results from the interaction of light with some medium for every surface normal direction. The use of a Stokes reflectance field enables direct mapping from a pixelwise measured Stokes vector to a surface normal.

The complexity of scattering mechanisms at the interface between two different media can be difficult to precisely model. That is, it is difficult to find a Mueller matrix that fully captures this behavior. One way to go around this is to experimentally determine the Stokes parameters for some known surface (e.g., a sphere) under unpolarized light. This

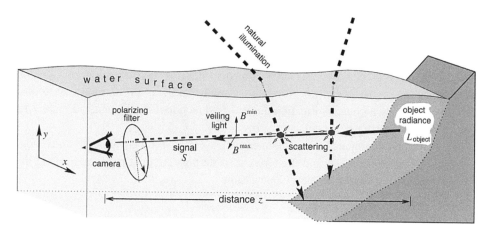

Figure 7.11 Image formation model for passive polarization imaging. Reprinted from (Schechner and Karpel, 2005).

data is then used as a reference for future data measurements. To account for differences in specular albedo and scattering, the maximum of

$$\sqrt{(S'_1)^2 + (S'_2)^2} \text{ and } S'_3$$

is scaled appropriately to match the maximum of the reference measurement. Note that the S' denotes the Stokes vector after interacting with the surface. Such first-order correction works well for surfaces with a wide range of surface normals. The surface normal can be calculated using the degree of polarization. The improvement occurs because other methods calculate the degree of polarization by assuming $S'_3 = 0$ (Guarnera et al., 2012).

7.4 Imaging through Scattering Media

Imaging through scattering media such as fog, haze, and water is crucial for a number of commercial and scientific applications, including autonomous driving, marine exploration, navigation, and photography. Backscatter quickly degrades the visibility of images located far from the camera in such environments. It is also desirable not to have to model the scattering mechanisms causing these degradations, as they can be complex and highly dependent on factors like time of day, weather, and location. These scattering mechanisms often have a polarization signature, which can be used to filter away these unwanted components (Treibitz and Schechner, 2008).

7.4.1 Underwater Imaging

One key challenge with underwater imaging is that the image degradation caused by scattering is spatially varying, because objects are different distances from the camera. This makes image processing tools such as median filtering and histogram equalization ineffective, because these algorithms assume spatially invariant noise. One way to deal with this is by using an image formation model based on polarization images. Figure 7.11 depicts the passive imaging model.

In this passively illuminated setup, the measured signal is a composition of two components: the direct transmission and the forward scattering. As light propagates along the z-axis towards the camera, the direct transmission

$$D(x, y) = L_{\text{object}}(x, y) \, e^{-cz},$$

is light that will reach the camera without being scattered or absorbed. L_{object} is the intensity that would be measured at the camera if the signal is unattenuated; and $c = a + b$ is the attenuation coefficient, where a is the absorption coefficient and b is the total scattering coefficient of the medium (i.e., water). The scattering component is forward-scattered light that deviates from the light's axis of propagation by some angle θ. This causes blur, which we can model as a convolution

$$F(x, y) = D(x, y) * g_z(x, y),$$

where $g_z(x, y)$ is the point spread function (PSF) of the blur. Note that the PSF is z-dependent, because light will scatter more the farther it has to travel. Therefore, we can model the total measured signal

$$\mathbf{S} = \mathbf{D} + \mathbf{F} = e^{-cz} \left(\mathbf{L}_{\text{object}} + \mathbf{L}_{\text{object}} * \mathbf{g}_z \right).$$

Interestingly, however, image degradation from underwater images are more affected by *veiling light* than by image blur. Veiling light is ambient light scattered toward the camera. It is not light considered to be part of the image-forming process. We use the fact that veiling light is partially polarized to algorithmically remove this component.

Under ordinary lighting conditions, the veiling light (i.e., the sun) would be unpolarized. However, as shown in Figure 7.12, the illumination source is above water, so that there exists an *optical manhole*, such that an observer can see only a small portion of the scene above water, caused by total internal reflection. Hence the angle at which the source reaches the scattering particles in the line of sight of the camera is restricted. Due to this anisotropy of irradiance, the veiling is partially polarized. The deeper the image is captured in the water, the more the veiling is polarized.

Using two different polarizers, we can now measure two images at polarizer orientations corresponding to the minimum and maximum intensity. From these measurements, we can apply image processing algorithms to recover the veiling light and remove it from the

7.4 Imaging through Scattering Media

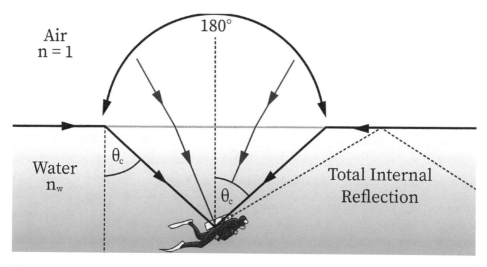

Figure 7.12 Snell's window (optical manhole). Total internal reflection past the critical angle creates only a small window visible from underwater.

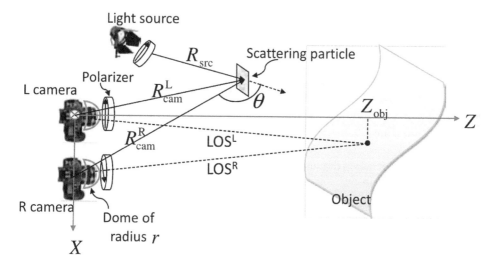

Figure 7.13 Stereovision and polarization for underwater imaging. The use of stereo enables video rate capture of polarization images underwater. Reprinted from (Sarafraz et al., 2009).

image (Schechner and Karpel, 2005). This reconstruction process is similar to that done in dehazing problems, which we discuss in section 7.4.2.

We now consider the integration of stereo vision and polarization imaging for underwater imaging, as shown in Figure 7.13. Each of the two cameras captures a different polarization state. The use of stereo cues enables the computation of distance to objects, and the polarization allows filtering of backscatter. We assume that in the absence of polarizers the cameras will capture a linear superposition of the object signal $S\left(\mathbf{x}_{obj}\right)$ and the backscatter $B\left(\mathbf{x}_{obj}\right)$

$$I\left(\mathbf{x}_{obj}\right) = S\left(\mathbf{x}_{obj}\right) + B\left(\mathbf{x}_{obj}\right).$$

The measured signal at a point $\mathbf{x}_{obj}$ in a scene is

$$S\left(\mathbf{x}_{obj}\right) = L_{obj}\left(\mathbf{x}_{obj}\right) F_{obj}\left(\mathbf{x}_{obj}\right)$$

where $L_{obj}\left(\mathbf{x}_{obj}\right)$ is the attenuated intensity of $\mathbf{x}_{obj}$, and $F_{obj}\left(\mathbf{x}_{obj}\right)$ is the *falloff function*, which is a function of an attenuation constant c, the distance to the object R_{src}, and inhomogeneities $Q\left(\mathbf{x}_{obj}\right)$ in the illumination source. It provides us with insight into the irradiation attenuation as light propagates from the source to the object and back to the camera. $B\left(\mathbf{x}_{obj}\right)$ is similarly a function of c, θ (as shown in Figure 7.13) and $Q\left(\mathbf{x}_{obj}\right)$, but is different from $S\left(\mathbf{x}_{obj}\right)$. Using expressions for $S\left(\mathbf{x}_{obj}\right)$ and $B\left(\mathbf{x}_{obj}\right)$, derived with the geometries of the left and right cameras, we obtain expressions for the input intensities for each of the cameras (L and R)

$$I^L\left(\mathbf{x}_{obj}^L\right) = S^L\left(\mathbf{x}_{obj}^L\right) + B^L\left(\mathbf{x}_{obj}^L\right),$$
$$I^R\left(\mathbf{x}_{obj}^R\right) = S^R\left(\mathbf{x}_{obj}^R\right) + B^R\left(\mathbf{x}_{obj}^R\right).$$

The use of polarization in this setup also enables backscatter modulation by removing *degeneracy*. Because the backscatter is partially polarized, and the object radiance is unpolarized, the polarization filters are able to modulate the backscatter without modulating the signal (Sarafraz et al., 2009). This backscatter can be removed by using coordinate adjustments (needed due to the stereo setup) and image models described in section 7.4.2.

7.4.2 Imaging through Haze and Fog

As mentioned previously, image degradation effects caused by haze are very strongly a function of the distance to the object being imaged. The more distant the object, the more light that gets scattered along the optical path between the object and the camera. In a hazy image, an image consists of *airlight* (natural illumination scattered toward the observer by aerosol particles) and *direct illumination*, which is the scene radiance that would be observed in the absence of haze. To enhance an image in haze, we seek to remove the airlight and correct for the attenuation caused by absorption and scattering. Similar to the imaging through water problem, the dehazing problem has an image formation model, as shown in Figure 7.14.

7.4 Imaging through Scattering Media

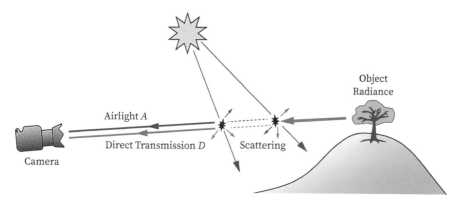

Figure 7.14 Image formation model for dehazing. The airlight has certain polarimetric properties, which are leveraged to be removed from the image.

We start by defining the plane of incidence here as the plane formed by the vector from the source to the scattering particle and the vector connecting the particle and the camera. The airlight is divided into two polarization components parallel and perpendicular to this plane, $A_{\parallel}$ and $A_{\perp}$. In Rayleigh scattering, when the size of the particles is small with respect to the light, $A_{\perp} \geqslant A_{\parallel}$. When larger haze particles cause the scattering, $A_{\perp} < A_{\parallel}$. We therefore can model the airlight

$$A = A_{\infty}\left[1 - t(z)\right] = A_{\perp} + A_{\parallel}$$

where A_{∞} is the airlight radiance for an object infinitely far away, and $t(z)$ is the transmittance of the light a distance z away, defined by the Beer-Lambert law. The degree of polarization ρ of the airlight is a strong function of the viewing and illumination direction

$$\rho = \left(A_{\perp} - A_{\parallel}\right)/A.$$

Notationally, we define $A_{\parallel}$ to be the lowest intensity measured and $A_{\perp}$ to be the highest measured intensity with a linear polarizer. If $\rho = 1$, then the airlight can easily be optically filtered out. However, this occurs only in the restricted situation in which the scattering aerosols are small and the illumination is normal to the viewing direction. In spite of this, we can still algorithmically take advantage of the fact that the airlight is partially polarized. However, the degree of polarization must still be relatively high, so such an algorithm may not be as effective in fog in which multiple scattering will cause *depolarization*. Depolarization is the reduction of the degree of polarization.

Meanwhile, the direct transmission can be modeled

$$D = L_{\text{object}} t(z),$$

where L_{object} is the radiance that would be sensed by the camera if there were no attenuation. We also assume that D has negligible polarization (i.e., is unpolarized). Note that this assumption does not hold for specular surfaces, but if the object is distant enough, the specular component will still contribute a negligible polarization. This means that the polarization filters will modulate the airlight more than the direct component, as mentioned previously.

We know that the image (without polarizers) is a superposition of the direct transmission and airlight. When we add a polarizer at an orientation α that attenuates the most airlight, our measured intensity can be modeled

$$I_{\parallel} = D/2 + A_{\parallel}.$$

Similarly, the polarizer orientation that transmits the most airlight can be modeled

$$I_{\perp} = D/2 + A_{\perp}.$$

A depiction of this concept is shown in Figure 7.15. We refer to the most and least attenuation as the *best state* and *worst state*, respectively. Figure 7.15 depicts an example of two images captured at the best and worst state. Note that these images are not significantly different in quality. We can estimate $A_{\perp}$ and $A_{\parallel}$ to be

$$A_{\parallel} = A(1-p)/2,$$
$$A_{\perp} = A(1+p)/2.$$

We know from Figure 7.16 that optical filtering does not suffice for such a scene due to partial polarization, so we now apply a spatially varying algorithm. Combining the above equations, we obtain an estimate for the airlight $\widehat{A}$

$$\widehat{A} = \left(\widehat{I}_{\perp} - \widehat{I}_{\parallel}\right)/p.$$

Using this, we obtain an estimate for an image removed of airlight $\widehat{D}$

$$\widehat{D} = \widehat{I}_{\perp} + \widehat{I}_{\parallel} + \widehat{A}.$$

Finally, we account for the spatially varying attenuation by approximating the transmission $\widehat{t}$ as

$$\widehat{t} = 1 - \widehat{A}/A_{\infty}.$$

Our dehazed image $\widehat{L}_{\text{object}}$ is

$$\widehat{L}_{\text{object}} = \frac{\widehat{D}}{t} = \frac{\widehat{I}_{\perp} + \widehat{I}_{\parallel} - \widehat{A}}{1 - \widehat{A}/A_{\infty}}.$$

The dehazed image is shown in Figure 7.16 (Schechner et al., 2003).

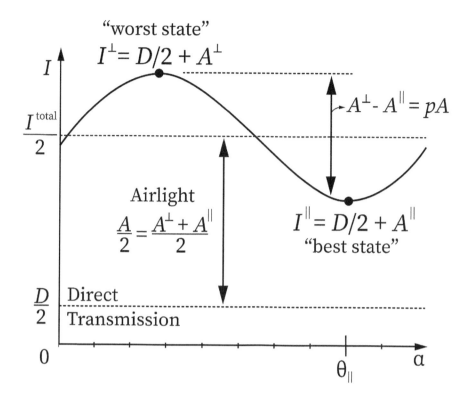

Figure 7.15 Contributions of airlight and direct transmission intensities. The polarization filter modulates the airlight and scattered light, but not the directly transmitted light. We leverage this fact to remove scatter and enhance the image.

So far, we have considered the case in which the airlight is polarized, but not the scene radiance. This assumption does not always hold. In reality, the polarizations of both the airlight and the object radiance contribute to the overall polarization of the scene. To account for both these polarizations, we consider the full Stokes vector at each pixel. For a partially polarized beam, we can consider its Stokes vector to be a superposition of a Stokes vector of a polarized beam and a Stokes vector of an unpolarized beam.

For the outlined method, we neglect S_3 because circular polarization is not prominent in natural scenes. The degree ρ^λ and angle ϕ^λ of polarization can easily be determined for

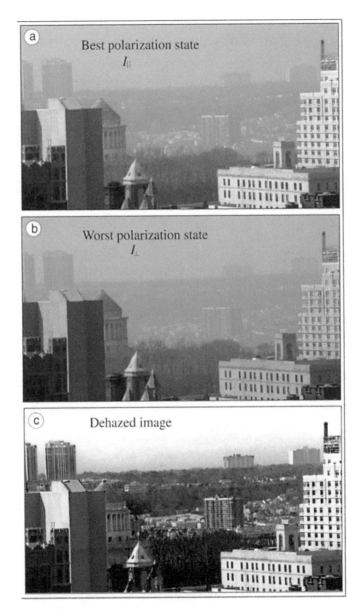

Figure 7.16 Image dehazing using polarization and physics-based models. Reprinted from (Schechner et al., 2003).

each color channel in $\lambda = \{\lambda_R, \lambda_G, \lambda_B\}$ by

$$\rho^\lambda = \frac{\sqrt{(S_1^\lambda)^2 + (S_2^\lambda)^2}}{S_0^\lambda},$$

$$\phi^\lambda = \frac{1}{2} \arctan\left(\frac{S_2^\lambda}{S_1^\lambda}\right).$$

To account for both the polarizations of the airlight and the radiance, we modify the transmission function earlier to instead obtain

$$t(x, y) = 1 - \frac{\Delta I(x, y) - \rho_D(x, y) S_0(x, y)}{A_\infty [\rho_A(x, y) - \rho_D(x, y)]},$$

where $\Delta I(x, y) = I_{\max} - I_{\min}$ is the *polarization-differenced image* (Fang et al., 2014).

7.4.3 Polarization-ToF Fusion for Depth Maps

As we discussed in 5.2, we can obtain useful information from time-resolved images, particularly scene depth. Scene depth is often measured based on a time-of-flight principle, in which the time between an emitted pulse and a measured spike can be used to infer distance. However, in scattered media this becomes nontrivial because of the mixing between scattered light from particles and reflected light from surfaces. The nonuniformity of polarization orientations and degree of polarizations with respect to space and time must be accounted for. We will build on our mathematical foundation for steady-state imaging and apply it to adaptive descattering imaging in a time-resolved manner.

Once again, we rely on the Stokes formulation to deal with this problem, but only the linear components. We measure the Stokes components here by measuring intensities of three different polarization orientations $\alpha = \{\alpha_1, \alpha_2, \alpha_3\}$. We obtain measurements for each orientation

$$I(t, \alpha) = \frac{1}{2}[S_0(t) + S_1(t)\cos(2\alpha) + S_2(t)\sin(2\alpha)].$$

From this, we can compute the degree of polarization for each pixel by

$$\rho(t) = \frac{\sqrt{S_1^2(t) + S_2^2(t)}}{S_0(t)}.$$

Note that the measured intensities now are taken as a function of time, because polarization states can change between frames. Using a Taylor approximation of Beer-Lambert's law, we can extract the direct component

$$D = S_0(t)\left[1 - \frac{\rho(t)}{\rho_{S_\infty}(t)}\right],$$

where $\rho_{S_\infty} = I_{S_\infty}^{-1}\left(I_{S_\infty}^{\max} - I_{S_\infty}^{\min}\right)$. Although this term does not account for attenuation, it still leads to reasonably good depth estimates. To obtain the depth, we consider a camera at position O, a light source at position S, and a scene point at X. We know that

$$|\overline{SX}| + |\overline{OX}| = c\tau(x),$$

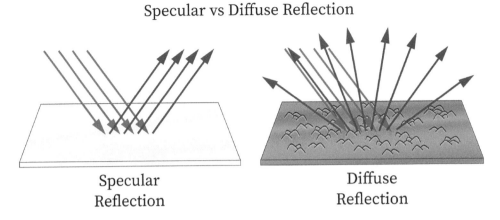

Figure 7.17 Specular versus diffuse reflection.

where c is the speed of light, and $\tau(x)$ is the time it takes for the reflection to be detected at pixel x. We can extract the depth for each pixel

$$d(x) = |\overline{OS}| = \frac{|\overline{OS}|^2 - (c\tau(x))^2}{2|\overline{OS}|\cos\theta - 2c\tau(x)},$$

where θ is the angle between $\overline{OS}$ and $\overline{OX}$ (Wu et al., 2018). For an example on how polarization can be used to obtain depth via polarized interferometry, the reader is directed to (Maeda et al., 2018).

7.5 Reflectance Decomposition Using Polarimetric Cues

7.5.1 Specular vs. Diffuse Reflection

When light bounces off a surface, it can reflect in either a specular or diffuse manner. *Specular reflections* occur when light bounces in a mirrorlike fashion, in which the reflected light is at the same angle as the incident light, as governed by Snell's Law. *Diffuse reflections*, however, go through several layers of interreflections and subsurface scattering before re-emerging at the surface of the material. The two concepts are illustrated in Figure 7.17. The problem with specular reflections is that the reflected light is spatially concentrated, which produces a strong highlight in brightness. This causes a glaring effect, which affects both the visual aesthetic of the image and the ability of a machine to perform vision tasks such as object detection. There is great benefit in being able to separate an image into its specular and diffuse components.

Non-polarization-based methods leverage aspects of the color intensity profile to separate specular and diffuse components. For dielectrics, the *dichromatic reflectance model* predicts that specular reflections have a similar spectral profile as the illumination spectrum, whereas

7.5 Reflectance Decomposition Using Polarimetric Cues

the spectrum of diffuse components is affected by the surface medium. Other approaches invoke a *Lambertian constraint*, in which the observed intensity of the diffuse components is isotropic or close to isotropic. Thus the observed diffuse reflection varies slowly (if at all) with observer position, unlike specular reflections, which vary rapidly with position. Intensities violating this Lambertian criteria are detected as specular. However, in their color dependence, these approaches do not consider the complexity of color profiles of real-world scenes.

Assuming that a scene consists primarily of dielectric materials, we can assume (1) that a dichromatic model is applicable and (2) that the specular component is polarized, while the diffuse is not. We can formulate the intensity I at every pixel

$$I = I_d + I_s,$$

where I_d is the diffuse intensity, and I_s is the specular intensity. If we place a linear polarizer in front of the sensor, we know that I_d will be approximately constant as a function of the polarizer orientation, because the diffuse component is unpolarized. However, I_s should vary as a function of the orientation angle θ. Therefore, we can now write the specular component as

$$I_s = I_{sc} + I_{sv} \cos\left[2\left(\theta - \alpha\right)\right],$$

where α is the orientation of the polarizer, I_{sc} is a constant specular offset, and I_{sv} is the amplitude of the cosine variation. In a situation in which the illumination can be controlled, this creates a straightforward specular removal process by using linearly polarized incident light. In such a case, we can take advantage of the fact that smooth surfaces preserve the polarization states of specularly reflected light. This means that we can place a polarizer phase shifted by $\pi/2$ from the illumination to effectively filter out the specular components and leave us with a diffuse image.

However, we cannot always control the source illumination, so we also consider specular/diffuse separation under passive illumination, which we assume to be unpolarized. I_{sc} and I_{sv} are dependent on the complex index of refraction η and the incidence angle ψ.

$$\frac{I_{sc} + I_{sv}}{I_{sc} - I_{sv}} = \frac{F_\perp(\eta, \psi)}{F_\parallel(\eta, \psi)} = q.$$

We know that for a filter with orientation θ_i, the measured intensity can be written

$$I_i = I_c + I_{sv} \cos\left[2\left(\theta_i - \alpha\right)\right]$$

where $I_c = I_d + I_{sc}$. We can reformulate this into a linear system of equations for M different filters by expressing the equation as a dot product

$$I_i = \langle \mathbf{f}_i, \mathbf{v} \rangle$$

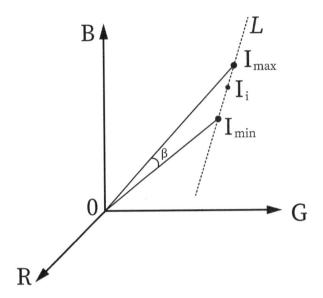

Figure 7.18 Color constraints on specular/diffuse decomposition.

where $\mathbf{f}_i = [1, \cos(2\theta_i), \sin(2\theta_i)]$ and $\mathbf{v} = [I_c, I_{sv}\cos(2\alpha), I_{sv}\sin(2\alpha)]$. For $M = 3$ independent filters, we can solve the linear equation to obtain $\mathbf{v}$, but there remains an ambiguity with α, because α can be either $\alpha + \pi$ or α. We need additional information to resolve this ambiguity. Note that we could solve for I_d by using q, but this is usually not known a priori.

Although we cannot rely solely on color information, it provides a way to constrain our image decomposition. For a set of collected polarized images, we know that the diffuse component will not change but the specular component will vary as a function of $\cos(\theta_i)$. This means that the measurements I_i will lie along a line L in color space, as shown in Figure 7.18. We can easily see from the expression for I_i that $I_{\max} = I_c + I_{sv}$ and $I_{\min} = I_c - I_{sv}$. We can then extract the *degree of polarization* via the analytical equation

$$\rho = \frac{I_{\max} - I_{\min}}{I_{\max} + I_{\min}}.$$

The degree of polarization gives us a measure of how polarized a light is, with $\rho = 0$ suggesting the light is unpolarized and $\rho = 1$ suggesting that the light is completely polarized along an axis. If ρ is below a certain threshold, we can assume that the light is unpolarized and mark the entire intensity of that pixel as diffuse. If the pixel is sufficiently polarized, and the colors of the specular and diffuse components are sufficiently different,

7.5 Reflectance Decomposition Using Polarimetric Cues

we can take advantage of the dichromatic model. We quantify the similarity of the colors by the angle β in Figure 7.18. Using color information from neighboring pixels, we can extract I_d from the measurements I_c (Nayar et al., 1997). In this case, we saw how polarization properties of diffuse and specular components further allow us to constrain color-based decomposition.

The specular and diffuse components can also be calculated if we can capture the full Stokes vector under circularly polarized spherical illumination. Recall from previous discussion that the Stokes vector of an incoming wave is transformed at the interface of another medium. This change can be modeled by the linear transformation

$$\mathbf{s}' = \mathbf{C}(\phi)\mathbf{D}\left(\delta;\vec{N}\right)\mathbf{R}\left(\theta;\vec{N}\right)\mathbf{C}(-\phi)\mathbf{s},$$

where $\mathbf{s}$ is the original Stokes vector, $\mathbf{s}'$ is the transformed Stokes vector, $\mathbf{C}$ is the Mueller rotation matrix, $\mathbf{R}$ is the Mueller reflection matrix, $\mathbf{D}$ is the retardation Mueller matrix, θ is the incidence angle, ϕ is the angle between the plane of incidence and the x-axis, $\vec{N}$ is the surface normal, and δ is the phase shift. This formulation holds for pure specular reflections. Using such a model, we can infer the reflectance behavior of the surface. First, we compute the degree of polarization using the measured Stokes parameters as

$$p = \frac{\sqrt{s_1^2 + s_2^2 + s_3^2}}{s_0}.$$

The specular intensity can be extracted as $\rho_s = s_0 p$, and the diffuse intensity can be extracted as $\rho_p = s_0(1-p)$, because specular light is polarized and diffuse is not. We can then couple these values with the measured Stokes parameters to determine the parallel and perpendicular reflection coefficients, by solving the following equations:

$$s_0 = \frac{1}{2}\left(R_\parallel + R_\perp\right) + \rho_d,$$

$$s_3 = \pm\sqrt{R_\parallel R_\perp}.$$

If $\theta \leqslant \theta_B$, then we take the positive value of s_3, where θ_B is the Brewster angle. Otherwise, we take the negative value of s_3. Using the reflection coefficients we can determine the per-pixel index of refractions from Fresnel equations, which would greatly improve the SfP task as discussed earlier, where refractive distortion was problematic. In practice, this holds only for dielectrics, because metals have a complex index of refraction. However, approximating them as real values still works relatively well (Ghosh et al., 2010). (Ghosh et al., 2011) and (Ma et al., 2007) used a multiview capture system with spherical gradient illumination to capture the geometry and reflectance of a human face, which contains a diverse mixture of specular and diffuse reflection.

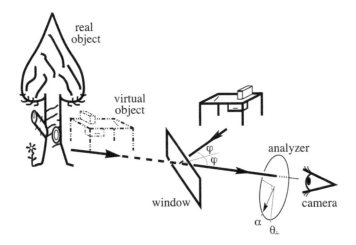

Figure 7.19 Image affected by semireflector. Reprinted from (Schechner et al., 1999).

7.5.2 Virtual vs. Real Image Decomposition

When a transparent material is present between a camera and the scene, we visually notice that a partial reflection is superimposed on the observed scene. Such semireflections negatively impact image aesthetic and the performance of vision tasks. We denote this transparent material as the *transparent layer*, the scene as the *real object*, and the semireflections as the *virtual object*, as shown in Figure 7.19. Although the presence of the semireflected image is unwanted in the image, it may still be desired to keep this image. Therefore, it is of great value to be able to decompose an image into a virtual and semireflected image.

One way we could deal with this is simply by placing a polarization filter in front of the camera to block the virtual image. However, this filtering is effective only when ϕ is at the Brewster angle, in which case we orient the polarizer parallel to the plane of incidence to filter out reflected perpendicularly polarized light. Furthermore, we may still want to keep the virtual image as a source of information about the scene. We also want to be able to identify which image is the virtual image and which is the real image.

Light incident on a semireflector can be separated into parallel and perpendicular components (relative to the plane of incidence), $I_\parallel$ and $I_\perp$. The reflectivities for a single surface medium are given by the Fresnel coefficients of reflection

$$R_\parallel = \frac{\tan^2(\phi - \phi')}{\tan^2(\phi + \phi')}, \quad R_\perp = \frac{\sin^2(\phi - \phi')}{\sin^2(\phi + \phi')},$$

where ϕ is the angle of incidence (from the virtual scene onto the semireflector), and ϕ' is the angle of the refracted ray (as governed by Snell's Law). Usually the semireflector, for example, glass of a car window, has a finite thickness. As the light enters the first air-glass interface, part of it will be transmitted and part of it will be reflected. At the second

7.5 Reflectance Decomposition Using Polarimetric Cues

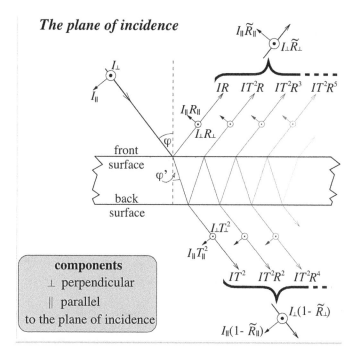

Figure 7.20 Multiple reflections and refractions in a semireflector. Reprinted from (Schechner et al., 1999).

interface, part of the light will again be reflected, and part of the light will be refracted into the air, and the process repeats, as shown in Figure 7.20. The total reflectivity is thus

$$\widetilde{R} = R + T^2 R \sum_{l=0}^{\infty} \left(R^2\right)^l,$$

assuming that the absorption within the medium is negligible and that spatial shift of the rays caused by refraction is negligible relative to the variations in the image. Thus, the total reflectivities and transmitivities are

$$\widetilde{R}_\| = \frac{2}{1+R_\|} R_\|, \quad \widetilde{R}_\perp = \frac{2}{1+R_\perp} R_\perp,$$

$$\widetilde{T}_\| = 1 - \widetilde{R}_\|, \quad \widetilde{T}_\perp = 1 - \widetilde{R}_\perp.$$

If unpolarized light is incident on the semireflector, and a polarizer with orientation α is placed in front of the sensor, we can determine the measured intensities for the reflected scene (virtual image) and transmitted scene (real scene) to be

$$f_R(\alpha) = \frac{I_R}{2}\left[\widetilde{R}_\perp \cos^2\left(\alpha - \theta_\perp + \widetilde{R}_\| \sin^2(\alpha - \theta_\perp)\right)\right],$$

$$f_T(\alpha) = \frac{I_T}{2}\left[\widetilde{T}_\perp \cos^2\left(\alpha - \theta_\perp + \widetilde{T}_\| \sin^2(\alpha - \theta_\perp)\right)\right],$$

where I_R and I_T are the true intensities of the reflected and transmitted scene, and $\theta_\perp$ is the angle at which $\widetilde{T}_\perp$ is maximized. We can then formulate the total measured intensity as

$$f(\alpha) = f_R(\alpha) + f_T(\alpha) = \left(\frac{f_\perp + f_\|}{2}\right) + \left(\frac{f_\perp - f_\|}{2}\right)\cos[2(\alpha - \theta_\perp)],$$

where $f_\perp = f(\theta_\perp) = \left(I_R\widetilde{R}_\perp/2\right) + \left(I_T\widetilde{T}_\perp/2\right)$ and $f_\| = f(\theta_\perp + 90°) = \left(I_R\widetilde{R}_\|/2\right) + \left(I_T\widetilde{T}_\|/2\right)$. We should note that $f_\perp - f_\| = 0.5\left(\widetilde{R}_\perp - \widetilde{R}_\|\right)(I_R - I_T)$. Therefore, if I_R and I_T are equal, the light coming out of the semireflector (toward the camera) is unpolarized. Generally, however, we assume that $I_T > I_R$, so that the $f(\alpha)$ is minimal at $\alpha = \theta_\perp$. Therefore, we conclude that the polarization of the transmitted light primarily dictates the overall polarization of the measured light.

Solving the previous set of equations, we obtain approximations for $I_T(\phi)$ and $I_R(\phi)$, assuming that we know $\widetilde{R}_\perp(\phi)$ and $\widetilde{R}_\|(\phi)$.

$$\widehat{I}_T(\phi) = \left[\frac{2\widetilde{R}_\perp(\phi)}{\widetilde{R}_\perp(\phi) - \widetilde{R}_\|(\phi)}\right]f_\| - \left[\frac{2\widetilde{R}_\|(\phi)}{\widetilde{R}_\perp(\phi) - \widetilde{R}_\|(\phi)}\right]f_\perp,$$

$$\widehat{I}_R(\phi) = \left[\frac{2 - 2\widetilde{R}_\|(\phi)}{\widetilde{R}_\perp(\phi) - \widetilde{R}_\|(\phi)}\right]f_\perp - \left[\frac{2 - 2\widetilde{R}_\perp(\phi)}{\widetilde{R}_\perp(\phi) - \widetilde{R}_\|(\phi)}\right]f_\|.$$

Again, if ϕ is at Brewster's angle, then we can easily approximate I_T by looking at $f_\|$. However, we do not know ϕ a priori. To find an estimate for ϕ, we assume that the reflected and transmitted intensities are uncorrelated, which is reasonable because they come from unrelated scenes. We can then solve for the angle of incidence by solving

$$\widehat{\phi} = \left\{\phi : \text{Corr}\left[\widehat{I}_R(\phi), \widehat{I}_T(\phi)\right] = 0\right\}.$$

The estimated images will be based on several captured polarization orientations α, because the polarizer sinusoidally modulates the intensities of the partially polarized light. The final images will have minimal information about each other, due to the decorrelated assumption about the two scenes (Schechner et al., 1999).

Chapter Appendix: Notations

7.5 Reflectance Decomposition Using Polarimetric Cues

Notation	Description
$(\hat{\mathbf{x}}, \hat{\mathbf{y}})$	Direction vectors along x- and y-axis
(E_x, E_y)	Electric field along x- and y-axis
(Γ, τ)	Reflection and transmission coefficients, respectively
(η_1, η_2)	Impedances of the media
θ_i, θ_t	Incident and transmitted angles of the wave
$\Gamma_\perp, \Gamma_\parallel, \tau_\perp, \tau_\parallel$	Fresnel coefficients
$\vec{N}$	Surface normal
(S_0, S_1, S_2, S_3)	Stokes vector [(S_0, S_1, S_2) describe linear polarization, and S_3 describes the circular polarization]
(I_0, I_1, I_2, I_3)	Light intensity measurements corresponding to Stokes vector
$E_i(z, t)$	Incident electric field
$E_t(z, t)$	Transmitted electric field
$(\mathbf{S}_i, \mathbf{S}_t)$	Stokes vectors corresponding to incident and transmitted electric fields
φ_{pol}	Polarization orientation
ϕ	Angle of polarization
ρ	Degree of polarization
n	Refractive index
$\mathbf{N}^{\text{depth}}$	Depth map
$\mathbf{N}^{\text{polar}}$	Matrix containing surface normals from polarization
$\mathcal{A}$	Binary operator
$\mathbf{N}^{\text{corr}}$	Corrected normal image
$\mathcal{R}$	Rotation operator
$\mathbf{M}_{x_i, y_i}$	Binary mask
$\{\theta^{\text{depth}}_{x_i, y_i}, \theta^{\text{corr}}_{x_i, y_i}\}$	Zenith angles obtained from $\mathbf{N}^{\text{depth}}$ and $\mathbf{N}^{\text{corr}}$, respectively
$D(x, y)$	Direct transmission
L_{object}	Unattenuated object intensity
$g_z(x, y)$	Point spread function of the blur
$\mathbf{x}_{\text{obj}}$	Coordinates of object to be scanned
$S(\mathbf{x}_{\text{obj}})$	Object signal
$B(\mathbf{x}_{\text{obj}})$	Backscatter signal
$L_{\text{obj}}(\mathbf{x}_{\text{obj}})$	Attenuated intensity
$F_{\text{obj}}(\mathbf{x}_{\text{obj}})$	Falloff function
$Q(\mathbf{x}_{\text{obj}})$	Illumination source
I^L, I^R	Input intensity from left and right camera, respectively
$(A_\parallel, A_\perp)$	Airlight polarization components parallel and perpendicular to the plane

A_∞	Airlight radiance for an object infinitely far away
$t(z)$	Transmittance of the light a distance z away
$(A_\parallel, A_\perp)$	Lowest and highest intensity measured with a linear polarizer, respectively
c	Speed of light
$\tau(x)$	Time it takes for the reflection to be detected at pixel x
I_d	Diffuse intensity
I_s	Specular intensity
I_{sc}	Constant specular offset intensity
I_{sv}	Amplitude of cosine variation
$\mathbf{C}$	Mueller rotation matrix
$\mathbf{R}$	Mueller reflection matrix
$\mathbf{D}$	Mueller retardation matrix

Exercises

Maxwell's mathematical formulation of electric and magnetic fields paved the way to our understanding of electromagnetic wave propagation. From such formulations arose the Fresnel equations, a set of coefficients that describe the reflection and transmission of light with respect to the light's polarization state. Knowledge of this polarization-dependent light-matter interaction has spawned innovations in polarimetric imaging. In this problem set, we use our understanding of polarization to filter out artifacts from an image caused by haze particles in the air.

1. Cross-polarization imaging through haze

   Art photographers often use polarizing filters to improve the aesthetic appeal of their photographs. The reflection of shiny objects creates specular highlights on an image that can negatively impact the aesthetic quality of the image. These specular reflections are often strongly polarized. By orienting a linear polarizer orthogonal to the polarization direction of the glare, we can filter out glare and produce a clean image, as shown in Figure 7.16. This method is known as cross-polarization. However, under certain conditions, this reflection may not have a high degree of polarization (DoP), for example, imaging through haze. In such a case, cross-polarization on its own is insufficient. We instead use this theory derived in the chapter to dehaze an image.

   a) Capturing polarization images

   To capture our polarization images, all we need is a linear polarizer that can be mounted in front of the camera aperture. There are two polarization images we need to capture: $I_\|$ and $I_\perp$. $I_\|$ is the image captured when the linear polarizer is oriented such that the intensity passing through the polarizer is at a minimum, and $I_\perp$ is the image captured when the amount of light passing through the polarizer is at a maximum. This minimum and maximum can be determined by empirically rotating the polarizer orientation.

   b) Measuring and calculating DoP and A_∞

   The degree of polarization (DoP) is calculated as

   $$\rho = \frac{\widehat{I}_{\perp,\text{sky}} - \widehat{I}_{\|,\text{sky}}}{\widehat{I}_{\perp,\text{sky}} + \widehat{I}_{\|,\text{sky}}}, \tag{7.1}$$

   where $\widehat{I}$ denotes a measurement of the true image intensity I. The subscript *sky* denotes a measurement of the sky. This should be calculated for each color channel independently. Prior work has reported $\rho = [\rho_r, \rho_g, \rho_b] = [0.28, 0.25, 0.22]$ but this could depend on environmental conditions. You should find a larger DoP for longer wavelengths. A_∞ is the airlight radiance infinitely far from the camera. This is computed

$$A_\infty = \widehat{I}_{\perp,\text{sky}} + \widehat{I}_{\parallel,\text{sky}} \qquad (7.2)$$

c) Modeling airlight

The linear polarizer sinuisoidally modulates the measured intensity as a function of the polarizer orientation angle, as shown in Figure 7.15. Assuming that the direct transmission (D) is unpolarized, half of D will be transmitted to the sensor through the polarizer. Based on the relationships derived from this figure, we can approximate the airlight

$$\widehat{A} = \frac{\widehat{I}_\perp - \widehat{I}_\parallel}{p}. \qquad (7.3)$$

d) Computing de-hazed image

The final dehazed image can be computed for each color channel independently

$$\widehat{L} = \frac{\widehat{I}_\perp + \widehat{I}_\parallel - \widehat{A}}{1 - \widehat{A}/A_\infty}. \qquad (7.4)$$

8 Spectral Imaging

As humans, we can see only a small subset of all radiation surrounding us. While we can see visible light, we cannot perceive the radio signals originating from our cell phones or the thermal radiation emanating from blackbody sources like the human body. Each of these radiation types is characterized by differing wavelengths (or frequencies) and holds fascinating properties worthy of further investigation. In this chapter we continue our study of the plenoptic dimensions, focusing specifically on the spectral nature of light. In particular, we delve into the capture and analysis of spectral images, as well as its usefulness for imaging applications such as material classification, anomaly detection, and remote sensing.

8.1 Spectral Effects on Light-Matter Interaction

8.1.1 Formal Definition of Spectrum

Light is an electromagnetic (EM) wave, propagating through a medium with spatially and temporally oscillating electric and magnetic fields. One of the important features of light is its *wavelength*. Frequency (ν) and wavelength (λ) are inversely proportional, so that

$$c = \nu\lambda,$$

where c is the propagation speed of light, and ν and λ are measured in units [Hz] and [m] respectively. Wavelength and frequency are measures of spatial and temporal frequency, respectively, of the wave oscillation. The range of all frequencies of the wave is known as the electromagnetic *spectrum*. The EM spectrum consists broadly of radio waves, microwaves, infrared light, visible light, ultraviolet light, X-ray light, and gamma rays. Humans are able to perceive only the visible spectrum as a quality we refer to as *color*. Whereas spectrum is continuous in nature, color is a discretization of the visible spectrum. These concepts are illustrated in Figure 8.1.

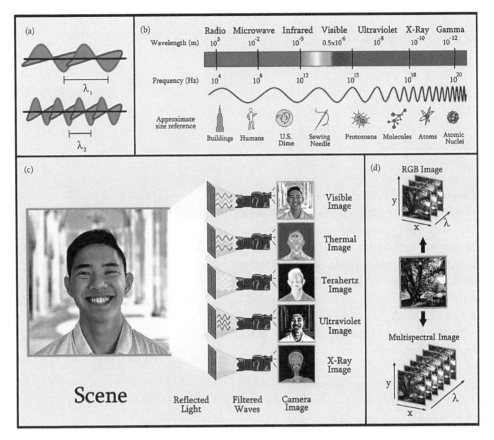

Figure 8.1 What is wavelength, and how do we use it in imaging? (a) Electromagnetic (EM) waves are characterized by a wavelength. (b) Electromagnetic spectrum. (c) A standard camera, similar to our eyes, captures visible light that is reflected by a scene, from which we extract photographs. However, images at different wavelengths capture different information about a scene. For example, a thermal image would be useful for heat seeking, while an infrared image would be useful for food quality inspection. (d) A spectral image samples scenes at a higher spectral frequency than normal RGB images.

8.1.2 Absorption, Reflectance, and Transmittance

The interaction between light and matter is highly wavelength dependent. Different materials have different crystalline structure. This crystalline structure can be better understood as a periodic arrangement of electrons, which can take on only a discrete set of energy states that satisfy Schroedinger's wave equation. The incident photon energy is directly

8.1 Spectral Effects on Light-Matter Interaction

proportional to frequency by *Planck's relation* (Hecht, 1998)

$$E = h\nu$$

where $h = 6.62607015 \times 10^{-34}$ J · s is Planck's constant. Because electrons can take on only a certain set of energy states, only certain photon energies will interact with the crystal structure of a medium. The frequency-dependent photon energy results in a frequency-dependent light-matter interaction. There are hundreds of wavelength-dependent interactions in imaging, but students should be familiar with four primitive types: absorption, reflection, scattering, and transmission.

Absorption dependence A medium will absorb a photon depending on its wavelength. Conceptually, this occurs when the energy of the incident light is near the activation energies of electrons in the material, enabling the electrons to absorb the light. This electron energy is later lost to lattice vibrations. Because the energy of light is related to its frequency, it follows that the absorption is dependent on frequency. Visually, we perceive absorption as the opacity of a material, such as cardboard or brick. Mathematically, this is expressed by a wavelength-dependent absorption coefficient in the context of the Beer-Lambert law (Swinehart, 1962).

Reflection dependence Reflections are also a wavelength-dependent phenomena. Conceptually, this process happens analogous to absorption. EM radiation strikes atoms in a material, and its electrons are excited. When the electrons return to a reduced energy state, they re-emit the absorbed light. The light can be emitted in either a mirror-like fashion (what we might call "reflected light.") or in a random direction (i.e., scattering). This process is extremely dependent on the frequency of light.

Scattering dependence In everyday life, we notice that some wavelengths penetrate fog (e.g., radio waves), whereas others become scattered (e.g., light). The scattering of light is a wavelength-dependent process. Concretely, light scatters differently in a medium depending on the size of the wavelength with respect to the particles. Refer to Figure 8.2 for an end-to-end example of how light scatters in our atmosphere, making the sky appear blue. This type of wavelength-dependent scattering, in which waves are scattered by particles much smaller than the wavelength, is known as *Rayleigh scattering*.

Transmission dependence If a light wave does not have the frequency corresponding to the activation energy of the electron, it will not be absorbed. Instead, the light will simply be able to pass through the medium. Visually, the more transparent an object, the more light that has been transmitted. We see the usefulness of transmitted wavelengths when imaging through barriers such as walls, as shown in Figure 8.4.

The above phenomena play an important role in the way our eyes, and cameras, perceive our surroundings. For example, the atomic structure of glass allows visible light to transmit through the medium, making it transparent to the human eye. However, the same glass material would be opaque to a camera sensitive to microwave wavelengths. From an

Figure 8.2 Why is the sky blue? The interaction of the broadband beam coming from the sun with particles in the atmosphere is highly wavelength dependent. Blue light's shorter wavelength causes it to undergo Rayleigh scattering in the atmosphere, which enables our perception of a blue sky.

algorithm standpoint, this wavelength-dependent behavior gives us critical insight into the material composition. A material's response (reflection, absorption, etc.) to different wavelengths of light is referred to as its spectrum and is an excellent indicator of what the material actually is. This makes spectral imaging one of the most powerful tools to date for material classification. Material classification is done by compiling a dictionary of different materials and their spectra. A material is then classified by measuring its spectrum and matching it to one of the dictionary entries.

8.1.3 Multispectral and Hyperspectral Imaging

Spectroscopy is the study of matter's interaction with different wavelengths of light. *Spectral imaging* combines the spatial resolution of ordinary imaging with the spectral resolution of spectroscopy by capturing a per-pixel spectrum. Let us consider a practical application of spectral imaging in food analysis. A common method to identify the soluble solids content

8.1 Spectral Effects on Light-Matter Interaction

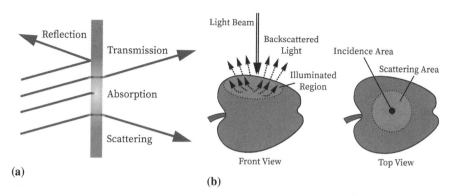

Figure 8.3 Interaction between light and matter. (a) When light interacts with an object, it will reflect off it, be absorbed by it, scatter through it, transmit through it, or do a combination of these. (b) Examining the interaction of light with an apple is a powerful, nondestructive method of analyzing the fruit's freshness. These interactions tend to be wavelength dependent, which is where spectral imaging is useful.

(SSC) in fresh fruits is to use a near-infrared spectroscopy (NIRS) approach. Consider the interaction of the fruit with light that is shown in Figure 8.3. Although the spectrum and SSC are correlated, scattering is most informative about the density, cell structures, and extracellular and intracellular matrices of the fruit tissue. Scattering information is ordinarily extracted from imaging. NIRS does, however, still provide valuable insight into the importance of certain wavelengths in fruit analysis. For example, the 680 nm band can be used for predicting chlorophyll content; the 880 nm, 905 nm, and 1060 nm bands are useful for predicting SSC content; and the 940 nm band is useful for predicting fruit firmness. By combining information from traditional imaging (scattering profiles) and spectroscopy (spectral measurements), we can robustly detect fresh fruit (Lu, 2004).

We now make the distinction between a *multispectral* and a *hyperspectral* image, although we refer to both of them generally as *spectral images* in this text. A spectral image is a 3D matrix, as depicted in Figure 8.1d, with spatial dimensions x and y along one plane, and the reflectance spectrum at each pixel along the third axis. A hyperspectral image captures the per-pixel spectrum over some spectral band, with uniform bandwidth for each spectral point. The band sensitivities must be continuous and cover the entire spectral band. Hyperspectral images also capture images with narrow band sensitivities, on the order of 1–10 nm. Meanwhile, a multispectral image captures only a subset of the spectrum in a spectral band, with potentially nonuniform bandwidths. The captured wavelengths in this case are usually handpicked based on the application. Also, the sensitivity of the bands does not necessarily have to be continuous in the spectral domain. Refer to Figure 8.4 for a comparison between the two.

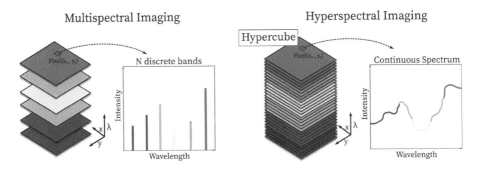

Figure 8.4 Multispectral versus hyperspectral imaging.

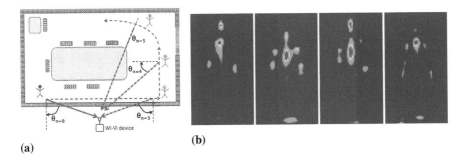

Figure 8.5 Seeing through walls with Wi-Fi. An interesting application of spectral imaging is in the use of non-traditional frequencies with Wi-Fi imaging (2.4 GHz) to image through walls. (a) Setup of Wi-Vi imaging module (Adib and Katabi, 2013) and (b) Wi-Vi image capturing different poses through a wall. Reprinted from online article by Adib.

8.1.4 Applications of Nonvisible Light

Most light is invisible to the human eye, but that does not make these wavelength regimes unimportant. We point to a few applications in which the use of nonvisible light is more effective than visible light. Such physical considerations are important to keep in mind in designing imaging systems (e.g., imaging through or around objects).

Wi-Fi imaging Wi-Fi wavelengths are on the order of several centimeters, allowing these waves to penetrate nonmetallic walls without interacting with the molecular structure, unlike visible light, which is significantly attenuated through walls. These Wi-Fi signals propagate through the wall, interact with objects in a scene, and reflect back to the receiver on the other side of the wall. Figure 8.5 shows the setup of such a system and the kinds of images we can extract with Wi-Fi imaging modules. Wi-Fi vision has been applied to see through walls, and even monitor heart rates and detect gestures through walls (Adib and Katabi, 2013).

Thermal imaging Visible light imaging of humans fails when the reflectance of visible light to the camera is weak. This can occur when (1) the scene is weakly illuminated, or (2) the light undergoes multiple reflections and scattering. In such contexts, measurement of thermal images is particularly useful. Thermal imaging solves (1) by making the person a light source, because humans can be approximated as blackbody emitters in the long-wave infrared (LWIR) regime. In doing so, (2) is also dealt with because we are no longer measuring visible light reflected off of a person. Instead, we are measuring light emitted from the human, which results in a stronger signal because no power is lost due to scattering and transmission into the skin. Thermal non-line-of-sight (NLOS) imaging makes use of this feature, and it is discussed in chapter 10.

X-Ray imaging Whereas visible light primarily reflects and thermal radiation is overwhelmingly emitted from human skin, X-rays transmit through skin. This makes them particularly useful for computed tomography (CT), projected radiography, and positron emission tomography (PET). All these are able to image through our skin to capture bone structure, monitor blood flow, and brain activity.

8.2 Color Theory

Color is the human perception of EM radiation in the visible spectrum, which falls in the range of 400–700 nm. As shown in Figure 8.1b, this range is extremely narrow—the human eye is sensitive to less than one-trillionth of the possible frequencies of light. Most of us are familiar with colors in everyday life as descriptive notions, such as blue, red, yellow, or green. The field of *color science* (also known as colorimetry), seeks to formalize these descriptive notions into a mathematical and psychological structure. In the following, we discuss two important keystones of color theory.

8.2.1 Retinal Color

The first keystone is *retinal color*, describing the human perception of color that corresponds to the activation of photoreceptor cells in the retina known as *cones*. There are three types of cones: *L-cones*, *M-cones*, and *S-cones*, corresponding to long, medium, and short. The nomenclature stems from sensitivity of each cone type to long, medium, and short wavelengths. An illustration of a cone cell and its corresponding spectral sensitivity curves is shown in Figure 8.6a. Note that the spectral sensitivity curves slightly vary between individuals. Cone cells are connected to neurons. The differential pattern of neurons that fire is sent to the brain as electrical signals to yield the perception of color. University courses in computational imaging often focus on retinal color, as one can mathematically express this using a vector space with three elements (activations of L-, M-, and S-cones).

Retinal Sensitivity to Color

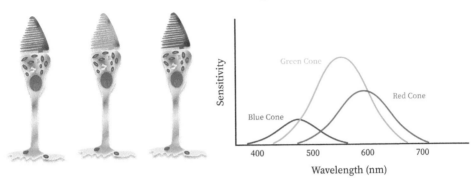

(a) Retinal Sensitivity to Color

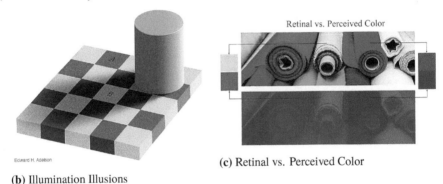

(b) Illumination Illusions

(c) Retinal vs. Perceived Color

Figure 8.6 (a) Retinal sensitivity to color. Our eyes have three types of cone cells: L-cones, M-cones, and S-cones (*left*). Each cone is optimized to sense light at different wavelengths. The spectral absorption of each cone is shown (*right*). (b) Illumination illusions. Our brain adapts to different illumination conditions to render a scene with spatial and color consistency. Reprinted from online artwork by Adelson. (c) Retinal vs. perceived color. Even with a blue overlay (*bottom*), our visual system is still able to correctly label each color.

Although retinal color is closer to a mathematical science (in contrast to perceptual color, discussed in the next paragraph), there are several nuances that are difficult to model, in general, by mathematics alone. For example, different individuals have biological variations in the structure and composition of L-, M-, and S-cones. For example, in canonical forms of *color blindness*, one type of cone is missing (e.g., the L-cone corresponding roughly to peaks in "red" wavelengths). However, one can say that "we are all color blind" (phrase courtesy of Professor Wojciech Matusik, MIT) as there are different spectral combinations that the majority of humans will perceive as the same color! These special colors are known as *metamers*. An example is the use of a mixture of $N > 1$ lights that, mixed together, yield a reddish color that most humans cannot distinguish from a single laser at 625 nm. In this case, the mixed color is metameric to the laser. Note that metameric does not mean mathematically equal to; it means that a controlled sample of humans perceive the color as the same.

8.2.2 Perceptual Color

The second keystone of colorimetry is *perceptual color*, which encompasses not only retinal color but also a remarkable adaptation by the human visual system to illumination and context cues. Perceptual color is perhaps best illustrated by example. Figure 8.6c shows a scene overlaid with a blue tint. Even with the blue overlay, most humans can see the photograph and ascertain that certain colors are "red" or "orange." However, when zooming into a subset of pixels, we see that all the colors in the tinted range are what one might ordinarily perceive as "blue." In this example, the human brain is adapting to variations in illumination. This same principle (not shown) applies to the reason that we can identify the colors of a hot air balloon in different illuminant conditions, ranging from the reds of sunset and sunrise to the blues of daylight. The human brain knows that light reflected from the same object can stimulate different retinal cones depending on the time of day. This principle, in which the human visual system is able to identify colors under different illuminants, is known as *color constancy*. Another example of perceptual color is the "Checker Shadow Illusion" published by Edward H. Adelson and reproduced in Figure 8.6b. Here, the squares A and B are exactly the same retinal color. However, the human brain is able to correctly account for illumination to render a checkerboard pattern.

Taken together, color theory is highly specific to individuals. A viral internet sensation in 2015, known as "The Dress," underscores the diversity in how individuals perceive color. In February of 2015, a bride and her friends in Scotland could not come to an agreement on the color of the dress. The photo was posted to social media and quickly found its way to mainstream media in New York City. Just under 60% of people perceived the dress as blue/black, 30% described it as white/gold, and 10% as other colors. What color do you perceive the dress to be? Color constancy is thought to be the cause of this difference in perception, in which individuals have a different "correction factor" for ambient illumination.

8.2.3 Information Loss in Human-Inspired Vision

Color theory has wide applications to the engineering of computational imaging and display systems. For example, to obtain color images, the ubiquitous Bayer filter is conventionally placed on top of a monochrome camera sensor. As illustrated in Figure 8.10, the Bayer

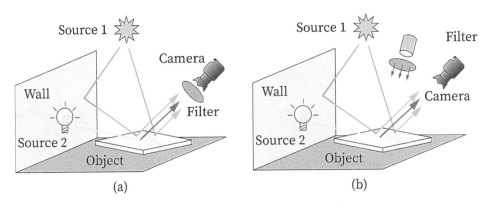

Figure 8.7 Capturing a spectral image. A multispectral image can be captured by either (a) passive illumination, or (b) active illumination. With active illumination, external spectral light sources are used (either by placing several filters in front of one broadband source, or by using several narrow band sources). Passive illumination setups place several narrow band filters in front of the focal plane array.

filter consists of a mosaic of red, green, and blue (RGB) filters. This enables camera pixels to record luminance projections that mimic human cone cells. Within each 2 × 2 cell are one red pixel, two green pixels, and one blue pixel. Why was green chosen as the color to duplicate? Hint: Study the cone response given in Figure 8.6a.[7]

Human perception plays a critical role in standards for how color images are converted to grayscale. Concretely, if one seeks to convert a color image in the CIE colorspace to grayscale, the following equation is used:

$$Y_{\text{linear}} = 0.2126 \cdot R_{\text{linear}} + 0.7152 \cdot G_{\text{linear}} + 0.0722 \cdot B_{\text{linear}},$$

where the coefficients are based on human perceptual sensitivity to these colors (i.e., green has the highest and blue has the lowest). (Reprinted from online article by Stokes et al.) The *linear* coefficients indicate that the electrical signal measured by the retina is approximated to be linearly proportional to the optical power. Note, however, that the only quantity that is measured by the retina is Y_{linear}. Therefore, $A = (R_{\text{linear}}, G_{\text{linear}}, B_{\text{linear}})$ is irretrievably lost information because infinite values of A can result in the same value of Y_{linear}, resulting in a loss of spectral information. This is a critical weakness of human-inspired imaging systems.

---

[7] Answer: Because the Bayer filter is a 2 × 2 repeating pattern, it was desirable to choose one of the filters to duplicate. Green was chosen because the eye has the highest sensitivity to green, and it overlaps in response to the M and S cones.

8.3 Optical Setups for Spectral Imaging

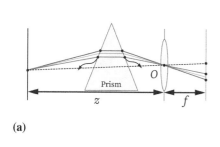
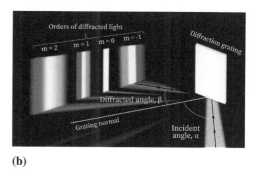

(a)

(b)

Figure 8.8 Wavelength separation by (a) prisms, and (b) diffraction gratings.

8.3 Optical Setups for Spectral Imaging

In this section, we restrict our scope to multispectral and hyperspectral imaging at *optical wavelengths*, defined as the wavelengths of light that conventional camera sensors can capture (300–1000 nm wavelengths). At the top level, there are two motifs used to obtain spectral images. Concretely, it is possible to use (1) a filter on the imager side, or (2) selective illumination with specific spectral sources. These two basic approaches to multispectral imaging are shown in Figure 8.7. Although the top-level idea may seem simplistic, there is richness in the design of filters. Systems that simultaneously capture both spectral and spatial features generally incorporate some form of wavelength separation through gratings or prisms. Other systems use selective filters that allow only a narrow band of light to pass through. These tools and principles are combined in different ways for different spectral applications, depending on the desired cost, acquisition time, and spatio-spectral resolution of the setup, as we analyze next.

8.3.1 Prisms, Gratings, and Scanners

Prisms and *diffraction gratings* both spatially separate incoming light into their constituent wavelengths but do so using different physical mechanisms. Prisms utilize the property of *dispersion*, a phenomenon in which light of different wavelengths propagating through a certain medium will bend at different angles. The wavelength-dependent angle of bending results is a wavelength-dependent index of refraction. The result of this is a spatially separated array of wavelengths. Diffraction gratings, on the other hand, use the property of *diffraction*. A diffraction grating is a periodic structure that operates based on Huygens-Fresnel principle, which states that every point on a wavefront is a secondary point source. The grating will produce diffraction angles at angles that result in constructive interference of light. Angles corresponding to constructive interference are determined by analyzing phase shifts caused by differences in optical path lengths (OPL). The phase shift induced by an OPL is wavelength-dependent, therefore light of different wavelengths construc-

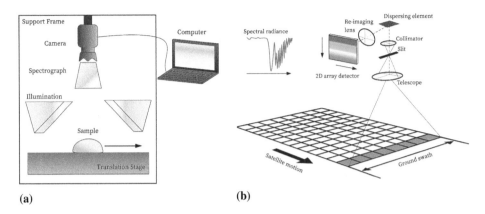

Figure 8.9 Spatiospectral scanning. (a) An example of a pixelwise scan of an image. The scanner captures a spectrum for each pixel, then iterates to the next pixel, and repeats. (b) Satellite hyperspectral imaging using a push-broom camera.

tively interfere at different diffraction angles. This causes a spatial separation of different wavelengths of light. A comparison of these principles is shown in Figure 8.8.

Consider the use of a prism in front of a DSLR camera, as shown in Figure 8.8a. In such a setup, it is common to account for factors like spatially variant dispersion by including collimating optics. A coded aperture mask would also provide a means of obtaining spectral cues, allowing for a constrained reconstruction (Baek et al., 2017). Similarly, the wavelength separation can also be performed by placing diffraction gratings in front of the sensor, as shown in Figure 8.8b. In such a setup, the spatial resolution has to be sacrificed for the spectral dimension (Alvarez-Cortes et al., 2016). The key takeaway, however, in both setups is that the wavelengths are *spatially* separated, so that their reconstruction algorithms will differ from non-separation-based methods.

Many standard spectral imaging setups incorporate some form of scanning to capture a spectral image. The scanning can be either spatial, in which a detector scans across each pixel of an image and collects the spectrum, or spectral, in which a 2D image is recaptured for each spectral point, as shown in Figure 8.1d. An example of such a scanning system is shown in Figure 8.9a. Although accurate, scanning is often too slow and requires a laboratory-like setting, making it infeasible for practical applications. Another example of a scanning spectral camera is the *push-broom camera*. A push-broom camera images one spatial line in a scene at a time. This corresponds to a measurement of a 1D array of pixels. Using either a dispersive or a diffractive element, the light is separated onto a 2D plane. Along one axis lies the wavelength, and along the other lies the spatial coordinate. A full hyperspectral image is obtained by "sweeping" across the scene. The measurement process

8.3 Optical Setups for Spectral Imaging

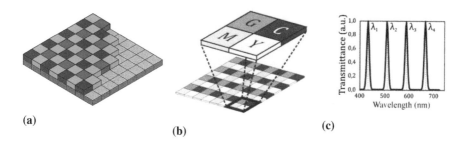

Figure 8.10 Color filter arrays. Side-by-side comparison of (a) Bayer filter and (b) multispectral filter array, specifically a CMYG CFA. (c) Spectral sensitivity of C, M, Y, and G channels with a QBPF (solid line) and without a QBPF (dashed line). Reprinted from (Themelis et al., 2008).

in which a push-broom camera in space images a point on Earth from space is illustrated in Figure 8.9b.

8.3.2 Multispectral Filter Arrays and Compound Imaging

Most cameras today employ a color filter array (CFA) placed in front of their focal plane array (FPA), the most common being the *Bayer filter*. The Bayer filter is an arrangement of red, green, and blue (RGB) filters, as shown in Figure 8.10a. It aims to mimic the sensitivities of the L-, M-, and S- cones of the human eye. A 2D grid of RGB values are measured, which are then demosaicked into a 3D RGB color image. The problem with such a system is that it integrates the product of a scene's spectral reflectance with the spectral response curves of just three filters, giving just a single output for three broad spectral ranges. This inevitably causes spectral information loss. However, the concept of a CFA can be extended to multispectral images for higher spectral resolution.

Consider Figure 8.10b, which shows a Cyan, Magenta, Yellow, Green (CMYG) multispectral filter array (MSFA). The C, M, Y, and G filters still have broad spectral sensitivity. However, by inserting another quadruple-bandpass filter (QBPF) on top of the CFA, the light is filtered twice before reaching the sensor. As shown in Figure 8.10c, this noticeably improves the spectral resolution of the camera. Although there are parallels between the optical hardware in CFAs and MSFAs, the spectral reconstruction and demosaicking process are slightly different, as we go over in detail in section 8.4. Also common nowadays is the use of a tunable filter. Such filters can change their spectral sensitivity without drastic changes in bandwidth or transmission. One example is the liquid crystal tunable filter (LCTF), which electronically modulates liquid crystals to transmit only certain wavelengths. The use of electronically controlled filters (rather than mechanical) enables compact optical hyperspectral setups that can be captured within seconds or even milliseconds with high spectral performance.

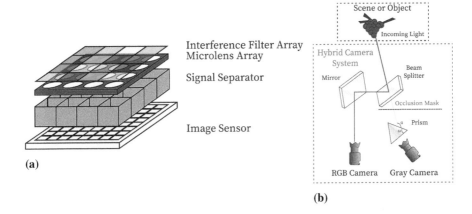

Figure 8.11 (a) Multispectral compound imaging setup. A compound imaging setup consists of several units, each capturing an image at different wavelengths. The units capture spatially offset versions of the same scene. (b) Hybrid capture. An optically parallelized setup to capture an RGB and hyperspectral image simultaneously.

Building on the use of CFAs, some setups use the design principles of a *compound imaging system* for multispectral image capture. A compound imaging system captures several images by using units within the optical setup. Each unit will capture a different aspect of the same scene (i.e., different wavelengths) at spatial offsets, based on the physical location of each unit. The set of these unit images form a compound image. The optical setup for such a system is depicted in Figure 8.11a.

The key idea in such a setup is that each point in the scene will be observed by all the units. This means that a spectrum can be measured for each spatial location in a scene, depending on the location of the observation plane. However, the main challenge is combining the unit images in postprocessing to account for spatial offsets (Shogenji et al., 2004). This can be done via cross-correlation between unit images, pixel rearranging, or other methods. The advantage of such a compound system lies in its compactness, compared with other setups that need to incorporate additional bulky hardware (e.g., tunable filters and spectral sources). However, these systems also require very precise measurement environments that may be hard to enforce in a practical environment. Multispectral image mosaicking operates on a similar principle. In such a setup, a camera with spatially varying spectral filters is panned across a scene to extract a wide-field-of-view multispectral image by observing each point multiple times (Schechner and Nayar, 2002).

8.3.3 Spectrum-RGB Parallel Capture

Meanwhile, other methods attempt to tackle the inherent low spatial resolution of spectral images more robustly and directly. Low spatial resolution is a consequence of difficulty in

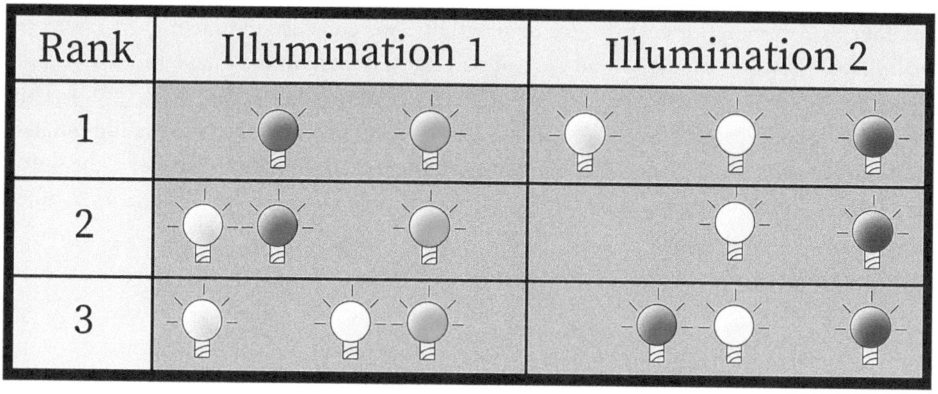

8.3 Optical Setups for Spectral Imaging

Rank	Illumination 1	Illumination 2
1		
2		
3		

Figure 8.12 Multiplexed illumination. Methodically illuminating the scene with more than one spectral source at a time can enable efficient data capture and higher reconstruction accuracies. An example of the top three optimal illumination patterns are shown with two allowed measurements.

spectrally multiplexing every point in a scene. In certain cases, the lower spatial resolution is also caused by diffraction limits in imaging at longer wavelengths. One widely explored optical setup to improve spatial resolution parallelizes the capture of a hyperspectral and RGB image. The incoming light beams are separated using a *beamsplitter*, which separates light into two directions equally. Along one optical path lies a high spatial resolution RGB camera, and along the other lies a broadband monochrome camera, as shown in the schematic in Figure 8.11b. The RGB and spectral image can be fused together by leveraging the spectral, spatial, and temporal (for videos) correlation of the pixels (Cao et al., 2011). We analyze a linear algebraic approach to combine an RGB image with high spectral resolution multispectral images in section 8.4.1.

8.3.4 Coded Spectral Illumination

So far, we have seen that a natural way to take a spectral image is to keep the scene illumination constant, and vary the camera's spectral sensitivity (via filters). We now consider the method of changing the illumination source, rather than the pixel's sensitivity. *Active illumination* is the process of using controlled light sources to illuminate a scene, in contrast to *passive illumination*, which relies exclusively on ambient light. The use of a controlled light source enables postprocessing algorithms to take advantage of spectral, spatial, and/or temporal features of the illumination pattern for better image reconstruction. In certain setups, multispectral illumination, rather than multispectral detection, can also enable faster image acquisition.

Consider a scene actively illuminated by Q narrow band sources (unlike Figure 8.7b, which depicts one broadband source occluded by spectrally discriminative filters).

Rather than turning on each source sequentially, we consider the method of *coded illumination*, in which we determine the optimal lighting pattern for image reconstruction. Figure 8.12 shows an example of such an optimized lighting pattern (Park et al., 2007). This particular lighting pattern uses *multiplexed illumination*, in which the scene is illuminated by multiple sources at a given time. By using a multiplexed illumination pattern, the number of needed measurements is also reduced. Every light source q has a mutually distinct spectrum $L_q(\lambda)$ known a priori. The illumination for frame n is given by a weighted sum of each source

$$p_n(\lambda) = \sum_{q=1}^{Q} d_{nq} L_q(\lambda)$$

where $0 \leqslant d_{nq} \leqslant 1$. The key idea is to find a basis **D** that minimizes the least-squares error of the image reconstruction, for a predefined number of allowed measurements. The example shown in Figure 8.12 assumes two allowed measurements per image. An ideal basis will fully utilize both light sources, have linearly independent illuminations within each frame, and keep total illumination power similar across frames.

Passive illumination setups are often negatively impacted by ambient noise, and struggle to extract reflectance properties of surfaces. A useful consequence of active illumination is that it provides a way to modulate incident light, enabling filtering of ambient light. Consider the setup shown in Figure 8.7a. In a passively illuminated scene, the radiometric response $\left(\rho_k^{xy}\right)$ of a given pixel P $= (x, y)$ can be modeled as

$$\rho_k^{xy} = t \int R_k(\lambda) S^{xy}(\lambda) E^{xy}(\lambda) \, d\lambda + N(\alpha t, \alpha t)$$

where t is the integration time; $R_k(\lambda)$ is the camera's response function at channel k; $S^{xy}(\lambda)$ is the reflectance spectral distribution at P; $E^{xy}(\lambda)$ is the spectrum of the incident flux at P; and $N(\alpha t, \alpha t)$ is additive Gaussian noise. Note that the mean and variance of the noise are proportional to the integration time. If we add an illumination source, the flux can be modeled as $E^{xy}(\lambda) = A^{xy}(\lambda) + L(\lambda) \Phi_i(\lambda)$, where $A^{xy}(\lambda)$ is the ambient illumination, $L(\lambda)$ is the spectrum of the light source (known a priori), and $\Phi_i(\lambda)$ is the transmission spectrum of filter i. If we take a measurement with filter i and j, each having integration time t_i and t_j, we find that their difference (after normalizing by t) is

$$t_j \rho_{k,i}^{xy} - t_i \rho_{k,j}^{xy} \approx t_j t_i \int R_k(\lambda) S^{xy}(\lambda) \Phi_i(\lambda) L(\lambda) \, d\lambda$$
$$- t_i t_j \int R_k(\lambda) S^{xy}(\lambda) \Phi_j(\lambda) L(\lambda) \, d\lambda + N(0, 2\alpha t_i t_j).$$

Note that we have effectively filtered out the ambient illumination $A^{xy}(\lambda)$. By taking several such measurements, we can construct a basis from which we can recreate the spectrum for each pixel. Such filtering cannot be done in a passive illumination setup,

8.3 Optical Setups for Spectral Imaging

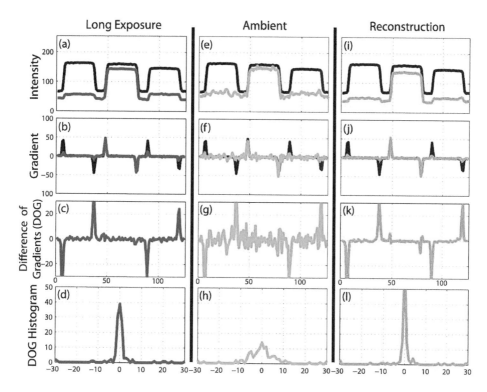

Figure 8.13 Dark flash photography. Capturing nonintrusive, high-quality images can be challenging in dimly lit environments. One way to get around this hurdle is by actively illuminating the scene with an infrared light source. We can then leverage the spectral proximity of red with infrared wavelengths to constrain the image reconstruction problem. Reprinted from (Krishnan and Fergus, 2009).

because the ambient illumination would be affected differently for each filter in the CFA in such a situation (Chi et al., 2010).

An interesting and unique application of active illumination is in dark photography (Krishnan and Fergus, 2009), where the scene is dimly lit. Measurements made at such low light intensities are corrupted by Poisson noise in the sensor. On the other hand, bursts of light caused by a flash unit can be disruptive to the scene. The innovation here is to illuminate the scene with infrared light so that the flash will be invisible to the human eye, hence leaving the scene undisturbed. By combining a dark flash photo (F) with an ambient image (A), a color image can be woven together in postprocessing. Contrary to flash/no flash photography (Petschnigg et al., 2004), this method by design has nonoverlapping ambient and flash illumination because the illumination sources are at different wavelengths. However, although the spectral intensities of A and F will be mostly different, the intensities will be correlated at the red and IR wavelengths due to their spectral proximity. We examine this spectral relationship by analyzing the measurements taken for a 1D scanline across three colored squares (blue, magenta, and yellow). In Figure 8.13a, the black lines correspond to measured IR intensities (F_1) and colored lines denote intensities from the red channel of the camera in a long exposure shot (L_1). Even though these measured intensities are rather different, note how the intensity drops are spatially aligned at the edges of the squares for both F_1 and L_1. This correlation at the edges is even more noticeable when analyzing the gradients ∇F_1 and ∇L_1, as shown in Figure 8.13b, and the difference of gradients (DOG) $\nabla F_1 - \nabla L_1$ in Figure 8.13c. We see that the DOG is sparse, because the DOG histogram shown in Figure 8.13d has a sharp peak centered at 0, so that most DOG values are 0. However, when the same measurements and calculations are made with ambient light as shown in Figure 8.13e–h, we see that the DOG is no longer sparse. Using what we have learned from Figure 8.13a–d, we can now reconstruct our final image R by (1) minimizing the differences in intensity between A and R; (2) making sure that $\nabla F_1 - \nabla R_1$ is sparse and (3) making sure $\nabla F_1 - \nabla R_1$ is sparse. Sparsity can be enforced by adding an L_1 norm to the cost function (Krishnan and Fergus, 2009). In this context, we use spectral measurements to act as a useful constraint in recreating an RGB image.

8.4 Computational Methods for Analyzing Spectral Data

8.4.1 Spatiospectral Matrix Representations

Linear algebra gives us the ability to represent images as a linear function of basis vectors. The expressive capability of a linear basis is a powerful tool at our disposal when we seek to express an image in terms of its spatial and spectral features. In spectral images, *mixed pixels* are often present. Mixed pixels contain more than one distinct substance, either because (1) the spatial resolution of the camera is too low, (2) the substance is a mixture of several different materials, or (3) both. *Spectral unmixing* is an inverse problem that aims to decompose each pixel into its constituent spectra (endmembers) and spectra intensities (abundances) (Keshava and Mustard, 2002). In other words, it aims to represent each pixel's spectrum as a linear combination of a set of basis spectra (corresponding to a set of a few fundamental basis materials). Whereas work has been done in *learning* representations of spectral images for hyperspectral image reconstruction (Choi et al., 2017), this chapter focuses on analytic methods for interpretable decomposition of spectral images.

Let us consider the hardware setup shown in Figure 8.10b, where we stack a QBPF on the CMYG CFA to increase the spectral sensitivity of our imager. A given pixel in the camera will have a sensitivity of w_{X_i}, where $i = \{1, 2, 3, 4\}$ refers to the QBPF filters, and $X = \{C, M, Y, G\}$ refers to the CFA filters. If I_{λ_i} is the intensity of light reaching the filter at wavelength λ_i, then the measured intensity (S_X) at the sensor for filter X is

$$S_X = w_{X1}(I_{\lambda 1}) + w_{X2}(I_{\lambda 2}) + w_{X3}(I_{\lambda 3}) + w_{X4}(I_{\lambda 4}).$$

8.4 Computational Methods for Analyzing Spectral Data

This can be further generalized for all filters as a matrix multiplication $\mathbf{s} = \mathbf{Wi}$.

$$\begin{bmatrix} S_C \\ S_M \\ S_Y \\ S_G \end{bmatrix} = \begin{bmatrix} w_{C1} & w_{C2} & w_{C3} & w_{C4} \\ w_{M1} & w_{M2} & w_{M3} & w_{M4} \\ w_{Y1} & w_{Y2} & w_{Y3} & w_{Y4} \\ w_{G1} & w_{G2} & w_{G3} & w_{G4} \end{bmatrix} \begin{bmatrix} I_{\lambda_1} \\ I_{\lambda_2} \\ I_{\lambda_3} \\ I_{\lambda_4} \end{bmatrix}.$$

The calibration matrix $\mathbf{W}$ can be experimentally found by measuring m different color samples with known reflection spectra $\mathbf{i}_1, \cdots, \mathbf{i}_m$, which can be concatenated into a matrix $\mathbf{I} \in \mathbb{R}^{4 \times m}$. We then obtain a series of measurements $\mathbf{s}_1, \cdots, \mathbf{s}_m$, which we concatenate into a matrix $\mathbf{S} \in \mathbb{R}^{4 \times m}$. We can then solve for the calibration matrix by $\mathbf{W} = \mathbf{SI}^{-1}$, where $\mathbf{I}^{-1}$ is the pseudoinverse of $\mathbf{I}$ and $\mathbf{W} \in \mathbb{R}^{4 \times 4}$. One condition is that $m \geqslant 4$. This ensures that $\mathbf{I}$ has linearly independent rows, which ensures that $\mathbf{I}$ has a right inverse. Once the $\mathbf{W}$ matrix is found, we can use $\mathbf{S}^{-1}$ as a transformation to go from sensor measurements to spectral intensities, by $\mathbf{i} = \mathbf{W}^{-1}\mathbf{s}$. This provides a conceptually straightforward way to increase the spectral resolution of a standard CMYG camera.

Now, we consider ways to digitally stitch together a low spatial resolution hyperspectral image $\mathbf{Y}_{hs} \in \mathbb{R}^{w \times h \times S}$ and a high spatial resolution RGB image $\mathbf{Y}_{RGB} \in \mathbb{R}^{W \times H \times 3}$ (typically $W \gg w$ and $H \gg h$), as captured by the setup shown in Figure 8.11b. Our goal is to extract a high-resolution hyperspectral image $\mathbf{Z} \in \mathbb{R}^{W \times H \times S}$. We try to extract a high-dimensional output from low-dimensional inputs, which means that the problem is under-constrained and we must make some assumptions about the scene. In this case, we assume that there is a very small number M of distinct materials in a scene. We also assume that scene radiance as a function of wavelength is a smooth function, and so can be expressed with fewer basis functions (Kawakami et al., 2011). Note that the mathematical assumptions made are derived from optical properties of the scene and may differ from application to application.

We first claim that $\mathbf{Y}_{hs}$ and $\mathbf{Y}_{RGB}$ are linear functions of $\mathbf{Z}$, such that $\mathbf{Y}_{hs} = \mathbf{P}_{hs}\mathbf{Z}$ and $\mathbf{Y}_{RGB} = \mathbf{P}_{RGB}\mathbf{Z}$. For a scene containing M different materials, we can express the spectrum $\mathbf{Z}(i, j)$ at point (i, j)

$$\mathbf{Z}(i, j) \approx \sum_{m=1}^{M} \mathbf{a}_m \mathbf{h}(i, j) = \mathbf{Ah}(i, j)$$

where $\mathbf{a}_m$ is the vector corresponding to the reflectance spectrum of material m, and $\mathbf{h}(i, j) = [h_1(i, j), h_2(i, j), \cdots, h_M(i, j)]$ are scaling coefficients. Now, we can express $\mathbf{Y}_{hs}(i, j, *)$ as a sum of pixels in a spatial window W_{ij}

$$\mathbf{Y}_{hs}(i, j, *) \propto \sum_{k,l \in W_{ij}} \mathbf{Z}(i, j, *) = \mathbf{A} \sum_{k,l \in W_{ij}} \mathbf{h}(k, l) = \mathbf{Aq}(i, j).$$

This formulation comes from the fact that $\mathbf{Y}_{hs}$ is a low spatial resolution version of $\mathbf{Z}$, so the spectrum measured at a point in $\mathbf{Y}_{hs}$ is actually a combination of different spectra in pixels surrounding it. Intuitively, this is analogous to colors in nearby pixels getting mixed (i.e., pixelated) in a low-resolution image.

Using this information, we can now express $\mathbf{Y}_{hs}$ as a matrix factorization, $\widetilde{\mathbf{Y}}_{hs} = \mathbf{AQ}$, where $\widetilde{\mathbf{Y}}_{hs} \in \mathbb{R}^{S \times wh}$; $\mathbf{A}$ is the reflectance spectra matrix, containing the spectra of M materials in its column space; and $\mathbf{Q}$ is the spatial matrix, containing the fractions of material found at every spatial point. A common method to solve for such a factorization is Gauss-Newton nonlinear optimization, with a constraint that the 1-norm of $\mathbf{Q}$ be minimized. The 1-norm constraint comes from the fact that we assume that there are very few materials in the scene.

We now search for $\mathbf{h}\,(i, j)$. We do so by using the previous assumption that

$$\mathbf{Y}_{RGB} = \mathbf{P}_{RGB}\mathbf{Z} = \mathbf{P}_{RGB}\mathbf{A}\mathbf{h}\,(i, j).$$

We use the $\mathbf{A}$ calculated in the previous step, and search for a $\mathbf{h}\,(i, j)$ that satisfies this equation and has the sparsest representation. Using our estimated value for $\mathbf{h}\,(i, j)$, we can calculate our high-resolution spectral image

$$\widetilde{\mathbf{Z}}\,(i, j, *) = \mathbf{A}\widehat{\mathbf{h}}\,(i, j).$$

We see that by making certain assumptions about the structure of the image data, we can decompose two images of the scene into their spatial and spectral features, and then recombine these features into a single high-resolution hyperspectral image by matrix multiplication (Kawakami et al., 2011). Related work has extended these principles to combine panchromatic (low spectral information) images with high spatial resolution for hyperspectral superresolution in satellite imagery (Wang et al., 2010; Nguyen et al., 2011; Li et al., 2013) in a field known as panchromatic superresolution.

The assumption of few distinct materials in a natural scene is a rather common one and is of great significance, considering how well it tends to model hyperspectral image spaces. It supports the idea that a hyperspectral image can be expressed accurately as a projection onto a low-dimensional subspace contained by few basis spectra. Optical hardware can also be developed in such a way that captured measurements are a projection of the spectral image onto this subspace, eliminating the need to capture a spectrum for each spatial point (Saragadam and Sankaranarayanan, 2019). Hyperspectral anomaly detection applications build on this principle by modeling the image as a sum of a low-rank background and sparse anomalies (Saragadam et al., 2017). These physical assumptions (anomalies → sparse, spectral background → low rank) can inherently inspire end-to-end design of image capture systems.

8.4.2 Dimensionality Reduction

We now consider linear algebraic principles in reducing the dimensionality of the hyperspectral data. As the amount of captured data increases, the data will likely not only contain

8.4 Computational Methods for Analyzing Spectral Data

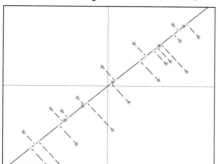
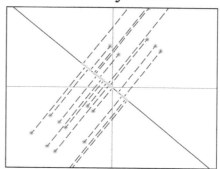

Figure 8.14 Principal component analysis (PCA). PCA seeks to represent data in a coordinate system as to maximize the variance of the data's projection onto each axis. Observe that by minimizing the least-squares error of the projection, the axis also maximizes the variance of the projections.

redundant (i.e., linearly dependent) information, but also become more difficult to computationally process. *Principal component analysis (PCA)* is a popular technique to reduce the dimension of hyperspectral data. In the context of spectral imaging, PCA searches for a transformed v-dimensional coordinate space, where v is the number of wavelengths and n is the number of pixels in the image (Baronti et al., 1998). The transformed coordinate system consists of axes that are highly uncorrelated (i.e., orthogonal) to each other. Each vector in the transformed v-dimensional basis is referred to as a *principal component (PC)*. An important result of PCA is that by projecting the hyperspectral data into its PCs, the data becomes more interpretable because it enables analysis of fewer, independent features of the data.

We now outline the steps to perform PCA on a dataset. We start off by subtracting the average spectrum from every spectral point (i.e., zero-mean data). Then, we determine an axis, which we denote PC1, such that the residuals of the projections of the data points onto this axis will be minimized. Mathematically, we formulate this

$$\min_{\|\mathbf{u}\|=1} \frac{1}{N} \sum_{i=1}^{N} \left\| \mathbf{x}_i - \left(\mathbf{u}^\top \mathbf{x}_i\right) \mathbf{u} \right\|^2.$$

We state without proof that this formulation is equivalent to finding a PC1 such that the variance of the projection is maximal. However, one can intuitively grasp this concept by contrasting two different projections, as shown in Figure 8.14, and noticing that the data projections are more spread out in the optimal projection. We repeat this process of projecting data onto an axis to maximize variance until we find PC1, PC2, $\cdots$, PCn. This idea of maximizing variance within each PC is at the heart of PCA, and is perhaps the most

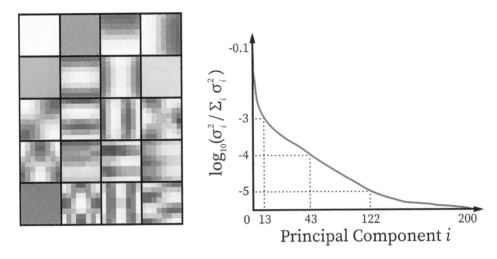

Figure 8.15 Statistical representation of spectral images. (*left*) PCA representation of patches in a hyperspectral image, and (*right*) log scale of variance of first 200 PCs.

important concept to understand. We again state without proof that all the PCs of the data can be found by extracting the eigenvalues and eigenvectors of the covariance matrix of the zero-meaned data. The magnitude of the eigenvalues indicates how great the variance is along the axis given by the corresponding eigenvector (Abdi and Williams, 2010).

Because we want to condense our image representation, we choose only the eigenvectors with the r largest eigenvalues to be our PCs ($r \ll v$). Although dropping principal components will reduce the reconstruction accuracy, this accuracy is not significantly affected if the eigenvalues are small for that PC. A PC with low variance indicates that most (or all) of the data share this feature, indicating that the feature is uninformative. This idea that a spectral image can be represented within a few principal components is also a testament to the idea that any pixel in a scene can be represented by a mixture of a few materials, that is a linear combination of a few spectra (Jaaskelainen et al., 1990).

Our new condensed data representation $\mathbf{Z}$ can be found by projecting measured data onto these PCs, which is done by a simple matrix multiplication $\mathbf{Z} = \mathbf{U}^T (\mathbf{Y} - \boldsymbol{\mu})$, where $\mathbf{Y}$ is the measured data, $\mathbf{U}$ is the matrix containing the r principal components in its column space, and $\boldsymbol{\mu}$ is a matrix containing the mean of the spectrum along every point.

We now use principles from PCA to gain a glimpse into the statistical representation of hyperspectral images, and interdependencies between spatiospectral features. Consider a real-world hyperspectral $\mathbf{X}$, separated into $P \times P$ spatial patches. Performing PCA on each patch separately yields a representation that looks something like that shown in Figure 8.15. As shown on the right side, the variance of the data contained in the PC rapidly decreases.

8.4 Computational Methods for Analyzing Spectral Data

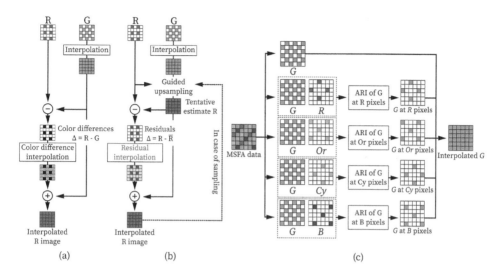

Figure 8.16 Image demosaicking using (a) color difference interpolation, (b) residual interpolation, and (c) adaptive residual interpolation.

In fact, 99% of the variance in the data is contained in the first 200 basis vectors (out of $\sim 2,000$).

Note that many patches share similar geometrical structure, suggesting similar intrapatch spatial relationships. Yet, they have different spectral features, evident through their different colors. This suggests that the data is separable into spectral and spatial components via matrix factorization. Further probabilistic analysis into the coefficients in the spectral and spatial matrices reveal that for a given spatial point, spectral components are nonindependent (Chakrabarti and Zickler, 2011). Such analysis gives us an idea of how PCA can be leveraged in research, as well as some insight into the underlying structure of a hyperspectral image. However, much research is still being devoted to both understanding these spatiospectral statistical representations and how to harness them effectively.

8.4.3 Multispectral Demosaicking

In 2.2.4, for CFA-based spectral setups we alluded to the need to *demosaic* a digital RGB grid to interpolate it into a color image. In this subsection, we make the distinction between demosaicking an RGB image and demosaicking a multispectral image using difference and residual interpolation. For an RGB image, a G image is first interpolated using only the G measurements. Then, the R and B channels are interpolated using either *difference interpolation*, as shown in Figure 8.16a, or *residual interpolation (RI)*, as shown in Figure 8.16b. Both are valid approaches for RGB demosaicking, but state-of-the-art work tends to use RI (Kiku et al., 2013), as smoother residuals often lead to higher interpolation accuracies.

For multispectral images, the CFA is replaced with a MSFA. In particular, we will examine the approach of *adaptive RI (ARI)* to demosaic outputs from the MSFA. ARI combines principles from *minimized-Laplacian RI (MLRI)* and *iterative RI (IRI)*. Unlike RI, which seeks to minimize the magnitude of the residuals, MLRI seeks to minimize the squared Laplacian energy **E** of the residuals, given by

$$E(a_{p,q}) = \sum_{i,j \in \omega_{p,q}} \left(M_{i,j} \widetilde{\nabla}^2 \left(R_{i,j} - \check{R}_{i,j} \right) \right)^2,$$

where **M** is a 2D binary mask, that has a value of 1 for red pixels and 0 for others (assuming that we are interpolating red pixels); $\omega_{p,q}$ is a window around a pixel (p,q); and $R_{i,j}$ and $\check{R}_{i,j}$ are the ground truth and estimated value, respectively, of the pixel (i,j). IRI is similar to traditional RI, but includes an iterative component as shown in the dashed line of Figure 8.16b. The interpolated red image in iteration k is used to guide the upsampling in iteration $k + 1$. The iterations are stopped based on the magnitude and smoothness of the residuals. The blue image is interpolated in a similar manner.

ARI first interpolates the G band, as shown in Figure 8.16c. The interpolation is broken down into $n - 1$ streams (excluding the G band), where n is the number of spectral bands being sampled. The G values are interpolated at the pixel locations of the other bands by adaptively choosing the iteration value for each pixel in IRI and methodically combining RI and MLRI estimates. All the other bands are similarly calculated iteratively with guided upsampling from the G image. Demosaicking a multispectral image is more challenging than demosaicking an RGB image because each spectral band is sampled at a lower spatial frequency (Monno et al., 2017). This is why more sophisticated algorithms are needed; one example is ARI.

Chapter Appendix: Notations

Notation	Description
c	Speed of light
ν	Frequency
λ	Wavelength
h	Planck's constant
E	Incident photon energy
$L_q(\lambda)$	Spectrum of light source q
$p_n(\lambda)$	Illumination for frame n

8.4 Computational Methods for Analyzing Spectral Data

ρ_k^{xy}	Radiometric response
$R_k(\lambda)$	Camera's response function at channel k
$S^{xy}(\lambda)$	Reflectance spectral distribution at (x, y)
$E^{xy}(\lambda)$	Spectrum of the indicent flux at (x, y)
$A^{xy}(\lambda)$	Ambient illumination
$\Phi_i(\lambda)$	Transmission spectrum of filter i
$\mathbf{Y}_{\text{hs}}$	Low spatial resolution hyperspectral image
$\mathbf{Y}_{\text{RGB}}$	High spatial resolution RGB image
$\mathbf{Z}$	High-resolution hyperspectral image

Exercises

Color images broadly capture the reflectance of a scene at three wavelength peaks, i.e., red, green, and blue (RGB). Multispectral images go one step further, capturing a scene at several peak wavelengths with narrow-band spectral sensitivities. This increased information capacity provides us with the ability to better understand a scene's composition. In this problem set, we learn to (a) sustainably process this increased information content, and (b) render scenes using hyperspectral information.

1. Spectral image compression using PCA

    Hyperspectral images densely store spectral information about every pixel in an image, making them highly memory intensive, computationally different to process, and uninterpretable. Raw spectral data often contains redundant information due to interdependent spatio-spectral features. The goal of this section is to apply principal components analysis (PCA) to a multispectral image and extract meaningful features from just a few independent principal components.

    a) Capturing the image

    Either capture a spectral image using filters, or use an online database such as the *Manchester Hyperspectral Image Database* for this section (Foster et al., 2006). If you have a relatively small number of captured wavelengths, you will find better results with less nuanced scenes.

    b) Extracting covariance matrix

    Our image data is represented as matrix $\mathbf{A} \in \mathbb{R}^{w \times h \times n}$, where w and h are the width and height, respectively, of the images, and n is the number of spectral points. Flatten this matrix such that all pixels are along one dimension. This should form a new matrix $\mathbf{A}' \in \mathbb{R}^{wh \times n}$. We then subtract the mean spectrum (calculated over all pixels), and subtract this mean spectrum pixelwise. This will give us the de-meaned matrix, $\mathbf{B}$. The covariance matrix $\mathbf{C}$ is then calculated

    $$\mathbf{C} = \mathbf{B}^T \mathbf{B} \qquad (8.1)$$

    where $\mathbf{C} \in \mathbb{R}^{n \times n}$.

    c) Extracting principal components

    By performing a symmetric eige-decomposition of $\mathbf{C}$, we can compute a factorization for $\mathbf{C}$

    $$\mathbf{C} = \mathbf{Q} \mathbf{\Lambda} \mathbf{Q}^T \qquad (8.2)$$

where $\Lambda \in \mathbb{R}^{n \times n}$ is a matrix containing eigenvalues sorted from greatest to least, and $Q \in \mathbb{R}^{n \times n}$ is the matrix containing the corresponding eigenvectors along its column space. The eigenvectors correspond to the principal components of the image, and the eigenvalues correspond to the variance of the projected data along the principal component.

d) Data projection

The key concept here is that certain spectral images can be reconstructed with high fidelity with just a few principal components. Select r number of principal components with the highest variance. We want to choose the principal components that account for large variances in the data, as these components account for higher-level features. We can project our image data onto a lower-dimensional subspace occupied by these principal components. To do this, construct a matrix $\mathbf{R} \in \mathbb{R}^{n \times r}$ with the r greatest principal components along its column space. We can reconstruct the original image from the low-dimensional subspace via the equation

$$\mathbf{A}_{new} = \mathbf{R}\mathbf{R}^T \mathbf{B}^T + \mu \qquad (8.3)$$

where $\mathbf{A}_{new} \in \mathbb{R}^{w \times h \times n}$ is our reconstructed image matrix, and $\mu \in \mathbb{R}^{w \times h \times n}$ is the mean spectrum (that we subtracted from previously). Choose a value of r that causes your reconstructed image to look similar to the original image. Place your reconstructed image, and report your chosen value for r.

2. Hyperspectral images under variable lighting

Scenes around us are illuminated by different types of sources (e.g., the sun, halogen, incandescent, and fluorescent). Each of these sources has a different known spectral illumination profile. If we know the reflectance profile at every point in a scene, then we can recreate an image of the scene under different illumination profiles by the given analytic equation

$$l(\lambda) = r(\lambda) e(\lambda) \qquad (8.4)$$

where $l(\lambda)$ is the observed spectral radiance at point P in the scene; $r(\lambda)$ is the spectral reflectance at P; and $e(\lambda)$ is the spectral profile of the light incident on P. This model assumes approximately uniform spectral illumination across all pixels in the scene, which is often a reasonable assumption. This problem will give us insight into the usefulness of knowing the spectral profile of a scene, particularly for image rendering.

a) Rendering a hyperspectral image

Choose any hyperspectral image from the *Manchester Hyperspectral Image Database*. These images contain the reflectance spectrum of the scene at each pixel sampled at

33 wavelength bands from 400 nm to 720 nm with a step size of 10 nm (Foster et al., 2006).

b) Rendering under illumination source

Extract the spectral profile of at least two different illumination sources from the *colorimetric data* available from the International Commission on Illumination de l'éclairage (n.d.). Ensure that the spectral profile is extracted at wavelength bands matching the hyperspectral image 400–720 nm with 10 nm step size). Using equation (8.4), determine the spectral radiance $l(\lambda)$. You can assume that the *Manchester database* gives us information about $r(\lambda)$ and the colorimetric database gives us information about $e(\lambda)$. Plot the observed radiance at point [2, 3] in the image.

c) Producing an XYZ image

$l(\lambda)$ tells us what the spectral radiance looks like for a given point in the scene. This contains all the information that we need to recreate a human-interpretable image of a scene. Recall that the human eye has three cones, with L-, M-, and S-cones sensitive to long, medium, and short wavelengths, respectively. Our eyes (and cameras via Bayer filters) process this light through these three color channels (red, green, and blue). What each cone observes is

$$o_i = \int s_i(\lambda) l(\lambda) d\lambda \tag{8.5}$$

where o_i is the scalar value that cone i will observe at a point P in the scene, $l(\lambda)$ is the spectrum of the light reaching our retina; $s_i(\lambda)$ is the spectral sensitivity of cone i; and $i \in \{R, G, B\}$.

Extract the spectral sensitivity for each of the three cones from the *CIE database*. This database contains CIE's 1931 color space sensitivity data, which is a quantitative link between the electromagnetic wavelength of colors and their corresponding physiologically perceived color in humans (Smith and Guild, 1931). Using the corresponding L-, M-, and S- cones, produce an RGB image by merging the three separate images.

d) Rendering into an RGB image

The RGB color space is contained inside the XYZ space, and hence the values of o_i can be converted to RGB. You can use built-in functions in Python such as *scikit.image.xyz2rgb* to perform the conversion and to produce an RGB image. Compare this image with another image under the same lighting. Do they look similar?

III SHADING AND TRANSPORT OF LIGHT

9 Programmable Illumination and Shading

In the previous two chapters, we discussed ways to leverage certain properties of light (i.e., spectrum and polarization) to obtain features of objects in a scene. In this chapter, we focus specifically on using intensity measurements for shape information and applications in graphics and rendering.

9.1 Scene Reflectance and Photometry

9.1.1 Albedo, Radiance, and Irradiance

Imaging, and our visual system, relies on information obtained from light reflected from objects in a scene. The field of optics has derived complex relationships between light and surfaces in the ways that they interact through single or multiple absorption, reflection, and scattering. However, modeling a scene using *only* these relationships is often an impractical task, due to the complexity of a real scene (i.e., unknown optics and noise) and the computational burdens it would impose, even assuming that we could derive a sufficiently complex model to mitigate model mismatch. In practice, the dependence of the scene reflectance on angle and spectrum of incident light cannot always be accounted for. Instead, some simplifying assumptions are used that in practice have yielded good results.

Over some spectral window, the reflectance is assumed to be constant. Meanwhile, the angular dependence is simplified by the separation of the measured light into its *diffuse* and *specular albedo*. Albedo is the proportion of incident light that is reflected, that is not absorbed, by a surface. Note that due to its spectral and angular dependence, albedo is not an intrinsic property of the surface. The specular component is light that directly bounces off a surface to the sensor, and the diffuse component is light that scatters randomly upon interacting with the surface. As discussed in chapter 7, the diffuse component is approximately isotropic, whereas the specular component is highly anisotropic. Decomposing an image into just two components allows for a convenient representation of the albedo while accounting for its dependence on the angle of incidence (Coakley, 2003). More simplifying assumptions are discussed in section 10.1.2.

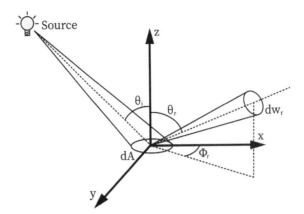

Figure 9.1 Lambert's law and foreshortening. When the incident light is at an angle with respect to the normal, the area of light incident on the surface is reduced, in what is known as foreshortening. This results in a reflected intensity proportional to the product of $\cos \theta_i$ and the incident intensity.

We now distinguish between two similar, but distinct, fields that study the measurement of light. *Radiometry* measures the absolute power of the light, and *photometry* attempts to mimic the radiant sensitivity of the human eye. Specifically, as discussed in chapter 8, three different cones in the retina have different spectral sensitivities, with peaks approximately centered at red, green, and blue light. Photometry is more common in imaging, whereas radiometry is more common for applications in astronomy, in which a faithful measurement of the spatial light intensity distribution is required. Radiometric quantities are typically converted to photometric quantities by cameras due to the spectral sensitivity of color filter arrays (e.g., Bayer filter) in front of the focal plane.

The *irradiance E* of a surface is defined as the incident radiant flux density, and is measured with unit $W \cdot m^{-2}$. Meanwhile, *radiance L* of a surface is emitted flux per unit of foreshortened area (from Lambert's law) per unit solid angle $W \cdot m^{-2} \cdot sr^{-1}$. Radiance in this context is useful because it enables understanding of how much irradiance will be received by the observer, depending on the observer's view angle. For some incident flux Φ_i, we can denote the irradiance

$$E = \frac{d\Phi_i}{dA},$$

where dA is the incident area. For $d^2\Phi_r$ flux radiated within a solid angle $d\omega_r$, the surface radiance is

$$L = \frac{d^2\Phi_r}{dA \cos \theta_r \, d\omega_r}.$$

9.1.2 Lambert's Law

A *Lambertian surface* is a surface that reflects light approximately equally across the azimuth. It obeys *Lambert's law*, in which the intensity is proportional to $\cos\theta$, where θ is the incident angle of illumination with respect to the surface normal. Assuming an ideal *point source*, the law can be expressed

$$I(\theta) = I(0)\cos\theta$$

where $I(\theta)$ is the light intensity reflected. The reflected light is maximum when the incident light is parallel to the surface normal, and is zero when the source is orthogonal to the surface normal. This reduction in intensity as the incident light deviates from the surface normal is due to *foreshortening*. Foreshortening is a phenomenon in which the area of light incident on a surface decreases at oblique incident angles with respect to the normal, resulting in an overall reduction in reflection, as shown in Figure 9.1. Note that there is no ϕ (azimuthal) dependence in Lambert's law. Any surface (typically diffuse) that approximately behaves as such is referred to as a Lambertian surface.

9.1.3 Bidirectional Reflectance Distribution Function

The ratio of the light incident on the surface and the irradiance measured by the observer is known as the bidirectional reflectance distribution function (BRDF). The BRDF is

$$f = \frac{L}{E}$$

and is a function of four variables: θ_i, θ_r, ϕ_i, ϕ_r, where subscript i denotes incidence angles, and r denotes reflected angles. Given the BRDF for a surface, we can determine the observed reflectance at some view angle as a function of incident angle.

The BRDF of an object can be empirically determined by densely measuring the reflectance of the object (Matusik, 2003). However, there is value in understanding some analytical BRDFs. Here are a few interesting, but fundamental, BRDF properties and ideas. The *Helmholtz reciprocity* states that, by the second law of thermodynamics, the appearance of the object does not change when the viewing and source directions are swapped. Mathematically, for any BRDF

$$f(\theta_i, \phi_i; \theta_r, \phi_r) = f(\theta_r, \phi_r; \theta_i, \phi_i).$$

(Zickler et al., 2002) showed that one can even exploit this reciprocity to constrain the solution of the surface normals of an object by surgically choosing the positions of the light sources and camera. If an object exhibits *rotational symmetry*, then the BRDF becomes a function of just three variables

$$f(\theta_i, \theta_r, \phi_i - \phi_r).$$

If a reflection is mirrorlike, or specular, then the BRDF becomes

$$f(\theta_i, \phi_i; \theta_r, \phi_r) = \rho_s \delta(\theta_i - \theta_r) \delta(\phi_i + \pi - \phi_r),$$

where ρ_s is the specular albedo, and $\delta(x)$ is the impulse function. This should make intuitive sense, because $\theta_i = \theta_r$, as dictated by Snell's law for a specular reflection. Furthermore, the reflected light should lie along the plane of incidence, so that $\phi_r = \phi_i + \pi$. Because all the specularly reflected light is contained precisely along these angles, the viewer angle must be aligned at $\theta_r = \theta_i$ and $\phi_r = \phi_i + \pi$ to detect a mirrorlike reflection. Otherwise, the viewer detects zero radiance. A depiction of this concept is shown in Figure 9.2. Note, however, that the double delta function is an idealized model. In practice, real surfaces are not perfectly smooth, so the highlights occur over some small patch rather than at a single point. It is, however, still instructive to understand these idealized BRDF models. One example of a BRDF accounting for specular reflections from rough surfaces is the Phong BRDF model. This model incorporates angular falloff of intensity into the idealized model by introducing a specular lobe rather than specular spike, shown in Figure 9.3 (Phong, 1975).

Let us consider the BRDF in the context of images taken in heavy snow conditions. In such scenes, a "white-out" effect is dominant, in which it is nearly impossible to distinguish the texture of the snow. Here is why. The sky is the illumination source for these images, with an approximately uniform radiance of $L^{\text{source}}(\theta_i, \phi_i) = L^{\text{sky}}$. The reflected radiance is

$$L^{\text{source}}(\theta_r, \phi_r) = \int_\Omega L^{\text{source}}(\theta_i, \phi_i) f(\theta_i, \phi_i; \theta_r, \phi_r) \cos\theta_i d\Omega,$$

$$= \int_{-\pi}^{\pi} \int_0^{\pi/2} L^{\text{sky}} f(\theta_i, \phi_i; \theta_r, \phi_r) \cos\theta_i \sin\theta_i d\theta_i d\phi_i,$$

where $f(\theta_i, \phi_i; \theta_r, \phi_r)$ is the BRDF for the snow, $\cos\theta_i$ is the term resulting from Lambert's Law, and $d\Omega = \sin\theta_i d\theta_i d\phi_i$ is the differential solid angle. For a diffuse surface, where the reflection is approximately isotropic, the BRDF will be a constant value given by ρ/π, where ρ is the albedo. Therefore, if we evaluate this integral, we will get $L^{\text{surface}} = \rho L^{\text{sky}}$. The surface reflectance at any point in the scene is simply a uniformly scaled version of the sky reflectance, which results in the uniform appearance of snow and the "white-out" effect.

9.2 Shape from Intensity

Shadows play an important role in human vision. Our two eyes enable binocular vision, which is particularly useful for depth perception. However, even if we cover one eye, we are able to perceive the shape of an object by observing its shading pattern. In computer vision, you can leverage local shading to obtain a concise representation of physically

9.2 Shape from Intensity

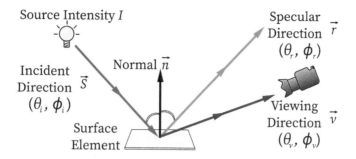

Figure 9.2 Specular or mirror-like reflection of light.

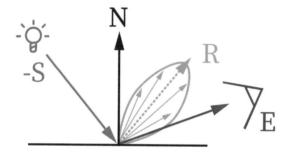

Figure 9.3 Phong BRDF model for specular highlights.

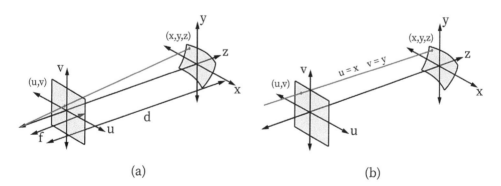

Figure 9.4 Geometry of image projection. (a) Perspective projection. (b) Orthographic projection.

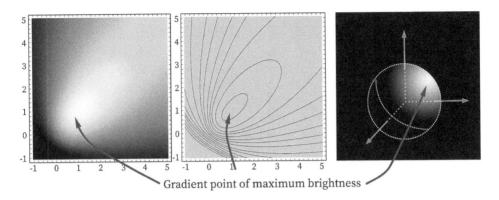

Gradient point of maximum brightness

Figure 9.5 Example of a reflectance map.

plausible shapes of objects in a scene (Xiong et al., 2014). In this section, we explore the fundamentals of shadow formation in the context of *shape from intensity*, a class of problems dedicated to extracting shape via raw intensity measurements from an imaging sensor. Note that the following sections serve as an introductory text. For a more rigorous treatment of image illumination and shadow formation, the reader is directed to (Horn, 1975).

9.2.1 Reflectance Maps and Gradient Space

A *reflectance map* maps a scene's reflectance as a function of the spatial gradients of the scene depth. It is particularly useful in understanding the image formation process because it is dependent on light source distribution and the shape of the object. Consider the object shape shown in Figure 9.4. The z axis lies along the direction connecting the camera to the object, with the x and y axis parallel to the image plane. The z value gives us the depth for each pixel. We can express the surface normals

$$\widehat{\mathbf{n}} = \frac{(p, q, -1)}{\sqrt{p^2 + q^2 + 1}},$$

where $p = \partial z/\partial x$ and $q = \partial z/\partial y$. For convenience, we assume that the camera is along the $-z$ direction. The reflectance map $R(p, q)$ is therefore expressed in terms of the surface gradients, and often plotted as a contour map in the p–q plane as shown in Figure 9.5. The p–q plane is known as *gradient space*, and is a convenient way to represent surface orientation. A given point in gradient space refers to a specific surface orientation with respect to the viewer. It is used in image analysis to relate the geometry of image projection to the radiometry of image formation. Every point along a contour has the same reflectance. The reader is referred to (Horn and Sjoberg, 1979) for details on how a reflectance map is obtained.

We can use a reflectance map to determine surface orientation. For this shape reconstruction task, there are two components: an offline job and an online job. The offline job consists of building a reflectance map and constructing a lookup table to convert reflectance measurements to surface normals, and the online job consists of measuring image intensities and determining surface orientations from the lookup table (Ikeuchi, 1981). A reflectance map provides us with a way to relate a fixed scene illumination, an object's photometry, and the imaging geometry together. However, this is insufficient for shape reconstruction because an intensity measurement provides one measurement but surface orientation has two degrees of orientation. Photometric stereo is useful because it adds another layer of information that enables shape information. The basic imaging equation, given the reflectance map, is

$$I_j(x, y) = R_j(p, q, x, y),$$

where $I(x, y)$ is the intensity output, or the image for a certain illumination j. The unknowns are p and q, while $I(x, y)$ is measured. Obtaining two images would constrain p and q, but often the solutions are nonlinear equations, yielding multiple possible values of p and q, which often makes it useful to have a third measurement. The benefit of using an *orthographic projection* is that $R_j(p, q, x, y)$ becomes only a function of p and q, $[R_j(p, q)]$, due to the spatial invariance of the reflectance field in far field. Although a *perspective projection* is a more accurate representation of the true geometry of image projection, an orthographic projection is a useful mathematical simplification in far field (see Figure 9.4). The orthographic projection also enables an easier measurement of the reflectance field. We can then directly match the reflectances of a given pixel to a point in gradient space, which we can then use to obtain the surface normal as given by $(p, q, -1)$. Figure 9.6 shows how we can obtain the surface orientation by determining the intersection points of multiple reflectance maps (Ikeuchi, 1981).

9.2.2 Calibrated Diffuse Photometric Stereo

Photometric stereo is a technique that observes a scene under different illumination conditions to extract the surface normals (and shape, by extension). Multiple illumination sources are placed at different locations while the camera is held in place. Note that this differs from traditional stereo imaging, in which the scene is captured at different camera positions. The increased information capacity provided by the location-dependent light source constrains the possible surface normal orientations and does not require additional image registration. Let us first consider the basic example of how we can use photometric stereo cues to extract surface normals for Lambertian surfaces. Refer to Figure 9.7 for the image formation model. For each illumination source, we assume that we know the vector $\mathbf{s} \in \mathbb{R}^3$ for each pixel, containing information about the direction and intensity of the ray for each pixel.

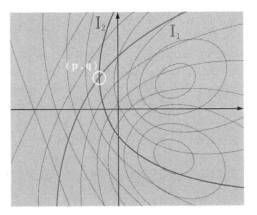

Figure 9.6 Mapping multiple intensities to surface orientations using a reflectance map.

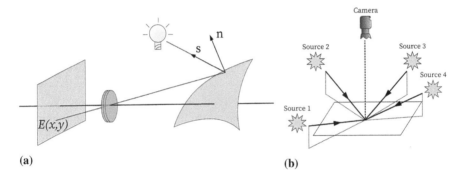

Figure 9.7 Photometric stereo for Lambertian surfaces. (a) Light from the illumination source is incident on the object, with the source vector **s** known for each pixel. The light reflected to the sensor is approximately independent of the sensor location, due to the Lambertian approximation. (b) Multiple spatially offset light sources are used in photometric stereo, with a fixed camera position.

There are three key photometric angles to understand in photometric stereo. The *incident angle* is the angle between the incident ray and the surface normal; the *view angle* is the angle between the reflected (i.e., observed) ray and the surface normal; and *phase angle* is the angle between the incident and the reflected ray. A depiction of these angles is shown in Figure 9.8. Lambert's law, as described previously, is an example of a reflectance model in terms of the incident angle (Woodham, 1980).

If we have k illumination sources, for each pixel we will have a measured intensity vector $\mathbf{i} \in \mathbb{R}^k$ and a measurement matrix $\mathbf{L} \in \mathbb{R}^{3 \times k}$ containing the **s** vector for each illumination source along its column space. Assuming the pixel has an albedo of ρ, we can then relate

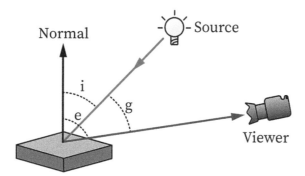

Figure 9.8 Important photometric angles: incident angle (i), view angle (e), and phase angle (g).

the pixel's surface normal $\hat{n}$ to our known quantities by the matrix product

$$\mathbf{i} = \mathbf{L}^\top \tilde{\mathbf{n}}'$$

where $\tilde{\mathbf{n}}' = \rho \hat{\mathbf{n}}$. For $k = 3$, we can easily invert and solve for $\hat{\mathbf{n}}$ and ρ, taking into account that $\hat{\mathbf{n}}$ has length 1. We can also have $k > 3$, in which case the linear system is overconstrained and the reconstruction will be more robust to noise. The reader is directed to (Shi, 2019; Zheng et al., 2020; Satkin et al., 2012) for surveys on data-driven photometric stereo methods. (Choy et al., 2016) also present a network that uses shape priors coupled with multiview image data to map images to their 3D shapes.

A subset of the photometric stereo problem is known as *shape from shading (SfS)*, where $k = 1$ and shading cues are used to obtain surface normals. This method is partly based on human vision. Incident light reflected by a surface is generally dependent only on the surface orientation with respect to the light source and the observer. Therefore, different points on a nonplanar surface will reflect different intensities of light back to the observer. This enables our brain to process these shadows and help us perceive shape through vision.

SfS is an underconstrained problem, so we have to make some assumptions to solve for the normals. One possible constraint is to enforce smoothness of gradients via the optimization

$$\min \iint_{\text{image}} \left(p_x^2 + p_y^2 \right) + \left(q_x^2 + q_y^2 \right) \, dx \, dy,$$

where $p_x = \frac{\partial p}{\partial x}$, $p_y = \frac{\partial q}{\partial y}$, $q_x = \frac{\partial q}{\partial x}$, and $q_y = \frac{\partial q}{\partial y}$. Another possible constraint would be to enforce the fidelity of the image intensity to the reflectance map by the optimization

$$\min \iint_{\text{image}} \left[I(x, y) - R(f, g) \right]^2 \, dx \, dy,$$

where $R(p, q)$ is the corresponding reflectance based on the normal at (x, y). A weighted combination of these optimization constraints is also possible, along with others. For a dedicated text on shape from shading, the reader is directed to (Horn and Brooks, 1989). For more work on numerical SfS, the reader is encouraged to review (Ikeuchi, 1981), and the reader is directed to (Zhang et al., 1999) for a comparison of different SfS algorithms, as well as code to implement them.

9.2.3 Uncalibrated Diffuse Photometric Stereo

The light source directions and intensities may not always be known, as we previously assumed. In such a case, we would have to resort to numerical methods to approximate our surface normal, even if we assume a Lambertian surface. Let us consider an image data matrix $\mathbf{I} \in \mathbb{R}^{p \times f}$, where p is the number of pixels and f is the number of frames captured. Based on the Lambertian constraint, we can express the image data

$$\mathbf{I} = \mathbf{RNMT},$$

where $\mathbf{R} \in \mathbb{R}^{p \times p}$ is the surface reflectance matrix (a diagonal matrix containing the albedo of each pixel along its diagonal); $\mathbf{N} \in \mathbb{R}^{p \times 3}$ is the surface normal matrix (containing surface normal vectors of each pixel along its row space); $\mathbf{M} \in \mathbb{R}^{3 \times f}$ is the light source direction matrix (containing the light source direction along its column space); and $\mathbf{T} \in \mathbb{R}^{f \times f}$ is the light source intensity matrix (a diagonal matrix containing the light intensities along its diagonal).

Using this formulation, we can express $\mathbf{I}$ as a matrix product $\mathbf{I} = \mathbf{SL}$, where $\mathbf{S} = \mathbf{RN}$ is the surface matrix and $\mathbf{L} = \mathbf{MT}$ is the light source matrix. If there are at least three surface normals in the image that do not lie in the same plane, then $\mathbf{S}$ will have three linearly independent rows, and will therefore have rank 3. Similarly, if the three light source directions do not lie in the same plane, $\mathbf{L}$ will have three linearly independent columns and also have rank 3. These are both reasonable assumptions, so we can assume that $\mathbf{I}$ will have rank 3 as well, provided our measurements are noiseless. If $p \geqslant f$, we can calculate a *singular value decomposition (SVD)* for our measured intensity matrix $\mathbf{I}$ in the nonideal case with noise

$$\mathbf{I} = \mathbf{U\Sigma V}^{\mathrm{H}}$$

where $\mathbf{U} \in \mathbb{R}^{p \times f}$, $\mathbf{\Sigma} \in \mathbb{R}^{f \times f}$, $\mathbf{V} \in \mathbb{R}^{f \times f}$, and $\mathbf{U}^{\top}\mathbf{U} = \mathbf{V}^{\top}\mathbf{V} = \mathbf{V}\mathbf{V}^{\top} = \mathbf{E}$ (where $\mathbf{E}$ is the identity matrix). $\mathbf{\Sigma}$ contains the singular values of the matrix $(\sigma_1, \cdots, \sigma_f)$ along its diagonals, sorted from greatest to least such that $\sigma_1 \geqslant \cdots \geqslant \sigma_f$. Because our measured intensities $\mathbf{I}$ are not noiseless, we assume that the first three singular values are above the

9.2 Shape from Intensity

noise threshold and decompose $\mathbf{U}$, $\mathbf{\Sigma}$, and $\mathbf{V}$

$$\mathbf{U} = \begin{bmatrix} \underbrace{\mathbf{U}'}_{3} & \underbrace{\mathbf{U}''}_{f-3} \end{bmatrix} \} p,$$

$$\mathbf{\Sigma} = \begin{bmatrix} \underbrace{\mathbf{\Sigma}'}_{3} & 0 \\ 0 & \underbrace{\mathbf{\Sigma}''}_{f-3} \end{bmatrix} \begin{matrix} \} 3 \\ \} f-3 \end{matrix},$$

$$\mathbf{V} = \begin{bmatrix} \mathbf{V}' \\ \mathbf{V}'' \end{bmatrix} \begin{matrix} \} 3 \\ \} f-3 \end{matrix}.$$
$\underbrace{}_{f}$

We then approximate a new denoised version of $\mathbf{I}$, given by $\widehat{\mathbf{I}}$, using the first three singular values from SVD, to get a rank 3 intensity matrix $\widehat{\mathbf{I}}$

$$\widehat{\mathbf{I}} = \mathbf{U}' \mathbf{\Sigma}' (\mathbf{V}')^H = \widehat{\mathbf{S}} \widehat{\mathbf{L}},$$

where $\widehat{\mathbf{S}} = \mathbf{U}' \left(\pm [\mathbf{\Sigma}']^{1/2} \right)$ and $\widehat{\mathbf{L}} = \left(\pm [\mathbf{\Sigma}]^{1/2} \right) (\mathbf{V}')^H$. Note that this approximation holds only if the third singular value is much larger than the fourth singular value. The sign ambiguity corresponds to ambiguity of the direction of the coordinate system. Without loss of generality, we can choose the right-handed coordinate system, corresponding to $+\mathbf{\Sigma}$. Note that $\widehat{\mathbf{I}} = \widehat{\mathbf{S}} \widehat{\mathbf{L}}$ is not a unique factorization. For some arbitrary invertible matrix $\mathbf{A}$,

$$\widehat{\mathbf{I}} = \left(\widehat{\mathbf{S}} \mathbf{A} \right) \left(\mathbf{A}^{-1} \widehat{\mathbf{L}} \right) = \widehat{\mathbf{S}} \widehat{\mathbf{L}}.$$

Therefore, we must find the matrix $\mathbf{A}$ that will yield the correct surface and light matrices, such that

$$\mathbf{S} = \widehat{\mathbf{S}} \mathbf{A},$$
$$\mathbf{L} = \mathbf{A}^{-1} \widehat{\mathbf{L}}.$$

To constrain the search for $\mathbf{A}$, we search for at least six pixels in which the relative value of the surface reflectance is constant or known a priori. We extract $p' \geq 6$ pseudovectors $\widehat{\mathbf{s}}$ from the row space of $\widehat{\mathbf{S}}$ that satisfy this condition. If all p' pixels have the same reflectance, then the following condition must hold:

$$\widehat{\mathbf{s}}_k^\top \mathbf{A} \mathbf{A}^\top \widehat{\mathbf{s}}_k = r, \qquad k = 1, \cdots, p',$$

where r is the magnitude of the surface reflectance. If this value is unknown, we can simply choose $r = 1$, because we are interested in relative surface reflectance. We first solve for $\mathbf{B} = \mathbf{A}\mathbf{A}^\top$ via a straightforward system of linear equations. Using the SVD of $\mathbf{B}$, we know that $\mathbf{B} = \mathbf{W}\mathbf{\Pi}\mathbf{W}^\top$, because $\mathbf{B}$ is symmetric. We can then determine that $\mathbf{A} = \mathbf{W}[\mathbf{\Pi}]^{1/2}$.

Once we solve for **A**, we can extract **S** and **L** using the given equations in terms of $\widehat{\mathbf{S}}$ and $\widehat{\mathbf{L}}$. The surface normals can be extracted by normalizing the magnitude of each row in **S** to be 1. The normalization factor is the albedo for that pixel. Note that the normals are represented in an arbitrary coordinate system. Once the surface normal of one pixel is determined with respect to the observer, the remaining surface normals can easily be determined via linear transformation (Hayakawa, 1994).

Another way to determine **A** is by taking advantage of the integrability constraint. From introductory multivariable calculus, we know that

$$\frac{\partial}{\partial x}\left(\frac{\partial z}{\partial y}\right) = \frac{\partial}{\partial y}\left(\frac{\partial z}{\partial x}\right).$$

Recall that $\widehat{\mathbf{S}}$ is a $p \times 3$ matrix, containing "pseudonormals" along its row space. Using the integrability constraint, we can set up a linear system of equations based on the derivatives of the normal vectors. This linear system of equations can be solved using linear least squares, to minimize

$$\left(\frac{\partial z}{\partial y \partial x} - \frac{\partial z}{\partial x \partial y}\right)^2$$

at each pixel. For more information on this integrability constraint, the reader is directed to (Basri et al., 2007). These methods provide us with a useful numerical method to extract shape information from multi-illumination when we do not have prior knowledge of the source direction or intensity, and the surface albedos.

9.2.4 Dichromatic Reflection Model

As we discussed in chapter 7, the highlights present in an image caused by specular reflections can often be distracting and affect computer vision tasks, including surface normal extraction. Photometric stereo is one way to deal with the specular reflections from glossy surfaces. We consider the case of images with purely specular reflections, because only 1%–2% of the reflections from metals are diffuse. This means that the Lambertian assumption does not work well here. For specular objects, we can also no longer illuminate the scene with point sources, as we previously assumed. Due to the single-bounce nature of specular reflections, only surfaces oriented in a specific direction will reach the observer's eye. Therefore, the observer will simply see virtual images of the point sources. Instead, we must use an *extended light source*, which can be obtained by uneven illumination from a diffusely reflecting planar surface. An extended light source can also be thought of as a series of multiple point sources.

Shadows cast by external objects can also negatively impact the performance of photometric stereo in shape analysis. Whereas reflectance properties of the object are illumination-independent, the surface relief is highly illumination-dependent and produces shadows depending on the location of the illumination. Ultimately, we want to be able to obtain an

9.2 Shape from Intensity

illumination-independent 3D characterization of shapes in a scene using 2D images. Note that both highlights and shadows are simply sudden, unexpected changes in pixel intensities. The only difference is that shadows are darker pixels, and highlights are brighter pixels.

We can handle challenges presented by external shadows and highlights by incorporating a *dichromatic reflection model*, in which the reflectance can be decomposed into a sum of a Lambertian and specular component. The Lambertian components can be dealt with similarly to the way we discussed previously. Meanwhile, we can detect specular components using spectral cues. Capturing an ordinary RGB image contains redundant information across channels, because the three color channels are linearly dependent. Instead, we can use a photometric stereo approach, in which we illuminate the scene from different angles, with each angle containing a different spectral source. This is known as *shape from color*. Note, however, that the spectral sources may have different intensities and will need spectral calibration accordingly. As discussed in chapter 8, the intensity measured by a sensor α is given by

$$I_\alpha = \cos\theta \int_{-\infty}^{\infty} \mu\, \epsilon(\lambda)\, R(\lambda)\, Q_\alpha(\lambda)\, d\lambda$$

where θ is the angle between the source direction and the normal; μ is a constant proportional to the strength of the light intensity; $\epsilon(\lambda)$ is the spectral distribution of the illumination source; $R(\lambda)$ is the reflectance at the pixel location; and $Q_\alpha(\lambda)$ is the spectral sensitivity of α. We first describe the process of *color photometric stereo*.

Recall that we can compactly express the intensity of a grayscale pixel in an image of a Lambertian object

$$\mathbf{I}_0 = \rho \mathbf{L} \mathbf{n},$$

where $\mathbf{I}_0$ is the intensity vector for light from three different directions $\mathbf{L} = [L_1, L_2, L_3]^\top$ and $\mathbf{n}$ is the surface normal for the pixel. In color photometric stereo, assuming that a surface patch has color $\mathbf{C} = [C_r, C_g, C_b]^\top$ and that we use an RGB camera capturing intensities $\mathbf{I}^k$, we can model the intensity for each pixel from illumination source k as

$$\mathbf{I}^k = [I_1^k, I_2^k, I_3^k]^\top = s^k \mathbf{C}$$

where $s^k = \langle \mathbf{L}^k, \mathbf{n} \rangle$ and $\mathbf{S} = (s^1, s^2, s^3)^\top$ is the *shading vector*. Note that

$$C_\alpha = \int_{-\infty}^{\infty} \epsilon(\lambda)\, R(\lambda)\, Q_\alpha(\lambda)\, d\lambda,$$

as alluded to earlier. If we combine the intensities for all three illumination sources, such that $[\mathbf{I}] = [\mathbf{I}^1, \mathbf{I}^2, \mathbf{I}^3]^\top$, we can express $[\mathbf{I}]$

$$[\mathbf{I}] = \begin{bmatrix} I_1^1 & I_2^1 & I_3^1 \\ I_1^2 & I_2^2 & I_3^2 \\ I_1^3 & I_2^3 & I_3^3 \end{bmatrix} = \begin{bmatrix} s^1 C_r & s^1 C_g & s^1 C_b \\ s^2 C_r & s^2 C_g & s^2 C_b \\ s^3 C_r & s^3 C_g & s^3 C_b \end{bmatrix} = \mathbf{S}\mathbf{C}^\top.$$

Each row of $[\mathbf{I}]$ is a different illumination, and each column is a different color. Note that $[\mathbf{I}]$ refers to measurements at one pixel, for three colors and three illumination directions. For a diffuse surface, the intensity vectors measured by each illumination source are linearly dependent, so they are all collinear in RGB space. In other words, these intensity measurements are scalar multiples (s^k) of each other. Noise in measurements may perturb this collinearity in RGB space. By applying principal components analysis (PCA) (discussed in detail in 8.4.2) on $[\mathbf{I}]$, we can determine the principal direction, which gives the chromaticity of the object. The principal direction is simply the principal component with largest variance. Each measured intensity in $[\mathbf{I}]$, which is in color space, is then projected onto this principal direction, to give us a new intensity measurement $[\mathbf{I}']$. $\mathbf{C}$ is simply the normalized unit vector along the direction of the principal direction, and $\mathbf{S}$ can be extracted by the known value of $\mathbf{C}$ and $[\mathbf{I}']$ by solving the linear system. Once $\mathbf{S}$ is determined, $\mathbf{n}$ can be recovered using standard grayscale photometric stereo.

This technique of color photometric stereo, however, will also falter under the presence of specular highlights. Assuming that there are no self-shadows, we can use four illuminations, and extract the albedos for each image separately using grayscale photometric stereo. If the albedos are significantly different, then the pixel has a specular highlight. In this case, the illumination with the smallest albedo is assumed to contain only the Lambertian constraint and used for surface normal recovery. Let us now formalize this physical intuition with mathematical definitions.

We will examine a method known as **4-*source color photometric stereo***. In this method, we capture four different illuminations instead of three as we normally do with Lambertian surfaces. Because we capture four images in 3D space, we will have four linearly dependent images. Therefore, for some vector $\mathbf{a} = [a_1, a_2, a_3, a_4]^\top$, a Lambertian surface will satisfy

$$\langle \mathbf{a}, \mathbf{I} \rangle = a_1 I^1 + a_2 I^2 + a_3 I^3 + a_4 I^4 = 0.$$

However, for specular highlights, the value will deviate from 0 by some factor ϵ. By thresholding this value for ϵ, we can identify specular regions. Then, we can interpolate specular regions from neighboring pixels. For more details on how the normals are extracted from this point, we direct the reader to (Barsky and Petrou, 2003). The reader is also welcome to study (Esteban et al., 2008) for use on how traditional stereo imaging can be combined with photometric stereo to extract shape from textureless, shiny objects.

9.2 Shape from Intensity

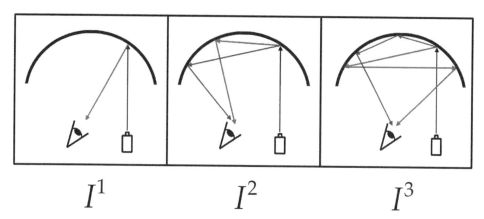

$$I^1 \qquad I^2 \qquad I^3$$

Figure 9.9 Scene interreflections. The most idealized model is the single-bounce model, in which light from the source bounces off the surface and directly reaches the sensor. However, the light can bounce off the surface n times, as shown for a two-bounce reflection and a three-bounce reflection. The total intensity measured at the sensor is the sum of the intensities for all possible number of bounces, from one to infinity. Reprinted from (Seitz et al., 2005).

9.2.5 Shape from Interreflections

In our idealized models so far, we assume that the light intensity measured at the sensor is light that propagates from the source, bounces off the surface, and directly reaches the sensor. This assumption generally works well for convex surfaces, in which no two points on a surface or visible to each other. However, for concave surfaces, the light will go through several *interreflections* before reaching the sensor, in which the light bounces off one point on a surface to another point, and repeats for a certain number of times before reaching the sensor. A comparison of the idealized image formation model and models containing interreflections is depicted in Figure 9.9.

Neglecting the effects of interreflections often results in incorrect image renderings. Refer to Figure 9.10 for an example of the poor reconstruction yielded by photometric stereo for a concave surface. In computer vision, it is desirable to remove the effects of interreflections to properly render an image with *color constancy*. Color constancy is the ability of an object to be approximately the same color under different illumination conditions. With interreflections, there are two known problems: the forward problem and inverse problem. The forward, or graphics, problem deals with trying to determine image intensity values from shape and reflectance information. The inverse, or vision, problem seeks to extract the shape and reflectance from intensity measurements. Often, inverse problems are challenging due to the nonlinearity of the forward model. (Shi et al., 2013)

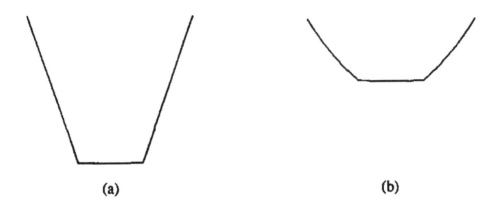

Figure 9.10 Concave shape reconstruction using photometric stereo: (a) original shape, and (b) shape reconstructed with standard photometric stereo. Reprinted from (Nayar et al., 1991).

present an example of a parametric bi-polynomial forward model for single reflections that is suitable for inverse problems.

The vision problem for interreflections is nontrivial due to interdependency between shape and interreflections. It resembles a "chicken and egg" problem, in which modeling interreflections requires prior knowledge of shape and reflectance, but our goal is already to determine the shape. However, assuming we work with Lambertian surfaces with continuous shape, we can extract "pseudoshapes" using standard photometric stereo techniques, and use those as a prior to understand interreflections in an object. These pseudoshapes can then be corrected to yield the actual shape of the object.

The pseudoshape of the object has certain interesting properties that we can exploit. The pseudoshape is unique for a given shape and is less concave than the actual shape. We now describe an interreflection model based on the forward problem. To be able to obtain a closed-form solution, we assume that all surfaces in the scene are Lambertian, but can have spatially varying albedos. Consider the concave surface in Figure 9.11a. Notice how the measured intensity from a given point on the surface is a superposition of direct illumination by the source and illumination from other points on the surface, shown in Figure 9.11b.

We want to determine the radiance of point $\mathbf{x}$ caused by point $\mathbf{x}'$, and we do so via a *visibility function*. The visibility function determines whether two points are able to "see each other" and is defined

$$V(\mathbf{x}, \mathbf{x}') = \left\langle \frac{\langle \mathbf{n}, (-\mathbf{r}) \rangle + |\langle \mathbf{n}, (-\mathbf{r}) \rangle|}{2|\langle \mathbf{n}, (-\mathbf{r}) \rangle|}, \frac{\langle \mathbf{n}', \mathbf{r} \rangle + |\langle \mathbf{n}', \mathbf{r} \rangle|}{2|\langle \mathbf{n}', \mathbf{r} \rangle|} \right\rangle.$$

Note that $V(\mathbf{x}, \mathbf{x}')$ is a binary function, with 1 indicating that the two points are oriented such that they can illuminate each other, and 0 indicating that they are not. We let $dE_m(\mathbf{x}, \mathbf{x}')$

9.2 Shape from Intensity

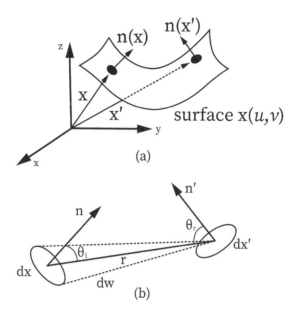

Figure 9.11 Direct and indirect illumination of surface points.

represent the irradiance of the surface element due to the radiance $L(\mathbf{x}')$ from element $d\mathbf{x}'$, as shown in Figure 9.11b. Using the geometry in Figure 9.11b, we can determine that

$$dE_m(\mathbf{x}, \mathbf{x}') = \left[\frac{\langle \mathbf{n}, (-\mathbf{r})\rangle \langle \mathbf{n}', \mathbf{r}\rangle V(\mathbf{x}, \mathbf{x}')}{\langle \mathbf{r}, \mathbf{r}\rangle^2}\right] L(\mathbf{x}' d\mathbf{x}').$$

Because we assume a Lambertian surface, we can model the radiance of $d\mathbf{x}$ due to $d\mathbf{x}'$

$$dL_m(\mathbf{x}, \mathbf{x}') = \frac{\rho(\mathbf{x})}{\pi} dE_m(\mathbf{x}, \mathbf{x}').$$

Recall from the "white-out" idea that the BRDF of a Lambertian surface is given by ρ/π. Meanwhile, the irradiance of the point due to a point source can be modeled

$$L_s(\mathbf{x}) = \frac{\rho(\mathbf{x})}{\pi} E_s(\mathbf{x}),$$

where $E_s(\mathbf{x}) = k\langle \mathbf{n}, \mathbf{s}\rangle$, analogous to what was mentioned in discussions of Lambert's Law and diffuse surfaces. The total radiance at a point is then expressed

$$L(\mathbf{x}) = L_s(\mathbf{x}) + \int dL_m(\mathbf{x}, \mathbf{x}'),$$

where the integration is done across the surface. This equation for the forward model is referred to as the *interreflection equation*. Although it is rather difficult to solve the integral equation, approximating a uniform reflectance $\rho(\mathbf{x}) = \rho$ would enable an iterative

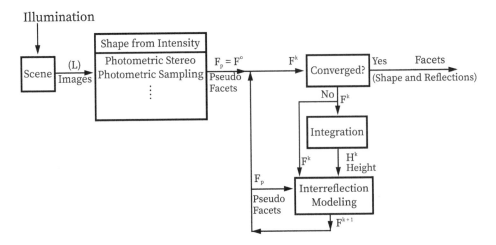

Figure 9.12 Iterative algorithm for extracting shape from objects with interreflections. Adapted from (Nayar et al., 1991).

solution by the Neumann series. Another possible method is by discretizing the surface into facets. Both these methods are out of scope for this text. Solutions to this integral form are somewhat analogous to ray tracing in computer graphics. The forward interreflection model is used in an iterative manner, as shown in Figure 9.12, to obtain the correct *facets*, which are simply discretized regions on the surface. The algorithm starts by estimating facets using traditional shape from intensity measurements. Using the properties of the pseudosurfaces described previously, the algorithm is able to make adjustments to the facets, based on the forward model, that eventually converge to the ground truth (Nayar et al., 1991).

9.2.6 Example-Based Photometric Stereo

Orientation consistency is a useful cue for interpreting scenes with arbitrary reflectance profiles. Orientation consistency is based on the premise that any two points with the same surface orientation must have a similar appearance in image, provided that they are under the right conditions. In other words, two surfaces with the same orientation reflect the same amount of light back to the observer. This assumption holds when the BRDF of both points are the same, the light sources are directional, the camera is orthographic, and there are no shadows or interreflection effects unaccounted for by the BRDF. Orientation consistency is particularly useful because if we know the surface orientation of some parts of a scene, we can infer the orientations of other points in a scene.

We leverage orientation consistency by matching a reference object to the imaged object, in what is known as *example-based photometric stereo*. By capturing several images with

9.2 Shape from Intensity

different illuminations, we can enforce orientation consistency across all images, using the fact that orientation consistency is independent of illumination direction. Per color channel, n reference images $\mathbf{V}^r = \left[\mathbf{I}_1^r, \cdots, \mathbf{I}_n^r\right]^\top$ and n target images $\mathbf{V}^t = \left[\mathbf{I}_1^t, \cdots, \mathbf{I}_n^t\right]^\top$ are captured. The best matching reference point can be determined by finding the pixel in the reference object that minimizes the ℓ_1 norm $\|\mathbf{V}^r - \mathbf{V}^t\|$ across all color channels and images. Note that any object shape can be used for the reference, provided that the shape is known and it has a sufficient distribution of surface orientations. The reference images contain known surface orientations. We then match the intensities of the target image to the reference and extract the surface orientation from the reference image.

We now generalize this method of example-based photometric stereo to spatially variant BRDFs. Assuming we have a homogeneous diffuse reference object, we can use a uniform albedo constant described in Lambert's law previously. For a given pixel p in the target object corresponding to pixel q in the reference object, we must account for the differences in albedo. We do so by

$$\mathbf{V}_p^t = \frac{\rho_p^t}{\rho^r}\mathbf{V}_q^r.$$

This relationship arises out of Lambert's Law, as previously discussed. Once this albedo calibration is done, the pixel correspondence for orientation can be done as described. Now, we consider surfaces composed of multiple materials. A common technique is to assume that all materials can be approximated by a linear combination of k basis *fundamental materials*, which can be accounted for in k homogeneous reference images. The intensity of the target image can therefore be expressed

$$\mathbf{I}_p^t = \sum_{i=1}^{k} \rho_{i,p}^t \mathbf{f}_i\left(n_p, v, L\right),$$

where $\mathbf{f}_i\left(\mathbf{n}_p, v, \mathbf{L}\right)$ is the reflectance map; $\mathbf{n}_p$ is the surface normal at pixel p; v is the direction to the viewer; and $\mathbf{L}$ is the incident illumination. Each material also has its own albedo $\rho_{i,p}^t$. Each of the reference images $\mathbf{V}^r = \left[\mathbf{I}_1^r, \cdots, \mathbf{I}_n^r\right]^\top$ can be expressed in a similar manner. If we have k linearly independent reference observation vectors $\left[\mathbf{V}_q^{r_1}, \cdots, \mathbf{V}_q^{r_k}\right]$, then they form a k-dimensional subspace. Therefore, we can express a pixel in the target image $\mathbf{V}_p^t$ as a linear combination of a corresponding pixel in $\mathbf{V}^r$

$$\mathbf{V}_p^t = \sum_{j=1}^{k} m_{j,p} \mathbf{V}_q^{r_j},$$

where $m_{j,p}$ is the material index for point p. We can express this concisely as a matrix multiplication by stacking the reference observation vectors into a single matrix $\mathbf{W}_q = \left[\mathbf{V}_q^{r_1}, \cdots, \mathbf{V}_q^{r_k}\right]$

$$\mathbf{V}_p^t = \mathbf{W}_q \mathbf{m}_p.$$

Note that $\mathbf{m}_p$ will differ for each color channel. A point p on the target image is considered orientation-consistent with a point q in the reference images if there exists a $\mathbf{m}_p$ that satisfies the matrix multiplication. The candidate q is chosen by minimizing

$$\left\| \mathbf{W}_q \mathbf{m}_p - \mathbf{V}_p^t \right\|_2^2.$$

Note that this could also also work well for Lambertian surfaces, at the expense of run time and potentially more captured images. Recall that Lambertian surfaces require only three different illuminations to constrain the reconstruction (Hertzmann and Seitz, 2005).

It is, however, desirable to be able to perform this generalized shape estimation without an example-based approach. If we know the BRDF of an object, photometric stereo methods enable shape reconstruction as described previously using example-based approaches. If shape is known, then some BRDF estimation methods can be applied to estimate the material properties of the scene. However, if both the material and the shape are unknown, the problem becomes more challenging, due to the high dimensionality of the optimization space.

As mentioned previously, we can leverage the fact that most objects, manmade or natural, can be decomposed into approximately just two materials. Using this constraint, it is possible to alternate between optimizing global parameters and optimizing per-pixel weights and normals. Then, using an isotropic Ward BRDF model, the normals and material weights for each pixel can be jointly optimized. For more details on this optimization process, we refer the reader to (Goldman et al., 2009). The reader is also welcome to study (Shi et al., 2016) for more work on non-Lambertian photometric stereo and (Ackermann and Goesele, 2015) for more information on photometric techniques and research challenges not discussed in this text.

9.3 Multiplexed Illumination

We discussed in chapter 8 the concept of *multiplexed illumination* in the context of spectral imaging, in which we illuminated a scene with sources of different spectral profiles to efficiently capture the spectral signature of a scene. In a similar spirit, we consider scenes simultaneously, rather than sequentially, illuminated by multiple sources to improve the dynamic range of the image (by appropriately dealing with specularity and shadows). Methods related to multiplexing make note of the fact that specular highlights are often concentrated in small image regions, that shadows and dark albedos typically coexist with bright albedos, and that low power illuminations may not yield good reconstructions.

Let us consider a multiplexed illumination setup with three sources and denote the acquired image $\mathbf{a} = \left[a_{1,2}, a_{2,3}, a_{1,3} \right]^\top$, where the subscripts denote which lights are on, and the image irradiance under each illumination source separately as $\mathbf{i} = \left[i_1, i_2, i_3 \right]^\top$. In the trivial case depicted in Figure 9.13a, we can approximate $\widehat{\mathbf{i}} = \left[i_1, i_2, i_3 \right]^\top$. Based on the

9.4 Applications in Graphics

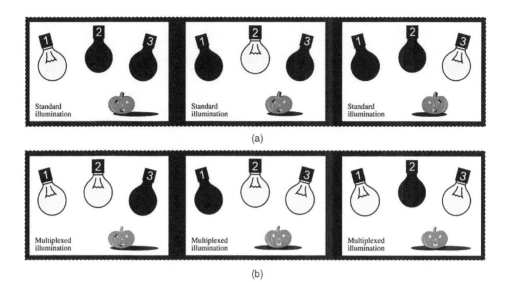

Figure 9.13 (a) Standard photometric stereo. (b) Multiplexed illumination. Reprinted from (Schechner et al., 2007).

scheme shown in Figure 9.13b, we can determine at a pixel that $\mathbf{a}\,(x, y) = \mathbf{W}\mathbf{i}\,(x, y)$, or

$$\begin{bmatrix} a_{1,2} \\ a_{2,3} \\ a_{1,3} \end{bmatrix} = \begin{bmatrix} 1 & 1 & 0 \\ 0 & 1 & 1 \\ 1 & 0 & 1 \end{bmatrix} \begin{bmatrix} i_1 \\ i_2 \\ i_3 \end{bmatrix},$$

where $\mathbf{W}$ is a weighting matrix for the intensities $\mathbf{i}$. This can be generalized to n light sources and n measurements. The rows of $\mathbf{W}$ correspond to a measurement and contain binary values indicating whether a source is turned on or not. The estimate for $\widehat{\mathbf{i}}$ in this case can be obtained by matrix inversion. The key benefit of the scheme shown in Figure 9.13b compared with that shown in Figure 9.13a is that using multiple sources of illumination raises the diffuse intensity relative to the specular intensity and illuminates shadows. The spatially offset illumination ensures that the specular components will not oversaturate, while increasing the brightness of the rest of the image (Schechner et al., 2007). Note that this is not necessarily a method for shape estimation but is still a useful way to robustly remove noise, specularities, and shadows in dimly lit environments. (Caorsi et al., 1994) also present an interesting approach for multi-illumination in microwave imaging.

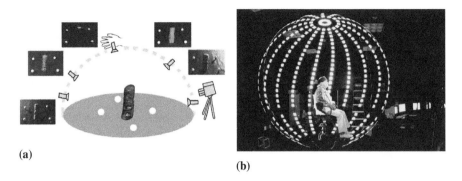

Figure 9.14 Light stage. (a) Light stage with a movable arm (Masselus et al., 2002). (b) Light stage based on several spatially offset light sources (Hawkins et al., 2001).

9.4 Applications in Graphics

9.4.1 Light Stage

The *light stage* is a fixed mechanical stage with several light sources at angular offsets, or with a single light source that can be manually moved around at fixed angular positions. Examples are shown in Figure 9.14. It enables capture of the reflectance field of a scene using a dense set of image measurements, captured at various illumination angles (Masselus et al., 2002). Being able to capture real-world scenes and render them as 3D computer models is useful for analysis of artifacts at archaeological sites. Standard methods for doing so, such as laser scanning, although often effective, struggle with complex materials that exhibit anisotropic or iridescent reflectance, self-shadows, or high specularity. To capture the scene with high fidelity, we can use a light stage to capture a dense set of reflectance field measurements. Note that a light stage is especially useful here, because a 3D model requires that we obtain accurate shape information and detail at any given angle. (Nam et al., 2016) used this multi-illumination idea at a smaller scale, building a microscope imager that captures microscale reflectance and surface normal information. The capture of the reflectance field also eliminates the need for prior knowledge of the scene geometry. The disadvantage is that it requires on the order of thousands of images, making them highly data intensive (Hawkins et al., 2001).

9.4.2 Image Rendering and Relighting

It is often desirable to be able to render a scene under different lighting conditions, particularly for graphics applications. Image relighting may be possible if we can obtain the BRDF for every pixel in a scene. With this information, we can determine the reflectance profile at every pixel in the scene for some arbitrary illumination, using the per-pixel BRDF. However, this information may not always be readily available with high accuracy, and it re-

9.4 Applications in Graphics

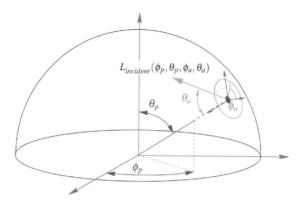

Figure 9.15 The coordinate system is defined such that a hemisphere completely contains the object of interest (Masselus et al., 2003).

quires prior knowledge of the geometry of the object. We consider some multi-illumination methods to relight a scene, using multiple images with varying illumination.

One way to perform image based relighting of real objects is by using 4D incident light fields. Recall that a light field is a 4D function that describes the illumination intensity leaving some 3D volume. They are used mainly in image-based rendering for displaying objects without knowledge of the scene geometry or material properties. To relight an image using a 4D incident light, we need to be able to characterize the scene from any incident illumination. The reflectance field is obtained by capturing the object under m basis incident light fields. These are later used for relighting, because any future lighting schematic can be expressed as a linear combination of the basis light fields. The total reflectance field is simply a linear combination of the individual basis reflectances. This is possible due to the linear nature of the illumination's interaction with objects.

For such an illumination setup, we define a coordinate system such that the object is completely contained by a hemisphere, as shown in Figure 9.15. The object is located at angle (ϕ_p, θ_p) and the illumination direction is parameterized in a local frame given by (ϕ_a, θ_a). We denote these parameters as $\Theta = (\phi_p, \theta_p, \phi_a, \theta_a)$ and Ω as the 4D space occupied by all values of Θ. We aim to separate Ω into N partitions such that the union of these partitions completely occupies Ω. Once we have done so, we can express any incident light field

$$\mathbf{L}_{\text{incident}}(\Theta) \approx \sum_{i=1}^{N} l_i \mathbf{B}_i(\Theta),$$

where $\mathbf{B}_i(\Theta)$ is a binary function determining whether $\Theta \in \Omega_i$, and l_i is the light field emanating from the space occupied by Ω_i. We can illuminate an object with different light fields using a projector and occluders, as shown in Figure 9.16. The projector is

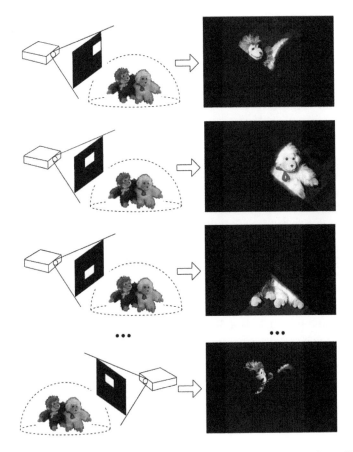

Figure 9.16 Relighting based on discretized 4D light fields. Reprinted from (Masselus et al., 2003).

mounted in place, while the object, the camera, and occluders are moved around using a turntable, enabling capture of the scene at different points with respect to the azimuth. The illumination is controlled with respect to the tilt angle by using the occluder. The object is then relit simply using a linear combination of the measured basis reflectances

$$\mathbf{L}_{\text{exitant}}(x, y) = \sum_{i=1}^{N} l_i \mathbf{R}_i(x, y),$$

where $\mathbf{R}_i(x, y) = \int_{\Omega_i} \mathbf{R}(\Theta, x, y) \, d\mu(\Theta)$ and $\mathbf{R}(\Theta, x, y)$ is the reflectance field of the scene (Masselus et al., 2003).

9.4.3 Local Shading Adaptation

If we can relight an image, we can also think about how we can increase the visual quality of an image by enhancing shape and surface details via shadows. We should understand that the goal of illustrators and photographers may not necessarily be to depict physically accurate lighting but instead to convey the shape and form of a scene. A real scene, as we have discussed, contains overexposed and underexposed pixels, shadows, and specular highlights. The principle of *local shading adaptation* is the idea that artists can increase local contrast in images and bring out fine shape details by manipulating shading. They can do so while ensuring that there are no sudden changes in intensity, keeping the dynamic range of the image high. We can leverage this idea by using multiple images illuminated from different angles. Each image will capture some features, which can be extracted via a multiscale edge-preserving decomposition using a bilateral filter. These features can then be recombined in a way to enhance the visual quality of the image. Note, however, that this may not necessarily yield a physically possible image. The intention is to convey features of the object, more so than the true optical nature of the scene.

The bilateral filter aims to decompose an image into a filtered image $\mathbf{I}^j$, preserving the edges in the image, and a detail image $\mathbf{D}^j$. Each filtered image can then be filtered repeatedly m times to obtain an m-level decomposition of the image such that

$$\mathbf{I} = \sum_{j=1}^{m} \mathbf{D}^j + \mathbf{I}^m,$$

where $\mathbf{D}^j = \mathbf{I}^{j-1} - \mathbf{I}^j$, $\mathbf{I}^j$ is a bilateral filtered image preserving the strongest edges in the image and $\mathbf{D}^j$ is a detail image containing smaller changes in intensity. The final enhanced image can be generated by

$$\mathbf{I}^{\text{result}} = \mathbf{I}^{\text{detail}} + \beta \cdot \mathbf{I}^{\text{base}}.$$

$\mathbf{I}^{\text{detail}}$ is a weighted sum of the difference image $\mathbf{D}^{(i,j)}$, where i corresponds to the image and j corresponds to the scale. Simply put, $\mathbf{I}^{\text{detail}}$ maximizes the detail at each scale j by choosing the value of $\mathbf{D}^{(i,j)}$ from image i that maximizes the detail at j. $\mathbf{I}^{\text{base}}$ is the base image, containing the coarsest level of shading information and contains high-level detail of the image. β is a scaling factor used to balance the tradeoff between the emphasis of the detail and the base image. For details on how the decomposition is performed and how individual images are extracted, the reader is directed to (Fattal et al., 2007). Note that even illustrators and photographers unfamiliar with the algorithm can still take advantage of visual shading effects by straightforward manipulation of β. This is a good example of how multiple images provide us with object features that can be easily extracted using numerical, rather than physics-based, models. (Akers et al., 2003) have also demonstrated an interactive easy-to-use application based on a weighted sum of multiple images for improved visual appeal.

Chapter Appendix: Notations

Notation	Description
E	Irradiance
L	Radiance
$\{\Phi_i, \Phi_r\}$	Incident and radiated flux, respectively
(θ_i, θ_r)	Incident and reflected zenith angles
(ϕ_i, ϕ_r)	Incident and reflected azimuth angles
$f(\theta_i, \theta_r, \phi_i, \phi_r)$	Bidirectional reflectance distribution function (BRDF)
$\delta(x)$	Dirac delta distribution/function
ρ_s	Specular albedo
p_x	Shorthand derivative notation of $\frac{\partial p}{\partial x}$
$\hat{\mathbf{n}}$	Surface normals
(p, q)	Gradient space
$R(p, q)$	Reflectance map
$\mathbf{i}$	Measured intensity vector
$\mathbf{L}$	Measurement matrix
$\mathbf{I}$	Image data matrix
$\mathbf{R}$	Surface reflectance matrix
$\mathbf{N}$	Surface normal matrix
$\mathbf{M}$	Light source direction matrix
$\mathbf{T}$	Light source intensity matrix
$\mathbf{S}$	Surface matrix
I_α	Intensity measured by a sensor α
$\epsilon(\lambda)$	Spectral distribution of the illumination source
$Q_\alpha(\lambda)$	Spectral sensitivity of sensor α
$\langle \mathbf{x}, \mathbf{y} \rangle$	Inner product between vectors $\mathbf{x}$ and $\mathbf{y}$

Exercises

The goal of this problem set is to reconstruct the shape of an object by observing it under varied lighting conditions. There are many different methods to reconstruct the shape of an object, including shape from polarization (SfP), stereo vision, and time-of-flight imaging (via depth maps). In this problem set, we explore the fundamentals of photometric stereo, a subset of shape from intensity.

(a) Object of interest.

(b) Computed normal map.

Figure E9.1 By observing an object under different lighting conditions, we can extract the surface normals of the object, which are used as a proxy for local shape.

1. Experimental component

    In this section, we capture the image data needed to perform photometric stereo analysis. First, choose some object in your house for which you would like to obtain the surface normals. You will find the best results with objects that are

    - approximately Lambertian (i.e., no glossy surfaces),
    - convex in shape,
    - homogeneous in material composition.

These constraints are advised because the algorithms we work with in this problem set do not account for specular reflections, interreflections, or spatially varying BRDFs. In addition, we will need

- a specular chrome sphere ball,
- four approximately point light sources (or one movable point source),
- a fixed camera.

a) Optical setup

Place your object in front of a uniform white background. Your four light sources should be placed around the object such that three source vectors are not coplanar, as shown in Figure E9.2. The source vector is the line connecting the origin to the light source, where the origin is defined as the center of the object. To approximate a point source, place your sources far away from the object. This also enables us to approximate a uniform incident lighting direction across all pixels. If you use one movable light source, be sure that you can move the source back and forth among the four locations with high precision. Eliminate all sources of light in the room except for the lighting sources used in the imaging apparatus. Keep your camera in a fixed location such that it is directly aligned to the front of the object. The line connecting the camera and the object will be defined as the z axis, while the x-y plane is orthogonal to it. This experiment assumes an orthographic projection model.

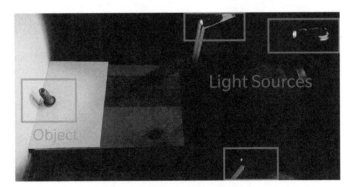

Figure E9.2 An example optical setup for photometric stereo.

b) Capturing images

Place your object in front of the camera. Keep the object and camera in place. Capture the four images, with each image corresponding to the lighting of one of the four sources. Keep track of which lighting direction corresponds to which image.

Exercises

Replace your object with a specular sphere, and repeat the image capture process. You should get a total of eight images that look similar to those in Figure E9.3.

2. Photometric stereo with known lighting directions

   a) Obtaining lighting directions

   We first use the specular sphere images to obtain vector representations of our four lighting directions (which are unknown at the moment). We do so by leveraging the known geometry of the sphere (i.e., known surface normals) and the known geometry of specular light reflection. You should use Python for the remainder of this assignment. Be sure to convert any RGB images to grayscale images.

   i. Detecting specular sphere

   Write a function that detects the edge of the specular sphere within each image. Then, draw a circle outlining the specular sphere. Note that the outline you obtain should be the same for all four images, because the camera is held in place. Keep in mind that the quality of your specular sphere data will affect your function's efficacy. Place the image of an outlined specular sphere in the box shown in Figure E9.5.

   (Hint: Use the Circle Hough Transform to detect the specular sphere.)

   ii. Detecting point of specularity on the sphere

   Assuming a point light source, there should be a small region on the sphere where incident light is directly reflected back to the viewer. Your goal is to write a function that detects this point of specularity. This point should be the pixel with maximum intensity within the sphere. We first blur the image, before attempting to detect bright pixels, so that outlier pixels do not affect your algorithm. Then, we find the brightest and largest region within the sphere, and average the pixel locations within that region to obtain a point (x, y) in the 2D image.

   iii. Calculating the lighting direction

   The point of specularity corresponds to the point at which light mostly behaves according to Snell's reflection law, from the viewer's perspective. The geometry of Snell's law allows us to compute the lighting direction. The normal vector N at a point (x, y, z) on the surface of a sphere is

   $$N = \frac{1}{\sqrt{x^2 + y^2 + z^2}}(x, y, z) \qquad (9.1)$$

Recall that a point (x, y, z) on the surface of a sphere with center $(x', y', 0)$, and radius r is parametrized by

$$(x - x')^2 + (y - y')^2 + z^2 = r^2. \tag{9.2}$$

Because we have already determined the x and y coordinates, we can easily solve for the z coordinate as well. Now, using the law of reflection, we can determine the light source direction L

$$L = 2(N \cdot R)N - R \tag{9.3}$$

where $R = (0, 0, 1)$ is the viewing direction (Debevec, 2008). You should scale your L vector such that its magnitude is the power of the lighting source. If this is unknown, assume unitary power.

b) Obtaining surface normals

Recall that for Lambertian surfaces, we can approximate the intensity of a single pixel as

$$\mathbf{I} = \rho \mathbf{L}^T \hat{\mathbf{n}}, \tag{9.4}$$

where $\mathbf{I} \in \mathbb{R}^4$ is the intensity vector containing the measured intensities from the four sources; $\mathbf{L} \in \mathbb{R}^{3 \times 4}$ is the illumination direction matrix containing the light source directions in its column space; $\hat{\mathbf{n}} \in \mathbb{R}^3$ is the surface normal at that pixel; and ρ is the albedo.

For an entire image containing p pixels, we can generalize this expression

$$\mathbf{I} = \mathbf{L}^T \hat{\mathbf{N}}, \tag{9.5}$$

where $\mathbf{I} \in \mathbb{R}^{4 \times p}$ and $\hat{\mathbf{N}} \in \mathbb{R}^{3 \times p}$. This equation resembles a least-squares problem, so we can obtain the surface normals by using the pseudoinverse of $\mathbf{L}^T$ (Woodham, 1980).

$$\hat{\mathbf{N}} = (\mathbf{L}\mathbf{L}^T)^{-1}\mathbf{L}\mathbf{I} \tag{9.6}$$

Note that the obtained normals have to be scaled appropriately to have unit length. The scaling factor for each normal describes the albedo.

c) Plotting normal map

Because the normals for each pixel have unit length, the values of each x, y, and z component will lie in $[-1, 1]$. Most plotting software accept either floats in range $[0, 1]$ or integers in range $[0, 255]$. Thus, for plotting purposes, we rescale them to be in range $[0, 1]$ by adding one to every coordinate value and dividing by 2. In the plotted RGB image, red corresponds to the $+x$ direction, green corresponds to the $+y$ direction, and blue corresponds to the $+z$ direction.

d) Error analysis

For data that we capture ourselves, we do not have ground truth surface normals for comparison. If we did have ground truth normals, we could determine the mean angular value (MAE) by calculating the pixelwise average of the angles between the reconstructed normals and the ground truth normals (via dot product). Instead, we reconstruct the shading images based on the computed surface normals. We can then compare these reconstructed shading images to the original shading images to determine the mean and median error. To reconstruct a shading image, we use the linear relationship from Lambert's law in Figure 9.5 to solve for the intensity matrix $\widehat{\mathbf{I}}$. Reconstruct all four shading images and determine the mean and median absolute difference between the pixels, across all four images. Be sure to scale the errors by the maximum pixel value (e.g., 1 on a $[0, 1]$ scale or 255 on a $[0, 255]$ scale). Note that the background pixels should be masked out. Place a reconstructed shading image in the box in Figure E9.7. Additionally, include the mean and median errors in the box.

3. Photometric stereo with unknown lighting directions

Now we consider the case in which we do not have the lighting directions readily available. We instead use a matrix factorization approach to decompose our intensity data into surface normals and lighting directions, as predicted by Lambert's law.

Please note the following:

1. The algorithm has a large time complexity that scales up with the resolution of the image.
2. The algorithm requires a prior knowledge of the surface normal of three points on the image. Specifically, each of these three points should have normals approximating $(1, 0, 0)$, $(0, 1, 0)$, and $(0, 0, 1)$. Ensure that you choose an object that allows you to provide these coordinates.

Provided that you have taken these factors into account, you can reuse the same images from the previous section (except that we will not use the specular sphere images).

a) Factorization of intensity matrix

Recall that we can express the intensity matrix $\mathbf{I} \in \mathbb{R}^{p \times f}$ of a scene with p pixels and f frames ($f = 4$ in our case)

$$\mathbf{I} = \mathbf{RNMT}, \tag{9.7}$$

where $\mathbf{R} \in \mathbb{R}^{p \times p}$ is the surface reflectance matrix; $\mathbf{N} \in \mathbb{R}^{p \times 3}$ is the surface normal matrix; $\mathbf{M} \in \mathbb{R}^{3 \times f}$ is the light source direction matrix; and $\mathbf{T} \in \mathbb{R}^{f \times f}$ is the light intensity matrix. This can also be expressed as a simple factorization

$$\mathbf{I} = \mathbf{SL}, \tag{9.8}$$

where $\mathbf{S} = \mathbf{RN}$ is the surface matrix and $\mathbf{L} = \mathbf{MT}$ is the light source matrix (Hayakawa, 1994). Our simplified objective, therefore, is to find the correct factorization of $\mathbf{I}$.

i. Singular value decomposition

Singular value decomposition (SVD) is particularly useful because it organizes the intensity data into its singular vectors and corresponding singular values, which we exploit here. First, flatten each image into vectors and create $\mathbf{I}$ by concatenating these vectors columnwise. We now decompose $\mathbf{I}$ using SVD. Python has libraries to help you extract the SVD of a matrix, such that

$$\mathbf{I} = \mathbf{U\Sigma V} \tag{9.9}$$

where $\mathbf{U} \in \mathbb{R}^{p \times p}$, $\mathbf{\Sigma} \in \mathbb{R}^{p \times f}$, and $\mathbf{V} \in \mathbb{R}^{f \times f}$. Note that calculating the SVD can be computationally expensive. You may need to downsample your images so that the algorithm executes in a reasonable time.

ii. Denoising the image

We inherently assume that there is some noise in our image data. The singular values we obtain from SVD are a measure of how much of the total variance in the data is captured by the corresponding singular vectors. We assume that the first three singular vectors are all tied to the image formation process, and that all subsequent singular vectors are the result of noise introduced into the measurement process. Therefore, we keep the singular vectors that have the three greatest singular values and drop the remaining vectors. We now approximate our true intensities as

$$\widehat{\mathbf{I}} = \mathbf{U'\Sigma'V'} = \widehat{\mathbf{SL}} \tag{9.10}$$

where $\widehat{\mathbf{S}} = \mathbf{U'}(\pm[\mathbf{\Sigma'}]^{1/2})$, $\widehat{\mathbf{L}} = (\pm[\mathbf{\Sigma'}]^{1/2})\mathbf{V'}$, $\mathbf{U'}$ contains the first three left singular vectors, $\mathbf{\Sigma'}$ contains the three highest singular values along its diagonal; and $\mathbf{V'}$ contains the first three right singular vectors. Note, however, that this approximation is valid only when the ratio between the third and fourth largest singular value is large enough (Hayakawa, 1994).

b) Constraining the factorization

The factorization we obtained in the previous section, however, is not unique. We can easily see this fact if we consider any arbitrary invertible matrix **A**.

$$\widehat{\mathbf{I}} = (\widehat{\mathbf{S}}\mathbf{A})(\mathbf{A}^{-1}\widehat{\mathbf{L}}) \tag{9.11}$$

Therefore, we must find the **A** matrix such that

$$\mathbf{S} = \widehat{\mathbf{S}}\mathbf{A}, \quad \mathbf{L} = \mathbf{A}^{-1}\widehat{\mathbf{L}} \tag{9.12}$$

We refer to the incorrect normals obtained in $\widehat{\mathbf{S}}$ as *pseudonormals*. To constrain the **A** matrix, we must find at least $p' = 6$ points on the image with constant surface reflectance. Assuming the material is homogeneous, we can manually select points that have relatively different surface normal orientations. For each of the corresponding p' pixels, we obtain these pseudonormals $\widehat{s}_k$ from the row space of $\widehat{\mathbf{S}}$ and constrain **A** such that

$$\widehat{s}_k^T \mathbf{A}\mathbf{A}^T \widehat{s}_k = \mathbf{r}, \quad k = 1, ..., p' \tag{9.13}$$

where r is the relative surface reflectance. If the exact value of the reflectance is unknown, we can assume $r = 1$. We can first solve for $\mathbf{B} = \mathbf{A}\mathbf{A}^T$ via a simple linear system of equations. Note that because **B** is a symmetric matrix, there are only six unknown entries in the matrix, which is why $p' \geq 6$. Once we have obtained **B**, we can solve for **A** by computing the SVD of **B**. Because **B** is symmetric, its SVD is given by $\mathbf{B} = \mathbf{W}\Pi\mathbf{W}^T$. Therefore, we can determine that

$$\mathbf{A} = \mathbf{W}[\Pi]^{1/2} \tag{9.14}$$

From **A**, we can determine **S** based on the algorithm shown in Figure 9.12 (Hayakawa, 1994). Be sure to normalize the length of the normal vectors.

c) Transforming normals to viewer-centered coordinate system

The surface normals we obtained in the previous subsection are oriented in an arbitrary coordinate system. We want the normals to be oriented relative to the camera's position. Manually find three points in the scene whose surface normals are oriented along the positive x, y, and z directions. Construct the following change of basis matrix

$$B = \begin{bmatrix} | & | & | \\ v_1 & v_2 & v_3 \\ | & | & | \end{bmatrix}^{-1}$$

where v_1, v_2, and v_3 are the corresponding computed surface normal vectors at those three points. We can determine the correct orientation of all points in the scene by applying this linear transformation (i.e., change of basis) to every normal in the image.

d) Normal map and error analysis

Plot the normal map and a sample reconstructed shading image. Also, calculate the error as you did in the previous section. Comment on the difference in performance between the two methods for photometric stereo (with known and unknown lighting), and potential reasons for the difference in performance.

Exercises

Figure E9.3 Insert an object image and its corresponding specular sphere image.

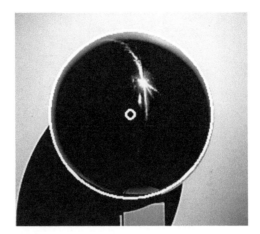

Figure E9.4 Insert your segmented specular sphere image.

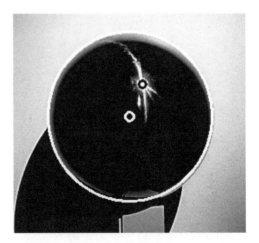

Figure E9.5 Label the point of specularity on your sphere image.

Exercises

Figure E9.6 Insert the normal map of the original object.

Mean normalized error = 0.01041, Median normalized error = 0.002481.

Figure E9.7 Insert the reconstructed shading image and the error statistics.

Exercises 355

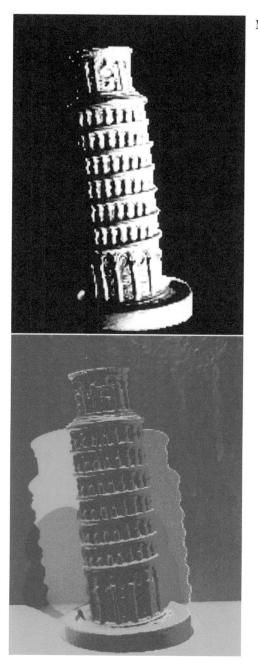

Mean Normalized Error = 0.1241, Median Normalized Error = 0.07291.

Figure E9.8 Insert the normal map and reconstructed shading image obtained with unknown lighting conditions.

10 Light Transport

In this chapter, we discuss light transport, an analytical framework that aims to describe the interaction between light and matter. Although light transport may seem to have a similar scope as the field of optics, there are distinctions. In particular, light transport involves the use of simplified representations of light (e.g, ray-based when possible, instead of wave-based) for the benefit of computational tractability. For this reason, researchers in vision and graphics sometimes add the additional adjective *computational*, referring to computational light transport (CLT).

10.1 Motivation

10.1.1 Curse of Dimensionality

Light transport is a simplification tool for the complexity of light. In an ordinary scene, there are trillions of light paths interacting with the scene. Imagine a scene that has been subdivided into three patches of space, each consisting of 100×100 scene points. From the lens of geometry alone, there are 1 trillion possible light paths that connect each pixel ($100^2 \times 3$). From the lens of color, polarization, and transient variation, there would be even more light paths. Unfortunately, an imaging system is tasked to sample this space with a camera that operates in megapixels (MP). This introduces a curse of dimensionality: trillions of light paths, sensed only by millions of pixels. To address this curse, light transport makes two simplifications. The *first simplification* is to reduce the optical complexity of the scene to the minimum amount of dimensionality that we feel is necessary to address an applied problem. For example, if trying to see around corners in gray scale, we might not need to consider color variations. The *second simplification* is to reduce high-dimensional light paths into a representation space that maps to the millions of pixels in imaging hardware, such as projectors, cameras, or other pieces of computational imaging hardware. In this way, light transport forms a bridge between unobtainable information (trillions of light paths) and obtainable measurements (cameras or projectors).

10.1.2 Light Transport Addresses Curse of Dimensionality

To overcome the high-dimensional nature of light transport, it is possible to relax the problem. Let us now dive deeper into the two simplifying reductions (from 10.1.1), which form the core of research advances in light transport. The first simplification was to use the minimum amount of optical degrees of freedom as needed. As we have seen in chapter 6, the 7D plenoptic function under certain assumptions and relaxations simplifies to a 4D light field, which is easier to sample and operate on. The second simplification is identifying a representation space for light paths. section 10.2.2 lays the foundation for this idea by decomposing light transport of a scene into a sum of different interreflection components, but this decomposition is not realizable in practice. In section 10.3 we discuss approaches that relax light transport in different ways. We start with a binary approach in 10.3.1 that separates light transport into either *global* or *direct* light paths under a smoothness assumption in the frequency domain. section 10.3.2 discusses finer separation of light transport, by separating the global component into near and far global components. In section 10.3.2 we also discuss how combinations of light transport and optical techniques such as interferometry can be used to achieve even finer decomposition of light transport. These techniques are useful in addressing various imaging applications like descattering, skin imaging, and time-of-flight imaging.

10.1.3 Forward vs. Inverse Light Transport

Light transport involves capturing the image of scene (I) illuminated by a source (S) using a sensor (P). For simplicity P can be thought of as a camera sensor, and we do not need to factor in the lens. There are two broad forms of light transport:

- **Forward light transport** Forward light transport addresses how a given scene appears under certain illumination conditions. Rendering approaches used in the field of computer graphics can be thought of as forward light transport. Ray tracing is a popular graphics technique for rendering scenes that models image formation by tracing the path of light from the illumination source to the sensor pixels, and simulating the interaction of the light ray with the objects in the scene. section 10.2.1 examines the forward light transport approach, but instead of using ray tracing we model image formation using a *light transport matrix*.

- **Inverse light transport** Most of the work in *imaging* addresses the problem of inverse light transport. Given a photograph of I acquired on the sensor P, inverse light transport aims to decompose it into multiple components, where each component records the contributions from certain groups of light paths in the scene. The second simplification in section 10.1.2 corresponds to the components we expect to recover as a solution of the inverse light transport.

10.1.4 Chapter Organization

The chapter is organized as follows. We begin with a discussion of forward light transport in section 10.2.1 by introducing the light transport matrix and associated concepts of superposition and Helmholtz reciprocity. section 10.2.2 describes inverse light transport and its relaxations are discussed in section 10.3. In section 10.4 we have a detailed look at one of the most popular problems in computational imaging: non-line-of-sight imaging. We conclude the chapter in section 10.5, summarizing additional real-world applications of light transport.

10.2 Light Transport Matrix

The light transport matrix describes the relationship between lighting, scene, and sensing. Here, we offer two views of the light transport matrix, from a forward (10.2.1) and inverse (10.2.2) perspective.

10.2.1 Light Transport Matrix: Forward Perspective

The light transport matrix, $\mathbf{T}$, is a part of the light transport equation: $\mathbf{p} = \mathbf{T}\mathbf{l}$ (where $\mathbf{p}$ is an $n \times 1$ vector of irradiance measurements, also known as camera pixels, and $\mathbf{l}$ is an $m \times 1$ vector of independent source/illumination pixels). It is an $n \times m$ matrix, with n being the irradiance measurements (e.g., camera pixels) and m being the independent illumination degrees of freedom (e.g., scene pixels). In modeling the propagation of light between a projector and a camera, the light transport matrix holds crucial information about the scene being illuminated. Using the light transport matrix, we can relight a scene to create novel pictures through the manipulation of various bounces of light. The transport matrix adheres to the superposition principle in that it allows for the creation of images under different lighting conditions. Mathematically, if the first and second light sources are

$$\mathbf{p}_1 = \mathbf{T}\mathbf{l}_1, \quad \mathbf{p}_2 = \mathbf{T}\mathbf{l}_2,$$

respectively, then their summation allows for new images due to the ability to manipulate lighting conditions. It is clear that a sum of the two equations would lead to a scene in which the illumination is the superposition of the two original illuminations:

$$\mathbf{p}_1 + \mathbf{p}_2 = \mathbf{T}(\mathbf{l}_1 + \mathbf{l}_2)$$

Another important application involves using the transpose of the transport matrix to capture two different perspectives using a fixed camera and projector. Suppose we had a scene with a playing card in it as shown in Figure 10.1, with the projector facing the front of the card and the camera facing the back, along with an object for light to reflect off (in this example, a book).

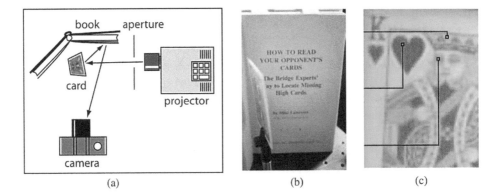

Figure 10.1 Dual photography leverages the light transport matrix and Helmholtz reciprocity to swap camera and projector viewpoints. (a) The setup, with the projector viewing the card's face and the camera viewing its back. (b) Live photo of the setup. (c) The image produced using dual photography. Reprinted from (Sen et al., 2005).

It is possible to image the card such that its front is visible, even though the camera is facing its back. The *Helmholtz reciprocity principle* enables this, stating that the light transport will be the same regardless of the flow of light. It relies foundationally on the conservation of energy. Because by this principle the same light is measured whether it starts at the projector or the camera, the transport matrix T can be transposed in order to produce a *dual image* as per the equation $\mathbf{p} = \mathbf{T}^\top \mathbf{l}$ (Sen et al., 2005).

The dual image, as shown in Figure 10.2, is synthesized and shows the projector's point of view, with illumination as if it were coming from the position of the camera. Through this concept, referred to as *dual photography*, the viewpoints of the camera and projector can be swapped. In summary, by the light transport equation, the picture at the camera (the primal image) is given by $\mathbf{l} = \mathbf{T}\mathbf{p}$, and the picture at the projector (dual image) is given by $\mathbf{p} = \mathbf{T}^\top \mathbf{l}$ (as shown in Figure 10.3).

Although swapping the projector and camera viewpoints to see the playing card is an incredible result, it requires that the projector be positioned in the line of sight to view the front of the playing card. Later in this chapter, we will discuss methods that can see around corners without a gadget in the line of sight.

10.2.2 Light Transport Matrix: Inverse Perspective

From this section onward, we focus our discussion on inverse light transport. Whereas forward light transport focuses on the propagation of light through a known scene, inverse light transport aims to infer the path of light in an unknown environment. The path of light can be parametrized by bounces of light. Concretely, an ordinary image **I** of a scene can

10.2 Light Transport Matrix

(a)   (b)

Figure 10.2 Example of dual photography (a) The primal image. Lighting is from the perspective of the projector, and the photo has a resolution equal to that of the camera. (b) The dual image. Lighting is from the perspective of the camera, and the photo has a resolution equal to that of the projector. Reprinted from (Sen et al., 2005).

be decomposed into an infinite summation of images $\mathbf{I} = \mathbf{I}_1 + \mathbf{I}_2 + \cdots$ where $\mathbf{I}_i$ is the image containing i^{th}-order interreflections.

Consider a setup with a scene, light source, and camera. Consider any $\mathbf{x}'$, $\mathbf{x}$, and $\mathbf{y}$ to be any arbitrary set of points on the source, scene, and camera sensor, respectively. Let $\omega_{\mathbf{x}'}^{\mathbf{x}}$ denote rays originating from $\mathbf{x}'$ and directed to $\mathbf{x}$. Let $\mathbf{L}_{\text{in}}\left(\omega_{\mathbf{x}'}^{\mathbf{x}}\right)$ denote the radiance as a function of all incident light rays $\omega_{\mathbf{x}'}^{\mathbf{x}}$, where $\mathbf{x}'$ is the collection of source points that illuminates the surface points represented by $\mathbf{x}$. Similarly $\mathbf{L}_{\text{out}}\left(\omega_{\mathbf{x}}^{\mathbf{y}}\right)$ represents radiance as a function of all outgoing light rays $\omega_{\mathbf{x}}^{\mathbf{y}}$. Because the outgoing light field, $\mathbf{L}_{\text{out}}\left(\omega_{\mathbf{x}}^{\mathbf{y}}\right)$ is partially composed of light that has been reflected by other surface points before reaching $\mathbf{x}$, we can decompose this light field to

$$\mathbf{L}_{\text{out}}\left(\omega_{\mathbf{x}}^{\mathbf{y}}\right) = \mathbf{L}_{\text{out}}^{1}\left(\omega_{\mathbf{x}}^{\mathbf{y}}\right) + \mathbf{L}_{\text{out}}^{2,3,\cdots}\left(\omega_{\mathbf{x}}^{\mathbf{y}}\right),$$

where the direct component, $\mathbf{L}_{\text{out}}^{1}\left(\omega_{\mathbf{x}}^{\mathbf{y}}\right)$, is determined by how the surface material reflects light off of the points $\mathbf{x}$ on the surface (e.g., bidirectional reflectance distribution function [BRDF]); and the indirect component, $\mathbf{L}_{\text{out}}^{2,3,\cdots}\left(\omega_{\mathbf{x}}^{\mathbf{y}}\right)$ is the contribution of all the interreflections that strike $\mathbf{x}$ after hitting some number of surface points (Seitz et al., 2005). For k

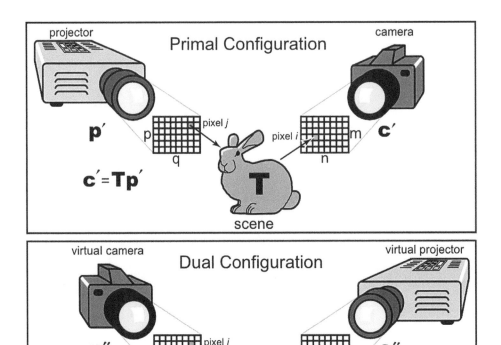

Figure 10.3 Primal and dual image matrices. The top diagram illustrates the primal setup in which light is emitted from the camera and captured by the projector. Helmholtz reciprocity, a consequence of conservation of energy, suggests that we can reverse this operation. For example, assume a ray from a projector pixel strikes the scene and is captured by a set of camera pixels. If those camera pixels were instead virtual projector pixels, the same amount of light would hit the scene and reach that single projector pixel (now a virtual camera). As illustrated in the bottom diagram, we can mathematically swap the location of the projector and the camera, in order to find out what the virtual camera would be capturing if it was in the projector's place. Reprinted from (Sen et al., 2005).

discretized surface points, we can rewrite this equation

$$\mathbf{L}_{out} = \mathbf{L}_{out}^1 + \mathbf{A}\mathbf{L}_{out}$$

where $\mathbf{A} \in \mathbb{R}^{k \times k}$ is the matrix that characterizes the proportion of irradiance (defined in chapter 2) from $\mathbf{x}'$ to $\mathbf{x}$ that is radiated toward $\mathbf{y}$. Rearranging this equation gives

$$\mathbf{L}_{\text{out}} = \left(\mathbf{C}^1\right)^{-1} \mathbf{L}_{\text{out}}^1,$$

which is indicative of relaxing the full light transport concept to an interreflection cancellation operator $\mathbf{C}^1 = (\mathbf{E} - \mathbf{A})$, where $\mathbf{E}$ is an identity matrix, that simply maps a direct illumination light field to a general light field (with direct and indirect components) (Seitz et al., 2005). $\mathbf{C} \in \mathbb{R}^{k \times k}$ is defined in the context of shape and reflectance properties (e.g., BRDF). This image formation process can be modeled by introducing the light transport matrix $\mathbf{T} \in \mathbb{R}^{k \times k}$, which can be measured as described in section 10.3.2. As a consequence of $\mathbf{T}$, we can say that

$$\mathbf{L}_{\text{out}}^1 = \mathbf{T}^1 \mathbf{L}_{\text{in}},$$

where $\mathbf{T}^1$ contains the components of $\mathbf{T}$ that are due to 1-bounce reflections (Seitz et al., 2005). Given this context, we can now extend our analysis to n-bounce reflections. Using the principles derived earlier, we can generalize

$$\mathbf{L}_{\text{out}}^n = \mathbf{C}^n \mathbf{L}_{\text{out}}.$$

Using Lambertian approximation for the surface, the cancellation operator can be expressed using $\mathbf{T}$

$$\mathbf{C} = \mathbf{T}^1 \mathbf{T}^{-1}$$

where $\mathbf{T}^1 \in \mathbb{R}^{k \times k}$ is the diagonal matrix containing the reciprocal diagonal elements of $\mathbf{T}^{-1}$ (Seitz et al., 2005). Thus, we practically decompose $\mathbf{L}_{\text{out}}$ into its component interreflections.

Ultimately, the existence of the cancellation operator shows that it is possible to compute different n-bounce reflections of a scene. However, capturing the light transport matrix is not efficient, which in turn makes the process of calculating the cancellation operator hard to implement. In the following sections, we show how to alleviate this problem by relaxing the inverse light transport problem allowing us to probe and manipulate light transport using the principles of light transport matrix, without having to acquire the full matrix. In the process we discover interesting relations between the scene constituents and the nature of the light transport matrix.

10.3 Relaxations of Inverse Light Transport

In the last section, our discussion of inverse light transport was conducted in the context of an infinite sum of n-order light bounces. When building a practical imaging system, it is nearly impossible to measure the number of bounces of a light path. In this section, we discuss two specific relaxations of light transport: global-direct separation in 10.3.1 and optical probing in 10.3.2.

(a) Scene   (b) Direct Component   (c) Global Component

Figure 10.4 Separation of global and direct for a complex scene. (a) Original image of a scene with many optically complex objects. (b) Decomposed direct illumination image, scaled up by a factor of 1.25. (c) Global illumination image that includes diffuse and specular interreflections (wall wedge and nut), volumetric scattering (milky water), subsurface scattering (marble), translucency (frosted glass), and shadow (fruit on board). Reprinted from (Nayar et al., 2006).

10.3.1 Global and Direct Separation

The light in an illuminated scene consists of two components: direct and global illumination. The *direct component* provides information with regard to how the material and local geometric properties of a scene interact with the light source and camera. The *global component* reveals the complex optical interactions within a scene, specifically between different objects and media. It also models interreflections and subsurface scattering, which indirectly illuminate a scene. The direct component is the light paths that reach the camera, from an object that has been illuminated directly by the source. In the framework of 10.2.2, it can be seen as a 1-bounce light path: from light source, to object, to camera. In contrast, the global component is formed from light paths that reach the camera from an object that has been illuminated indirectly from the source. Here, indirect means that the object is not illuminated directly by the light source, but by light that has bounced off other objects in the scene. In the framework of 10.2.2, the global component refers to the summation of all n-bounce light paths, where n is greater than 1. To illustrate the contrast between global and direct components, let us examine convex and concave shapes. Light will bounce directly off a convex object, whereas it may reflect within a concave object, hitting the inner walls before traveling to the camera.

Figure 10.4 depicts the separation of direct and global components in a real scene (Nayar et al., 2006). The authors have chosen this particular image for its remarkable effects of light transport. The V-groove where the two walls meet follows the logic just described about convex and concave objects. The marble appears bright in the global image due to subsurface scattering, which has a high global component. The surface of the curtain appears dark in the global image except for the fringe, which exhibits subsurface scattering.

10.3 Relaxations of Inverse Light Transport

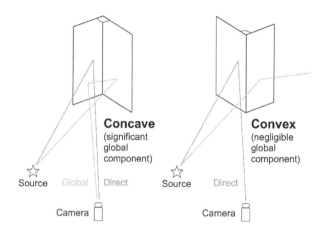

Figure 10.5 Direct-global decomposition of concave and convex surfaces. Concave surfaces are curved inward, and convex structures are curved outward.

The milky water appears dark in the direct image, but bright in the global image due to the scattering of milk.

Another example that illustrates direct and global components is shown in Figure 10.5. Here, we examine both the concave and the convex geometry of the V-groove. For the direct component of both of these different shapes, we obtain the same 2D image because the direct image takes into account only the light that travels directly back from the V-groove and not the interreflections. In contrast, whereas the concave geometry exhibits a global component because of the scattering of light, the convex geometry has a flat global image.

In separating the original image into the direct and global components, we gain the knowledge required to perform image manipulation and reveal more information about a scene. This makes direct-global separation a useful technique that has a lot of exciting applications such as novel scene synthesis, mitigating multipath interference in time-of-flight imaging, skin imaging, and many more. Refer to section 10.5 for more details on the applications of direct-global separation.

One can theoretically measure global light by illuminating one point of the scene at a time and capturing an image to determine the contribution of this point to all other points (Seitz et al., 2005). However, this method is too expensive to have practical applications.

However, through the use of high-frequency illumination, it is theoretically possible to separate direct and global light using just two images (Nayar et al., 2006). This approach requires certain assumptions. Each point in the scene must be illuminated by at most one source element. In other words, only a single light source is used. Additionally, the global contribution of each scene point is assumed to be a smooth function with respect to frequency of the lighting. In terms of equipment, the camera is assumed to have infinite resolution, and the projector is idealized to create patterns devoid of light leakage. With a practical camera of finite resolution and a projector with leakage and defocusing, around twenty-five images are required to produce the desired separation.

In a high-frequency binary illumination pattern (alternating lit and unlit patches along the scene's surface) the lit patches contain both direct and global light whereas the assumption is that patches unlit by the light source consist of only global light. This *smoothness assumption* in light transport makes the separation feasible, as patches may be subtracted to achieve separation (recall that $\mathbf{L} = \mathbf{L}_d + \mathbf{L}_g$). Dividing the surface into N sections, with M of the N sections directly corresponding to a pixel of the source, we can define $\mathbf{L}[c, i]$ as radiance of a patch i, measured by a camera c. Defining $\mathbf{A}[i, j]$ as the reflectance distribution over the set of patches $i, j \in \mathsf{P}$ (where P is the set of patches in the scene), we derive (Nayar et al., 2006),

$$\mathbf{L}_g[c, i] = \sum_{j \in P} \mathbf{A}[i, j]\, \mathbf{L}[i, j],$$

$\mathbf{L}_g[c, i]$ can be further decomposed into $\mathbf{L}_{gd}[c, i]$ and $\mathbf{L}_{gg}[c, i]$. $\mathbf{L}_{gd}$ represents the direct component of radiation from scene patches (i.e., applicable to the light travel scenario: source → other patch → patch of interest). $\mathbf{L}_{gg}$ represents the global component of radiation from scene patches (i.e., applicable to the scenario: source → occurrence(s) → other patch → patch of interest). In a pattern where only a fraction of the source's patches are lit (a checkerboard pattern), with a good distribution for high frequency, α represents this fraction. Let $\mathbf{L}^+$ be the image of the scene lit with high-frequency illumination with a fraction of the activated source pixels α and $\mathbf{L}^-$ be the image of the scene lit with a complementary illumination pattern, that is with a fraction of the activated source pixels $1 - \alpha$. We define a new $\mathbf{L}^+_{gd}$

$$\mathbf{L}^+_{gd}[c, i] = \alpha \mathbf{L}_{gd}[c, i],$$

because only the lit patches have a direct component and therefore contribute. Likewise,

$$\mathbf{L}^+_{gg}[c, i] = \alpha \mathbf{L}_{gg}[c, i].$$

The combination of $\mathbf{L}^+_{gd}$ and $\mathbf{L}^+_{gg}$ as well as $\mathbf{L}_d$ results in the first image, which is represented by

$$\mathbf{L}^+[c, i] = \mathbf{L}_d[c, i] + \alpha \mathbf{L}_g[c, i].$$

We represent the complementary illumination with $\mathbf{L}^-$. It has $(1 - \alpha)$ activated pixels. It forms the second image and is written as

$$\mathbf{L}^-[c, i] = (1 - \alpha)\, \mathbf{L}_g[c, i].$$

These equations hold under the assumption that the deactivated source pixel does not generate any light. If this assumption is false, we can assume the brightness of the

10.3 Relaxations of Inverse Light Transport

deactivated source to be a fraction of the activated element. We denote this fraction by b, where $0 \leqslant b \leqslant 1$. Taking b into account changes these equations as follows:

$$\mathbf{L}^+ [c,i] = \mathbf{L}_d [c,i] + \alpha \mathbf{L}_g [c,i] + b(1-\alpha) \mathbf{L}_g [c,i], \tag{10.1}$$

$$\mathbf{L}^- [c,i] = b\mathbf{L}_d [c,i] + \alpha b \mathbf{L}_g [c,i] + (1-\alpha) \mathbf{L}_g [c,i]. \tag{10.2}$$

If α is close to either 0 or 1, the scene would be very dimly lit in one of the illumination conditions. Choosing $\alpha = 0.5$ is hence a justified choice as it maximizes the sampling frequency of illumination in both the images. Hence if α and b are known, separation can be done using two images. In practice however, it is difficult to obtain ideal complementary patterns, as the lit and unlit regions may have brightness variations. Limited depth of field of the projector also causes defocusing of certain scene regions.

We now describe how one can separate an image into its direct and global components by solving the two equations. First, we estimate the value of b. To do so, we project a white pattern on the screen and capture its image. Similarly we obtain an image of the black pattern. The white pattern corresponds to $\alpha = 1$, and the black pattern corresponds to $\alpha = 0$.

We can then estimate b for each pixel by dividing the black pattern image and the white pattern image. We call this matrix $\mathbf{b}_{\text{mat}}$ as it stores the value of b for each pixel. For the illumination pattern, we use a checkerboard pattern with squares of size 8×8 pixels. The checkerboard pattern ($\alpha = 0.5$) is shifted 5 times (by 3 pixels each time) along the axes to obtain a total of twenty-six images. Using this data, the process of extraction of the global and direct components from the scene shown in Figure 10.4a can be expressed as follows using a MATLAB style pseudocode:

```
1  %% For a given scene:
2
3  % Estimating b
4  bmat = Black ./ White;
5
6  % Rescaling b from 0 to 1
7  bmat = bmat ./ max(bmat(:));
8
9  % Initializing Lplus to be a zero matrix
10 Lplus = zeros(size(White));
11
12 % Initializing Lminus to be Inf
13 Lminus = Inf * ones(size(White));
14
15 for i = 1:26
16    % img = image captured with ith checkerboard pattern
17    Lplus = max(img, Lplus);
18    Lminus = min(Lminus, img);
19 end
```

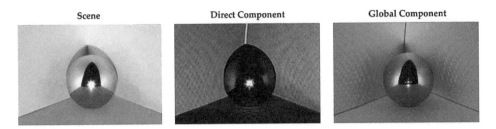

Figure 10.6 Failed direct-global decomposition. Failed separation due to the violation of the smooth global function assumption, when the checkerboard pattern is shifted. The highly specular reflections cause residual checkerboard patterns in each component. Reprinted from (Nayar et al., 2006).

```
20
21  % Direct Component
22  Ld = (Lplus - Lminus) .* (1 ./ (1 - bmat));
23
24  % Global Component
25  Lg = 2 * (bmat .* Lplus - Lminus) ./ ...
26                      ((bmat - 1) .* (bmat+1));
```

The twenty-six images obtained by shifting the checkerboard pattern help in accurately estimating L^+ and L^-. Using these obtained estimates, the last two steps of the pseudocode obtain the direct and global components Figure 10.4b–c by solving equations (10.1) and (10.2). The theory of direct and global separation discussed here is also applicable to other high frequency illumination patterns such as sine waves, and under certain assumptions can achieve direct-global separation from a single image as well. For sources like the Sun, which cannot be made to generate high frequencies, occluders like moving shadows in a scene can be used. If the shadow is thin enough, such as in the case of a line occluder, global-direct separation is possible using the method discussed here by approximating $\alpha = 1$ and $b = 0$. In practice, however, moving a line occluder throughout the entire scene is time consuming, therefore mesh occluders are more commonly used (Nayar et al., 2006).

We now consider a scenario in which global-direct separation fails. Although the high-frequency binary illumination pattern works very well for many cases, attempting to perform separation for a scene that contains mirrors violates the assumption that the global function is smooth in comparison with the illumination frequency due to the high-frequency off-diagonal structures that mirror reflections produce. This results in failed direct and global images containing residual checker patterns, as shown in Figure 10.6.

In the next section, we introduce a method that allows us to perform global-direct separation by optically probing the light transport matrix while capturing the scene. This approach also succeeds for cases such as that shown in Figure 10.6, which violate the smoothness assumption of the global component.

10.3 Relaxations of Inverse Light Transport

10.3.2 Optical Probing of the Light Transport Matrix

Motivations: Acquiring a light transport matrix. So far in this chapter, we have focused on forming images of different bounce orders of light. Although the light transport matrix is an important tool in this discussion, we skirted around acquiring the light transport matrix. Now, we assume that we do not know the light transport matrix and seek to capture it.

There are, of course, ways of naively capturing the light transport matrix; for example, (Sen et al., 2005) discussed one such method in their paper on dual photography, which used a projector and a light sensor as the setup. Here, the transport matrix is captured by using the projector to display a variety of patterns with one element lit up at a time. Of course, this is very time inefficient because it requires so many different measurements to acquire all components of the transport matrix. For example, it took ninety minutes to complete the brute-force pixel scan for a 3×3 pixel pattern.

There are other techniques to address this shortcoming, such as *fixed pattern scanning*. Fixed pattern scanning allows multiple elements to be lit at the same time. However, it fails to capture global illumination effects, which are crucial to understanding how an image is seen. Adaptive multiplexed illumination fixes this problem by illuminating multiple pixels at once, but controlling which pixels are illuminated so that each camera pixel is not affected by more than one projector pixel simultaneously.

To find which pixels cannot be illuminated at the same time, we start with one large block that contains all projector pixels. This block is then subdivided whenever conflicts between two blocks are found. This iterative procedure continues until each block represents a pixel. Whereas the brute-force method took ninety minutes for a 3×3 pattern, adaptive multiplexed illumination enables capture of the **T** matrix in just over two hours for a 578×680 pixel pattern.

However, when the scene is dominated by diffuse interreflections or subsurface scattering, adaptive multiplexed illumination may degenerate into the time-costly, brute-force pixel-by-pixel method. Also, it lends itself to incorrectly culled blocks: if a block happens to contribute a small amount of energy to a certain scene, and this amount of energy is below the noise threshold contributed by the other blocks, it may be unfairly culled and its energy lost. This becomes an issue especially in capturing diffuse-diffuse interreflections. To fix this problem, we can use a hierarchical approach in which the energies are recorded at the last possible level at which they can be measured, ensuring that we do not accidentally lose a block to noise.

Even a compressive sensing approach to capturing the transport matrix, although it improves the efficiency of the actual capture, takes an inordinate amount of time to postprocess the image (Peers et al., 2009). For a 512×512 pixel scene, this method yields a postprocessing time of fifteen hours. The general trade-off is that one must either dedicate a huge amount of effort to taking all the images under many different illuminations as necessary or minimize the effort of capturing the images at the cost of a huge postprocessing time.

Clearly, we cannot always easily and efficiently capture the transport matrix. However, *optical computing* implements numerical algorithms directly in optics, which allows replacing matrix products in a numerical algorithm with capturing a scene under a certain illumination. For example, suppose that we want to find the eigenvector of a certain transport matrix without figuring out the matrix itself. In this case, a numerical algorithm called power iteration can be used optically to find the eigenvector. The numerical domain power iteration operation $\mathbf{p}_i = \mathbf{T}\mathbf{l}_i$ can be implemented by project and capture operations in the optical domain. This is an iterative process that will converge to the principal eigenvector of the scene. Apart from this simple example, there are many other iterative algorithms, such as *Arnoldi iteration* and *generalized minimal residual*, that can be used to bypass the need for the exact knowledge of the transport matrix (O'Toole and Kutulakos, 2010).

Primal-dual coding. We have already seen how to use the transport matrix to analyze and manipulate photos to our advantage, using the light transport equation $\mathbf{p} = \mathbf{T}\mathbf{l}$. However, we introduce another matrix, known as the probing matrix $\mathbf{\Pi}$, which allows us to develop a generalized imaging method for active illumination, operating entirely in the optical domain. The light transport equation can then be modified using the probing matrix, yielding the *transport probing equation*:

$$\mathbf{p} = (\mathbf{\Pi} \odot \mathbf{T})\, \mathbb{1},$$

where the symbol $\odot$ denotes the elementwise multiplication of two matrices, and $\mathbb{1}$ denotes a vector of ones. The light transport equation contains only the degrees of freedom equal to the number of elements in the illumination vector, whereas the transport probing equation yields a much greater degree of freedom—equal to the number of elements in the probing matrix.

The primary use of the probing matrix is in *optical probing*, more specifically in a process that we call *primal-dual coding*. This technique is closely related to many methods in microscopy that are used to enhance microscope performances by eliminating out-of-focus light.

Primal-dual coding allows simultaneous control over two different aspects of image formation: first, the scene's illumination, which we consider the primal domain; and second, the modulation of the light coming into the camera, which we consider the dual domain. In physical terms, we project a certain pattern onto the scene we are photographing (practically, this can be done with a projector); then we use a secondary pattern which we insert between the scene and the camera, which modulates the light as it enters the camera's sensor (which can practically be done using an LCD mask, among other methods). In a sense, primal-dual coding consists of a combination of illumination coding and coded-exposure photography, but it is performed completely externally without any need to delve inside the camera.

10.3 Relaxations of Inverse Light Transport

We can also study primal-dual coding through the lens of the probing matrix and the transport probing equation. In the regular light transport equation (see section 10.2.1), we acquire information of the transport matrix **T** of size $L \times P$ through the illumination vector **i** of size $L \times 1$. However, with the transport probing equation, we can use the probing matrix Π of size $L \times P$ to control how **T** is mapped onto **i**, which gives us more control over how the resulting image is formed.

Figure 10.7 shows a summary of some of the possible transport probing equations that can be realized using structured probing matrices.

There are two main optical algorithms for using the probing matrix in practice. The first is known as *path isolation*, and the second is the *optical probing matrix*. We will begin by discussing path isolation. It is somewhat simpler conceptually to discuss as a starting point.

The idea of path isolation is to turn on one *single projector pixel* at a time while unmasking only a *single camera pixel* on the camera sensor. With only one pixel active at either end, we thus have an absolute guarantee that the only light paths contributed to the photo will be between those two points, because scene points are nonilluminated (on the projector's end) and photons are masked (on the camera sensor's end). To capture the full photograph, the naive path-isolation algorithm uses a time-multiplexing approach, in which we look at a certain time slice τ within the exposure period. During this time slice τ, every possible combination of the projector pixel n and sensor pixel m is accounted for, where $\Pi[m,n]$ captures the intensity of the projector pixel during τ.

However, path isolation is not a particularly efficient method of capturing the photograph. Many probing matrices are sparse matrices, so most of the elements have a value of zero. Because path isolation allocates equal time to every element, most of the exposure time is spent not integrating any light at all. A larger problem with path isolation manifests when we consider that normally millions of *other* pixels would, altogether, contribute a lot of noise on top of the light from the original isolated pixel; in fact, the light of the original pixel would be completely lost underneath the additive noise from all the other pixels combined.

The second algorithm, using optical matrix probing, involves more efficient acquisition through optical computing. We define $\{\mathbf{p}_k\}$ and $\{\mathbf{m}_k\}$ to be sequences of column vectors that correspond to the decomposition of the probing matrix into rank 1 matrices, so that

$$\Pi = \sum_{k=1}^{K} \mathbf{m}_k (\mathbf{p}_k)^\top .$$

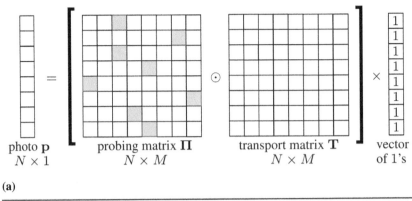

(a)

Transport Probing Equation	Expression for Element Π_{nm}	Expression for Photo Pixel p_n
$\times \mathbf{i}$		$\sum_m \mathbf{T}_{nm} \mathbf{i}_m$
$\odot \quad \times \mathbf{1}$	$\mathbf{i}_m$	$\sum_m \Pi_{nm} \mathbf{T}_{nm}$ $= \sum_m \mathbf{i}_m \mathbf{T}_{nm}$
$\odot \quad \times \mathbf{1}$	$\delta(n-m)$	$\sum_m \Pi_{nm} \mathbf{T}_{nm}$ $= \mathbf{T}_{nn}$
$\odot \quad \times \mathbf{1}$	$\delta(n-m+w)$	$\mathbf{T}_{n(n+w)}$
$\odot \quad \times \mathbf{1}$	$\delta(\mathbf{a}_n-m)$	$\mathbf{T}_{n\mathbf{a}_n}$

(b)

Figure 10.7 Operations using probing matrix. (a) The light transport matrix can be rewritten as being multiplied elementwise by the probing matrix. This offers a greater degree of freedom in the light transport matrix. (b) This table outlines some potential probing matrix operations that we can do without knowing the full light transport matrix. Adapted from (O'Toole et al., 2012).

10.3 Relaxations of Inverse Light Transport

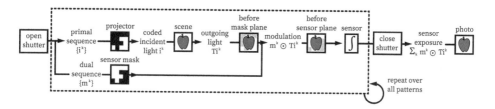

Figure 10.8 Optical probing pipeline. This diagram contains the full pipeline, with relation to the optical hardware, of the probing procedure. Reprinted from (O'Toole et al., 2012).

The probing equation can hence be expressed

$$(\Pi \odot T) \mathbb{1} = \sum_{n=1}^{P} \Pi[n] \circ T[n]$$

$$= \sum_{n=1}^{P} \left(\sum_{k=1}^{K} \mathbf{m}_k \mathbf{p}_k[n] \right) \circ T[n]$$

$$= \sum_{k=1}^{K} \left(\mathbf{m}_k \circ \sum_{n=1}^{P} T[n] \mathbf{p}_k[n] \right)$$

$$= \sum_{k=1}^{K} \mathbf{m}_k \circ T\mathbf{p}_k,$$

where $T[n]$ is the nth column of T; $\mathbf{p}_k[n]$ is the nth element of $\mathbf{p}_k$, and $\{\mathbf{p}_k\}$ is the rank 1 matrix of illumination patterns; and $\{\mathbf{m}_k\}$ is the rank 1 matrix of masks for optical probing. We illustrate this visually, with its correlation to the actual physical hardware, in Figure 10.8. A detailed description of the hardware setup to implement optical probing can be found in (O'Toole et al., 2012).

Figure 10.9 is useful for understanding our two alternate optical algorithms as we have presented them here: the naive path isolation method and the optical probing matrix.

Optical probing can be used for a variety of image manipulation and information extraction tasks such as enhancing direct components using a single photo, descattering using only two photos, separating the direct and indirect components under high-frequency indirect transport, and separating low- and high-frequency indirect transport. Optical probing is also successful in global-direct separation for scenes that violate the assumptions of the approach presented in section 10.3.1, as shown in Figure 10.10.

The primal-dual coding approach is a significant advancement, but it is confined specifically to a coaxial arrangement of the projector and camera. (O'Toole et al., 2014) showed that in a general noncoaxial formation, the dominant light paths are what we call *epipo-*

Path isolation:

In: exposure time E, probing matrix Π

Out: photo equal to $(\Pi \odot T)\mathbf{1}$

1: $\tau = E/NM$
2: open camera shutter
3:    **for** $n = 1$ to N
4:       unmask pixel n
5:       **for** $m = 1$ to M
6:          turn on projector pixel m for time τ with intensity Π_{nm}
7:       mask all pixels
8: close shutter
9: **return** captured photo

Optical matrix probing:

In: exposure time E, probing matrix Π, K illumination vectors $\{\mathbf{i}^k\}$

Out: photo equal to $(\Pi \odot T)\mathbf{1}$

1: $\tau = E/K$
2: open camera shutter
3:    **for** $k = 1$ to K
4:       apply mask $\mathbf{m}^k = \Pi\,\mathbf{i}^k$
5:       project vector $\mathbf{i}^k$ for time τ
6: close shutter
7: **return** captured photo

Figure 10.9 Optical probing algorithms. The two main algorithms used in the optical probing procedure: path isolation and optical matrix probing. Reprinted from (O'Toole et al., 2012).

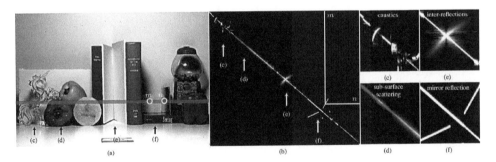

Figure 10.10 Light transport matrix of a scene. (a) An image of the scene, containing various objects that have complex optical interactions. (b) A slice of the light transport matrix for the single highlighted row in (a). A point, (n, m) in the image, represents the light paths that were emitted by pixel m of the projected image and captured by pixel n of the camera (in the highlighted row). The diagonality of the slice implies that light was transported between projector and camera pixels that were close to each other. (c–f) Various notable aberrations in the light transport matrix slice and their causes. Reprinted from (O'Toole et al., 2012).

10.3 Relaxations of Inverse Light Transport

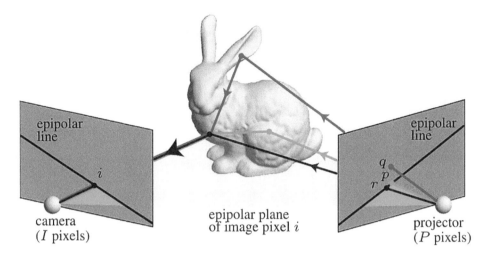

Figure 10.11 Stereo transport matrix using epipolar imaging. Diagram of the stereo light transport setup, where the matrix is subdivided into three groups of light: epipolar (green), non-epipolar (red), and direct (black). Reprinted from (O'Toole et al., 2014).

lar and *non-epipolar* paths. Epipolar paths contribute to a scene's direct image, whereas non-epipolar paths contribute to the indirect components of a scene.

To accommodate general configurations of the projector-camera setup, (O'Toole et al., 2014) introduced the *stereo transport matrix*. The stereo transport matrix, so named because the camera and projector form a stereo pair, contains three categories of matrix elements: *epipolar elements*, *non-epipolar elements*, and *direct elements*, as shown in Figure 10.11. Epipolar elements require the camera and projector pixels to be on corresponding epipolar lines. Non-epipolar elements do not have the camera and projector on corresponding epipolar lines and are by far the most common type of element in the matrix. In direct elements, the camera and projector are in stereo correspondence, in other words, equivalent to the direct light transport with which we have already worked. The image of the scene (denoted by **U**) can be expressed as combinations of these three elements

$$\mathbf{U} = \mathbf{T}^D \mathbf{p} + \mathbf{T}^{EI} \mathbf{p} + \mathbf{T}^{NE} \mathbf{p}$$

where $\mathbf{T}^D \mathbf{p}$ is the direct image, $\mathbf{T}^{EI} \mathbf{p}$ is the epipolar indirect image, and $\mathbf{T}^{NE} \mathbf{p}$ is the non-epipolar indirect image. In comparing the different types of matrix elements, we see that non-epipolar elements outnumber both epipolar and direct elements. Due to this, *non-epipolar dominance* is assumed, allowing for simplification of the matrix.

Table 10.1 summarizes the three kinds of light transport decomposition that we have examined so far.

Inverse light transport and interreflections	(Seitz et al., 2005)
Direct and global component separation	(Nayar et al., 2006)
Stereo light transport matrix	(O'Toole et al., 2014)

Table 10.1 Overview of light transport: three interrelated views of light transport. The first view separates an image into $1, ..., n$ bounces of light (Seitz et al., 2005). The second view uses a smoothness relaxation to reduce the separation into 1-bounce and n-bounce transport (Nayar et al., 2006). The third view shows the ability to discriminate global light transport based on the distance from the diagonal of the transport matrix (O'Toole et al., 2014).

Probing light transport using interferometry. Optical probing using the primal-dual coding approach allows us to implement a generalized probing pattern, but it does not account for path length resolution. Whereas interferometry has been widely used in other areas, such as in astronomy and physics, it is relatively unexplored in computational imaging. Optical interferometry allows for much higher precision in path length resolution, up to an order of $10\,\mu$m, but it can be applied to only small volume regions. The interferometric approach is thus useful for analyzing light transport at a micron scale, finding applications in microscopy and tissue imaging.

The light transport matrix can be decomposed as a sum of its constituents with varying path lengths τ

$$\mathbf{T} = \sum_\tau \mathbf{T}^\tau.$$

Following this decomposition, capturing images $\mathbf{i}$ of a static scene under uniform illumination vector of $\mathbf{I} = \mathbb{1}$ can be described

$$\mathbf{i} = \sum_\tau \omega(\tau) \left(\mathbf{M} \odot \mathbf{T}^\tau \right) \mathbb{1}.$$

Here, $\mathbf{M}$ is a binary matrix of dimension $P \times L$ where L is the number of points on the source and P is the number of pixels on the sensor. If $M_{pl} = 0$, the contribution of the path beginning at the point l on the source and reaching the pixel p on the camera is removed. Analogous to this, $\omega(\tau)$ is a binary function and can remove the contribution of paths of length τ. For a regular image, $\omega(\tau)$ and $\mathbf{M}$ are 1 everywhere as none of the paths are ignored.

Spatial decomposition techniques vary the contributions of spatial light paths by changing $\mathbf{M}$ but keeping $\omega(\tau)$ constant. On the contrary, interferometric approaches vary $\omega(\tau)$ to achieve path length resolution. Interferometric approaches for scene decomposition and optical probing are largely based on the classic Michelson interferometer setup.

Let d_r be the distance between reference arm and the beamsplitter, and d_s be the distance between the target mirror and the beamsplitter. The temporal coherence of the light source is the average correlation between its two instances delayed by τ, at any pair of times. If

10.3 Relaxations of Inverse Light Transport

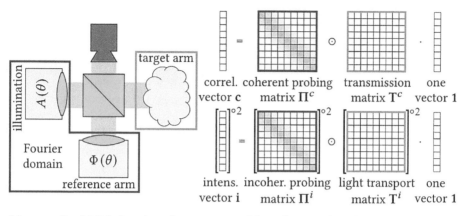

(a) generalized Michelson interferometer (b) implemented probing equations

Figure 10.12 Michelson interferometer light transport probing. (a) An input beam is split by a beam splitter into two copies that reflect off the two mirrors at differing distances from the source; the two copies then recombine at the beamsplitter before being imaged by the camera. One of the mirrors is the target arm (scene), and the other is the reference arm. (b) Varying the source coherence properties, light of different lightpath decompositions can be captured. Reprinted from (Kotwal et al., 2020).

$|d_r - d_s|$ is larger than the temporal coherence, the camera's measured image will be equal to the sum of the two images of the mirrors, that is no interference takes place. A $|d_r - d_s|$ smaller than the temporal coherence results in interference, with a fringe pattern recorded by the camera.

Using temporal coherence and detecting the extent of interference allows separation of light paths with different pathlengths down to a resolution equal to the temporal coherence length. Additionally, taking spatial coherence into account allows a precise spatial separation of light paths. An interference pattern is created if two light paths originate from points within the spatial coherence length of the source, and combine at the same camera pixel.

However, the method just discussed is somewhat constrictive, as it focuses almost solely on diagonal probing, which corresponds to the direct component of the image. (Kotwal et al., 2020) have introduced an interferometric approach using coded mutual intensity, which enables direct versus global decomposition and descattering (similar to the primal-dual coding method discussed previously).

The coded mutual intensity approach adds optical components to the original Michelson interferometer that enable amplitude and phase modulation (see Figure 10.12b). Modulating the amplitude and phase separately between the reference arm and source arm achieves two different effects: incoherent probing of the light transport matrices yields global versus direct separation, whereas coherent probing of transmission matrices allows us to perform descattering. Longer path lengths indicate indirect or global components, whereas shorter path lengths belong to direct components.

This is the first approach to perform both coherent probing of transmission matrices and incoherent probing of light transport matrices using complex probing patterns.

Coherent probing can be used for descattering by suppressing the scattering effect using a probing pattern that emphasizes the diagonal of the light transport matrix while subtracting the first few off-diagonals.

While the interferometry methods provide high resolutions, they are limited in that they are extremely sensitive to vibrations due the small path lengths they consider.

In this section we looked at a broad spectrum of approaches and relaxations to solve inverse light transport that allow separating light paths into global and direct components and are also capable of optically probing the light transport matrix. A combination of precise hardware setups and clever algorithms facilitates the use of light transport for a variety of practical applications. The remaining part of the chapter focuses on various complex imaging problems, which can be tackled using the principles of light transport that we have studied so far. To this end, the next section introduces the problem of non-line-of-sight imaging.

10.4 Non-Line-of-Sight Imaging

In section 10.2.1, we discussed the example of seeing the hidden playing card using dual photography, in which a projector was placed in the hidden scene. Here, we discuss methods to see around corners with no gadget in the line of sight.

Although the history of exploiting scattered radiation dates back to radar and seismic techniques, the concept of photography around corners may have originated with a technical report from (Raskar and Davis, 2008). This report conceptually formulated ideas of seeing around corners or non-line-of-sight (NLOS) imaging as a set of techniques aimed at recovering objects hidden around corners, with applications in disaster management, endoscopy, self-driving cars, and many more areas. NLOS imaging (hereafter, NLOS) can be expressed mathematically as the forward model

$$y = f(x) + \varepsilon$$

where x represents the parameters of the hidden scene such as albedo or class of hidden objects, and ε expresses the noise. $f(\cdot)$ is a map from the hidden scene to the measurement y. This map is determined by the illumination, the geometry of the hidden scene, and the sensor. The aim is to devise algorithms that invert the mapping $f(\cdot)$ such that x can be recovered when y is given, that is, properties of the hidden scene can be uncovered by taking some imaging measurements of the visible scene.

10.4 Non-Line-of-Sight Imaging

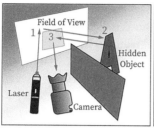 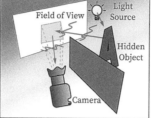 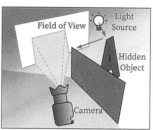

(a) Time-of-flight based methods exploit path-length constraints of photons

(b) Coherence-based methods exploit preserved coherence properties

(c) Intensity-based methods exploit shadows casted by depth discontinuities

Figure 10.13 Types of NLOS detection methods. We discuss (a) time-of-flight-based, (b) coherence-based, and (c) intensity-based methods in this chapter. Reprinted from (Maeda et al., 2019).

Advances in imaging technology have made NLOS possible through a variety of different methods (Figure 10.13). We focus on the following two popular classes of NLOS imaging methods:

- **Time-of-flight methods** (section 10.4.1): The time it takes a photon to traverse its optical path after being reflected from the hidden scene is analyzed to detect properties of the occluded elements.
- **Intensity based methods** (section 10.4.2): By exploiting the surface reflectance of a relay wall or object, this method recovers hidden scenes using typical (RGB) cameras, including smartphones, thus making NLOS accessible.

Our discussion about NLOS is centered mainly around time-of-flight-based and intensity-based approaches. Coherence-based approaches are beyond the scope of this text, but they are mentioned briefly in section 10.5.3 from an applications perspective.

10.4.1 Time-of-Flight Methods

Transient imaging. Each ray of light takes a distinct path through a scene. However, light travels extremely fast (approximately 1 foot/nanosecond). As a result, in a room-sized environment, a microsecond exposure (integration) time is long enough for a light impulse to fully traverse all the possible multipaths introduced due to interreflections between scene elements and reach steady state.

Traditional 2D cameras sample very slowly in comparison with the time scale at which the transient properties of light appear. Hence, they are able to capture only the final steady state sum of the rays at each camera pixel and express light intensity by $I(x, y)$. This is known as steady-state light transport. In this case, we assume that light takes no time to reach the final steady state, and hence the time parameter of light transport is ignored. This loss of multipath information is responsible for limitations in traditional imaging methods. Transient imaging overcomes these limitations by using cameras that are capable of sampling at subpicosecond scales along with ultrafast femtosecond lasers. A transient imaging camera can capture a 3D time image, expressing light intensity by $I(x, y, t)$. This allows us to directly observe the path of a ray traversing the scene as a function of time, and analyze this light transport to discover various properties, such as the geometry of a scene. The effects of multipath interference (MPI) can be captured in image space. (Marco et al., 2017) have used this fact to model MPI as a 2D convolution, because each pixel's light transport can be expressed as a linear combination of that of all other pixels. Using this observation, the effects of MPI can be corrected or leveraged using a convolutional neural network.

Transient imaging is one of the early methods to achieve NLOS. In contrast to the dual photography approach that required an illumination source to be in the line of sight, this method enables the camera to take a picture of an element occluded from both the camera and the illumination source by analyzing the light multipath information obtained through transient imaging.

Geometry of visible and hidden elements. We first understand how to use transient imaging along with the *space time impulse response* (STIR) to estimate the geometry of a visible scene. Dividing a scene S into M distinct patches, we assume that every patch is visible from all other patches, and that every patch has a nonzero diffuse component.

Consider a transient imaging system consisting of a pulse illumination source and a generalized sensor. Each sensor pixel (x_i, y_i) observes a unique patch p_i in the scene over time. (θ_i, Φ_i) represents the direction of a ray generated by the pulse source illuminating patch p_i. By synchronizing the sensor and illumination, the time difference of arrival (TDOA) of light is measured at regular intervals. This system is then used to form STIR (S) of the scene. The 5D function is given by STIR $(x_i, y_i, \theta_i, \Phi_i, t)$ and measured by (Kirmani et al., 2009), as follows. For every patch $p_i : i = 1, \cdots, M$

1. Illuminate p_i with an impulse ray (θ_i, Φ_i)
2. Capture a time image of every patch p_j visible to

$$p_i : \{I(x_j, y_j, t), \ j = 1, \cdots, M, \ t = 0, \cdots, T = \text{STIR}(x_j, y_j, \theta_i, \Phi_i, t)\}.$$

We now define $\mathbf{O}^1 = \{O_i^1 | i = 1, \cdots, M\}$ as the set of first onsets, that is the collection of all time instants O_i^1 when the pixel observing patch p_i receives the first nonzero response while the source illuminates p_i. As shown in Figure 10.14, O_i^1 is the time it takes for the light impulse ray that starts at p_0 and is directed toward p_i to go to p_i and come back, thus tracing the direct path $p_0 \rightarrow p_i \rightarrow p_0$. Analogous to this, the set $\mathbf{O}^2$ is the set of second onsets, and is defined by $\mathbf{O}^2 = \{O_{ij}^2 | i, j = 1, \cdots, M; i \neq j\}$. It is the set of time instances when the camera first receives a non-zero response at patch p_i while the illumination is directed at patch p_j. By Euclidean geometry, we have $O_{ij}^2 = O_{ji}^2$. With this collection of

10.4 Non-Line-of-Sight Imaging

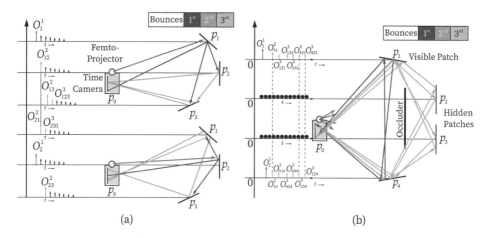

Figure 10.14 Measuring the space time impulse response (STIR). (a) A single patch is illuminated at a time (p_1 in the upper image and p_2 in the lower image), and the times at which reflected light reaches the camera (p_0) are recorded for each patch. Adapted from (Kirmani et al., 2009). (b) Onset data collected from illuminating visible patches can be used to calculate the locations of hidden ones (assuming third bounces arrive before fourth bounces, no interreflections, and a known number of hidden patches). Adapted from (Kirmani et al., 2009).

first and second onsets, we can compute the direct distance of each patch from the camera, as well as the relative distances between all patches, thus inferring the geometry of the scene (Kirmani et al., 2009). If $\mathbf{D} = [d_{ij}]$ is the matrix of pairwise Euclidean distances between all patches, including p_0 = camera, we define $\mathbf{d} = \text{vec}(\mathbf{D})$ and $\mathbf{T}_2$ to be the $[M(M+1)/2] \times [M(M+1)/2]$ matrix that contains the sum of possible pairings of path lengths between the M patches.[8]

We can then find the distance estimates $\widehat{\mathbf{d}}$ by solving the linear system $\mathbf{T}_2 \mathbf{d} = c\mathbf{O}$ where c is the speed of light. For example, for Figure 10.14, the system is

$$\begin{bmatrix} 2 & 0 & 0 & 0 & 0 & 0 \\ 1 & 1 & 0 & 1 & 0 & 0 \\ 1 & 0 & 1 & 0 & 0 & 1 \\ \hline 0 & 0 & 0 & 0 & 2 & 0 \\ 0 & 0 & 0 & 1 & 1 & 1 \\ 0 & 0 & 0 & 0 & 0 & 2 \end{bmatrix} \begin{bmatrix} d_{01} \\ d_{12} \\ d_{13} \\ \hline d_{02} \\ d_{23} \\ d_{03} \end{bmatrix} = c \begin{bmatrix} O_1^1 \\ O_{12}^2 \\ O_{13}^2 \\ \hline O_1^2 \\ O_{23}^2 \\ O_1^3 \end{bmatrix}.$$

---

[8] $M + (M-1) + (M-2) + \cdots + 1 = M(M-1)/2$

These pairwise distances $\widehat{\mathbf{d}}$ are then used to make conclusions about the geometry of the scene by using an isometric embedding algorithm (Kirmani et al., 2009).

The geometry of hidden scenes can be recovered similarly if some assumptions are made. We assume that we know the number of hidden patches, and that all third bounces of light arrive before higher order bounces, which is true when there are no interreflections among hidden patches.

For a hidden patch p_i, because the first and second onsets cannot be observed, we find the set of third onsets $\mathbf{O}^3$. Based on Euclidean geometry, $O^3_{ijk} = O^3_{kji}$.

Assume that as shown in Figure 10.14, patches p_2 and p_3 are hidden. We first find d_{01}, d_{04}, d_{14} because these are visible. Next, we apply a labeling algorithm to identify all third onsets. O^3_{141} and O^3_{414} are found from TDOA because p_1 and p_4 are visible, and hence distances are known. Additionally, $O^3_{124} = O^3_{421}$ and $O^3_{134} = O^3_{431}$. We use this equality relation to find these onsets. Furthermore, we make assumptions about the proximity of the hidden patches. For example, we may assume without loss of generality that p_2 is closer to p_1 than p_3, and hence $O^3_{121} < O^3_{131}$, which allows us to label these onsets as well.

We can then construct an operator $\mathbf{T}_3$ such that $\mathbf{T}_3 \mathbf{d}_H = c \mathbf{O}_h$ where $\mathbf{d}_H$ is the distances to the hidden patches and $\mathbf{O}_h$ is the third bounces of arrival times corresponding to hidden patches (Kirmani et al., 2009). In the given example, we can then solve the system:

$$\begin{bmatrix} 2 & 0 & 0 & 0 \\ 1 & 1 & 0 & 0 \\ \hline 0 & 0 & 2 & 0 \\ 0 & 0 & 1 & 1 \end{bmatrix} \begin{bmatrix} d_{21} \\ d_{24} \\ d_{31} \\ d_{34} \end{bmatrix} = c \begin{bmatrix} O^3_{121} - O^1_1 \\ O^2_{124} - (O^1_1 + O^1_4)/2 \\ O^3_{131} - O^1_3 \\ O^2_{134} - (O^1_1 + O^1_4)/2 \end{bmatrix}.$$

This enables us to reconstruct the geometry of the hidden scene using the same isometric embedding algorithm that is used for visible elements.

On a similar note, (Pandharkar et al., 2011) proposed an algorithm using the constrained least-squares model for estimating motion and absolute locations of NLOS moving objects in cluttered environments through tertiary reflections of pulsed illumination, using only relative time differences of arrival at an array of receivers. The authors also presented a method to estimate the size of NLOS moving objects by backprojecting extrema of their time responses.

10.4.1.1 3D Shape Recovery from Hidden Scenes

Streak cameras + femtosecond lasers. However, the method proposed in the previous section assumes well separated and isolated hidden patches with known correspondence between hidden patches and recorded pulses. In addition, the images recovered are 2D.

In this section, we discuss another method to recover the 3D structure of a hidden scene by extracting information from the multibounce path of light. The setup of this method

10.4 Non-Line-of-Sight Imaging

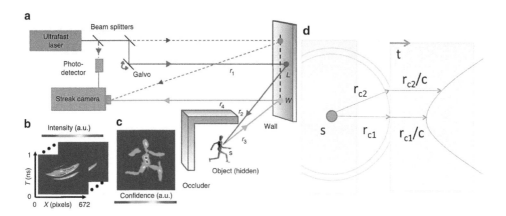

Figure 10.15 Image capture procedure and geometry. The laser is aimed onto the wall via galvanometer and mirrors (a), and the camera takes a series of images in time (b). A confidence map (c) of the hidden object can be constructed from the results (Velten et al., 2012a). (d) The hyperbolic curves in the individual camera images result from the varying distances (*left*) and thus times (*right*) light travels to reach the sensor (Gupta et al., 2012).

consists of a diffuser wall, an ultrafast pulsed laser, and a streak camera that is capable of sampling at extremely short time intervals to produce time images. The setup is shown in Figure 10.15a.

In this setup, a light impulse reflects off the diffuser wall and onto the hidden scene. It then reflects back onto the wall, carrying information about the 3D geometry of the hidden scene that the camera then captures. The streak camera has one spatial and one temporal dimension, and we focus it on the dashed line segment along the wall, as shown in Figure 10.15a. We capture images only of this line segment over time intervals as short as 2 picoseconds to create the space-time image. The streak image looks like the hyperbolic curve shown in Figure 10.15b. This is because the impulse of light has a spherical wavefront propagating from the hidden scene and it arrives at different points on the diffuser wall with different time delays (Figure 10.15). As different spots on the wall are illuminated, the hyperbolic curves vary according to the encoded information.

To analyze the light reflected from the hidden scene, it is important to ensure that no light is reflected directly off the wall without reaching the hidden scene. Hence, the laser illuminates only a spot above or below the dashed line. By pointing the laser to illuminate multiple spots on the wall and capturing multiple time images, we can uncover enough information about the hidden scene to reconstruct it.

However, although the images captured by the streak camera contain information about the hidden scene, they lack correspondence information. We do not know which light pulse

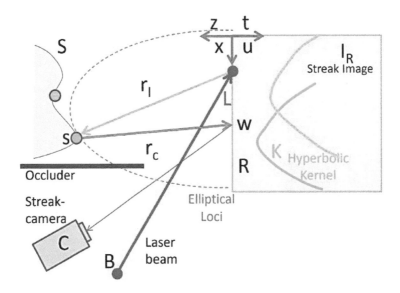

Figure 10.16 Backprojection geometry. The set of possible hidden object locations corresponding to an image pixel form an ellipse, as each image corresponds to a set distance that light has traveled. Reprinted from (Gupta et al., 2012).

is caused due to which surface point in the hidden scene. As demonstrated in Figure 10.16, this problem is solved using a backprojection algorithm (Velten et al., 2012a).

Consider any pixel p in a streak image with nonzero light intensity. The possible locations that could have contributed intensity to this pixel lie on an ellipsoid in Cartesian space. The focal points (L, w) of this ellipsoid correspond to the point on the wall illuminated by the laser and the point on the dashed line where it was reflected from the hidden scene (see Figure 10.16). In 2D, the intersection of the ellipses corresponding to some pixels p, q, r of a streak image would uniquely determine the location of the hidden surface patch contributing to these pixels. However, in practice, we do not know if light detected by two pixels came from the same 3D surface point, and hence we create a likelihood model (Velten et al., 2012a).

For this model, Cartesian space is discretized into voxels[9] and we compute the likelihood of each voxel being on the hidden surface (Velten et al., 2012a). Each pixel in the streak image is allowed to "vote" for every voxel that lies within its corresponding ellipsoid. Additionally, each pixel's vote is multiplied by the distance it travels between the wall and the hidden scene to account for distance attenuation ($r_2 r_3$ in Figure 10.15). Thus, the more

---

[9] Voxels represent values in a regular grid in 3D space. Unlike pixels, they do not have specific coordinates, and their position is inferred from their position relative to other voxels.

10.4 Non-Line-of-Sight Imaging

votes a voxel has, the more likely it is to lie on the hidden surface. The 3D scalar function on these voxels is called a *heatmap*. By summing the weighted intensities from all pixels of a single streak image, we can estimate a heatmap of the target patch. We can repeat this process by illuminating different points on the diffuser wall to get many streak images (Figure 10.15b), and hence better approximations to the heat map (Figure 10.15c).

The final step of the reconstruction algorithm is filtering. The second derivative of the heatmap along the depth (z) projected on the $x - y$ plane reveals the hidden shape contour as shown in Figure 10.17d.

Single photon avalanche diode. Compared to using femtosecond lasers and expensive streak cameras for non-line-of-sight imaging, a single photon avalanche diode (SPAD), used alongside a photon counter and laser, is a cheaper and more practical alternative. A SPAD is a type of p–n junction that responds electrically to incoming photons. It can be disabled for certain durations, thus ignoring first-bounce light.

Given a pair of positions on the wall where the SPAD and laser are focused, the photon counter generates a histogram of photon counts versus time (see Figure 10.18). By pointing the laser and detector at various locations on a grid of points on the wall, we can gather the data we need to run a backprojection algorithm similar to the one discussed in section 10.4.1.1 ("Streak Cameras + Femtosecond Lasers"), as follows

- Model the hidden scene as a 3D grid of voxels.
- Create a confidence map from the set of photon counts $N(t, x_i, y_i, x_o, y_o)$ – t for time, and (x_i, y_i) for laser coordinates, (x_o, y_o) for detector coordinates, to the set of voxels $V(x, y, z)$.
- Apply a Laplacian filter, and threshold the results.

To accurately model and recover occlusions within hidden scene parts in non-line-of-sight imaging, (Heide et al., 2019) developed both image formation and inverse methods. The nonlinear factorization method proposed was validated in simulation as well as physical measurements. The time-resolved imaging system built using an array of single photon avalanche diodes and a picosecond laser, provided superior quality reconstructions compared with other proposed methods.

10.4.1.2 ToF sensors for Real-World NLOS Imaging

Commercial ToF cameras. Commercial time-of-flight cameras can be used alongside nanosecond lasers to image hidden objects (Heide et al., 2014c). This approach requires considerably cheaper hardware, shortens the acquisition time of images, and is more robust to ambient lighting compared with methods using femtosecond lasers and streak cameras. Here, recovery of 3D shape of the occluded object is posed as an inverse problem that is solved using an optimization procedure supplied with appropriate structural priors on the data.

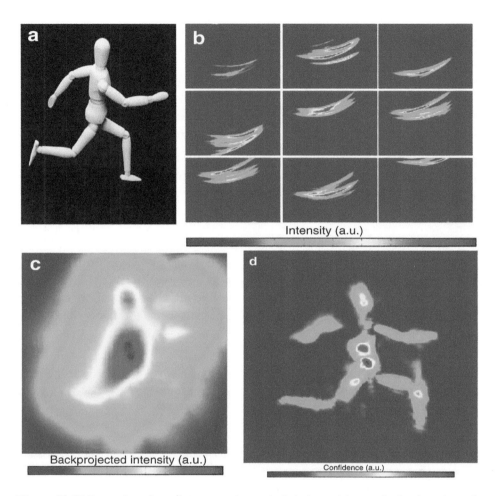

Figure 10.17 Examples of streak images. An occluded object (a) is probed indirectly with an ultrafast laser. (b) Many streak images of the hidden object are captured. The object can then be recovered via (c) backprojection and (d) filtering. Reprinted from (Velten et al., 2012a).

Using commercial time-of-flight cameras, we can define the hidden scene as a set of patches with certain heights and orientations, and assume a one-to-one mapping between wall patches and camera pixels. Under these assumptions, the radiance $L(w)$ at a wall patch w given an emitted radiance $L_e(l)$ hitting a wall patch l can be derived from the rendering equation:

$$L(w) = L_e(l) \rho(w) \int_V g(x) v(x) \, dx,$$

10.4 Non-Line-of-Sight Imaging

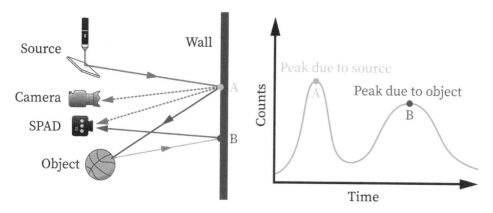

Figure 10.18 Data collection. Laser pulses bounce off a wall and hidden object to reach a single photon avalanche diode (SPAD) *(left)*, and a photon counter produces a graph of detector hits versus time *(right)*.

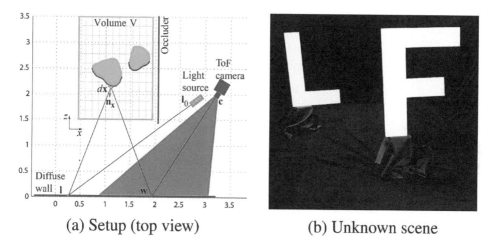

(a) Setup (top view) (b) Unknown scene

Figure 10.19 Image/camera setup. By now, this picture should seem familiar: (a) a relatively cheap laser and ToF camera (replace the faster lasers and expensive sensors of previous sections), with the goal of more accessibly capturing (b) the hidden scene. Reprinted from (Heide et al., 2014a).

where $\rho(w)$ indicates diffuse albedo, V denotes the the volume of the hidden scene, and x denotes the location of the hidden patches. Last, g, v are geometry terms correcting for hidden patch location and albedo/orientation, which can be expressed as functions of the positions and orientations of $w, l,$ and x.

These equations correspond to stationary light transport. To get to the transient version, we incorporate time as a dimension, ensuring that we include only the light reflected off the hidden scene. Because this light arrives at $t = t_0 + \tau(x)$, where $\tau(x)$ is the total travel time, the light arriving at the camera pixel c (corresponding to the wall patch w) can be calculated

$$L(c,t) = \int_0^T L_e(l,t_0) \rho(w) \int_V \delta(t_0 + \tau(x) - t) g(x) v(x) \, dx \, dt_0.$$

Representing the discrete hidden patch locations as $\mathbf{v}$, the transient image as $\mathbf{i}$, the light transport matrix as $\mathbf{T}$, and the correlation matrix corresponding to camera and image modulation as $\mathbf{C}$, the discrete version of this equation

$$\mathbf{h} = \mathbf{Ci} = \mathbf{CTv}$$

where $\mathbf{h}$ is the measurement from the ToF sensor. This equation can be expressed as an optimization problem and can be solved using a modified version of the alternate direction method of multipliers method (ADMM) augmented with several regularization priors: that the spatial gradients are smooth, that the hidden patches are sparse, and that the discretization results in each coordinate have at most one hidden patch. The priors can be combined into a single regularization term (in the given order) as follows:

$$\Gamma(\mathbf{v}) = \lambda \sum_z \|\nabla_{x,y} v_z\|_1 + \theta \|\mathbf{W}\mathbf{v}\|_1 + \omega \sum_{x,y} \text{ind}_C(\mathbf{v}_{x,y}).$$

(Heide et al., 2014a) demonstrated that this method recovers the shapes of cardboard letters at a resolution of about 5 cm, depending on material. At the cost of a difficult optimization problem, the paper demonstrates an approach that is more practical and deployable in real world scenarios than using a streak camera to reconstruct 3D images of hidden scenes. A novel density estimation technique was presented by (Jarabo et al., 2014) that allowed reusing sampled paths to reconstruct time-resolved radiance. Along with the introduction of a formal framework for transient rendering, in order to factor in the distribution of radiance along time in participating media, they also devised new sampling strategies.

Virtual sensor array. Much of the preceding work has been experimental in nature. (Kadambi et al., 2016) proposed a theoretical framework for ToF NLOS imaging by treating the intermediary wall as a virtual sensor array (VSA) and establishing a model for the importance of wall specularity by borrowing from the field of array signal processing.

The basic model considers a single occluded point light source. The phasor L representing light transport from the light source to the diffuse wall can be written as

$$L(u,v) = \frac{\cos\theta}{\phi_L^2(u,v)} e^{j\phi_L(u,v)},$$

10.4 Non-Line-of-Sight Imaging

Figure 10.20 Experimental results. The reconstructed depth *(left)*, albedo *(center)*, and hidden target *(right)* for both high *(bottom)* and low *(top)* ambient light. Reprinted from (Heide et al., 2014a).

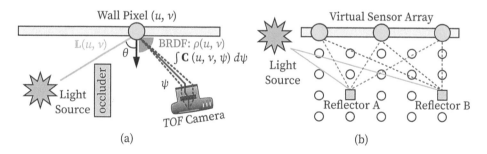

Figure 10.21 Scene geometry. The familiar image capture diagram *(left)* remains the same, but here the hidden target is interpreted as a set of point sources *(left)* or reflectors *(right)*, and the wall itself is modeled as a sensor array *(right)*. Reprinted from (Kadambi et al., 2016).

where (u, v) is the wall location, θ is the angle of incidence, and ϕ_L is the phase difference. Similarly, the phasor C representing transport between the wall and the camera can be written as

$$C(u,v,\psi) = \rho(u,v) \frac{\cos\psi}{\phi_C^2(u,v)} e^{J\phi_C(u,v)},$$

where ρ is albedo, ψ is the angle of reflection, and ϕ_C again represents phase difference.

Because the camera is focused on the wall by design, we can integrate across ψ to get the total transport phasor M

$$M(u,v) = A_0 L(u,v) \int C(u,v,\psi) \, d\psi,$$

where A_0 is the original amplitude. If the wall has roughly uniform albedo, the amplitude of C can be ignored and L can be approximated to L' as follows:

$$L(u,v) = \left(\frac{1}{A_0}\right)\left(\frac{M(u,v)}{C(u,v)}\right),$$

$$L'(u,v) = \left(\frac{1}{A_0}\right)\left(\frac{M(u,v)}{e^{J\phi_C(u,v)}}\right) = \frac{\cos\theta}{\phi_L^2(u,v)} e^{J\phi_L(u,v)} \int \rho(u,v) \frac{\cos\psi}{\phi_C^2(u,v)} d\psi,$$

where ϕ_C is known from scene geometry.

We now have an array of virtual sensors L', each a phasor with an amplitude and phase, and want the location of the original source. If we discretize the hidden scene as a grid of voxels, let $\mathbf{x} \in \mathbb{C}^N$ be the vector containing the confidence that the source is at each one of these N voxels and let $\mathbf{y}$ be our vector of measurements L', we can construct the matrix

$$\mathbf{D} = [\mathbf{s}(u_1, w_1), \mathbf{s}(u_2, w_2), \cdots, \mathbf{s}(u_R, w_Q)],$$

where each column $\mathbf{s}(u,w) : (u,w) \to \mathbb{C}^M$ represents the expected sensor measurements (of which there are M in total) for each voxel. Finally, the result, $\mathbf{y} = \mathbf{Dx}$, can be solved with either sparse solvers (assuming the number of target voxels is much less than the number of total voxels) or beamforming, which approximates an answer with

$$\mathbf{x} = \mathbf{D}^H \mathbf{y}.$$

In addition to this model, we can also explore the relationship between wall specularity and reconstruction accuracy (Kadambi et al., 2016): again borrowing from array signal processing, we can express the reconstruction resolution (FWHM$^\angle$)

$$\text{FWHM}^\angle = \arcsin\left(\frac{f_m \gamma^\angle}{\lambda + d\gamma^\angle}\right),$$

where $\gamma^\angle$ represents how diffuse the "sensor" or wall is, f_m corresponds to the modulation frequency of the camera, and d is the diameter of the virtual sensor array. This represents a nonlinear relationship between wall specularity and reconstruction accuracy, with very poor theoretical bounds (1 m) on the resolution given very diffuse walls. Experiments and data from the Mitsubishi Electric Research Labs BRDF database suggest that many

10.4 Non-Line-of-Sight Imaging

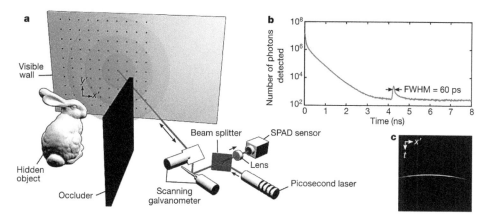

Figure 10.22 Confocal NLOS setup. Confocal NLOS involves (a) simultaneously imaging and sensing the same point on a wall. (b) For each point, photon counts are measured versus time. (c) These measurements are then combined into streak images (O'Toole et al., 2018).

real-world materials are indeed specular enough for this form of ToF imaging to be effective in practice (Kadambi et al., 2016).

10.4.1.3 Recent Advances in ToF-Based NLOS Imaging

Confocal NLOS. Many of the time-of-flight based NLOS imaging methods require extremely high processing power and memory. In addition, the flux of multiply scattered light is low. Hence, data needs to be acquired for long periods of time in dark environments. Confocal non-line-of-sight imaging (C-NLOS) seeks to solve this problem by aiding in the derivation of the light-cone transform to reconstruct hidden scenes.

Instead of illuminating and capturing every possible pair of distinct points on a diffuser wall, C-NLOS illuminates and captures the same point at one time and then raster scans[10] this point across the wall to obtain its transient image (O'Toole et al., 2018). Points (x', y') on the diffuser wall are confocally scanned at $z' = 0$. A 2D histogram of spatial and temporal dimensions is measured at this point, as shown in Figure 10.22b. The second spike corresponds to $2*$ distance from the hidden object. Many such images are put together across the row to form the streak image shown in Figure 10.22c. τ, the final 3D volume of

---

[10] Raster scanning is a technique in which the laser is pointed sequentially at every point in the row, for each successive row, thus effectively traversing every point on the wall.

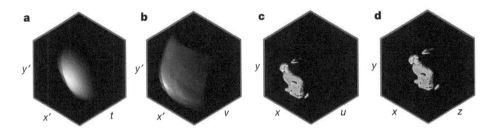

Figure 10.23 Object reconstruction. The steps of the reconstruction algorithm match the components of the convolution: (a)–(b) attenuation in time, (b)–(c) Wiener filtering, and (c)–(d) attenuation in space. Reprinted from (O'Toole et al., 2018).

measurements, is expressed as:

$$\tau(x', y', t) = \iiint_\Omega \frac{1}{r^4} \rho(x, y, z) \delta \left(2\sqrt{(x-x')^2 + (y-y')^2 + z^2} - tc \right) dx\, dy\, dz$$

where c is the speed of light, ρ is the albedo of the hidden scene at the given point (x, y, z), and the Dirac delta function δ represents a 4D spatiotemporal cone that models the propagation of light from the wall to the object and back to the wall. Using change of variables and substitution, this integral can be transformed into a 3D convolution expression

$$R_t\{\tau\}(x', y', v) = R_z\{\rho\}(x, y, u)\, h(x' - x, y' - y, v - u),$$

where $R_t\{\tau\}$ is τ multiplied with a constant; $R_z\{\rho\}$ is a function of ρ, and h is a shift-invariant transformation of the 3D solution kernel. The inverses of R_z, R_t both have closed-form solutions.

The discrete version of image formation can be represented by $\mathbf{R}_t \tau = \mathbf{H} \mathbf{R}_z \rho$ where τ is the vector form of the measurements, ρ is the vector form of the albedos, and $\mathbf{H}$ represents the 3D shift-invariant convolution operation. $\mathbf{R}_z$ is the transformation on the spatial domain and $\mathbf{R}_t$ is the transformation on the temporal domain. Because both these matrices operate independently, C-NLOS is both memory and power efficient.

We have now translated NLOS to a 3D deconvolution problem, and we can derive the closed-form solution. Based on the Wiener filtering method, the final geometry of the hidden scene is recovered as shown in Figure 10.23.

Because C-NLOS involves lesser data collection, and simpler processing due to independent spatial and temporal dimensions, it is computationally simpler. It can hence be applied in real-time NLOS tracking of a hidden scene.

Nonvisual NLOS. Most of the optical time-of-flight methods discussed so far require specialized ToF cameras and fast lasers. In addition, visible light signals for diffuse hidden elements fall off quickly, resulting in higher data acquisition time, and often failure to

10.4 Non-Line-of-Sight Imaging

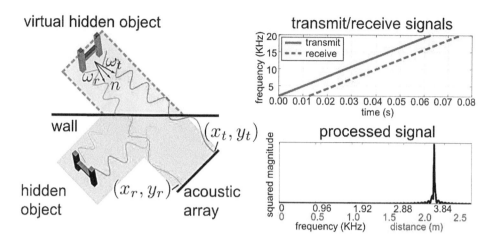

Figure 10.24 Acoustic NLOS. With sound, walls act in a much more specular manner than they do with light, which results in a clearer virtual object *(left)*. This can be quantified by measuring the time delay of the return signal *(upper right)*, and then conducting a Fourier analysis *(lower right)*. Reprinted from (Lindell et al., 2019).

uncover objects that are farther away. One of the ways to overcome these issues is using higher wavelength signals like radio and sound waves for NLOS. Walls have specular (mirrorlike) properties for *sound* and *radio* waves, and hence hidden scenes are revealed more easily than in the optical case. NLOS using sound employs off-the-shelf hardware like microphones and speakers, and is therefore more accessible. The acoustic NLOS setup is demonstrated in Figure 10.24.

A virtual hidden object is formed behind the wall in acoustic NLOS due to the specular reflection of acoustic waves. The received signal is a delayed version of the transmitted signal, and after undergoing Fourier transformation, these signals produce a sharp peak at a frequency proportional to the distance of the reflecting object. The wavefield for this model is given by a 5D function:

$$\tau(x_t, y_t, x_r, y_r, t)$$

where x_t, y_t are the spatial positions of the transmitter (speaker), x_r, y_r are the spatial positions of the receiver (microphone), and t is the time. For the signal transmitted from $(x_t, y_t, z = 0)$, a response is recorded at $(x_r, y_r, z = 0)$. After reductions in the Fourier domain, the preprocessed measurements can be approximated as functions of the spatially varying albedo of the hidden object and the acoustic BRDF.

There are two methods to capture such images. In the confocal method that is analogous to O'Toole's optical confocal imaging method, $x_t = x_r$, $y_t = y_r$. In this case, a closed-form solution can be developed for the reconstruction of the hidden scene. However, in the case

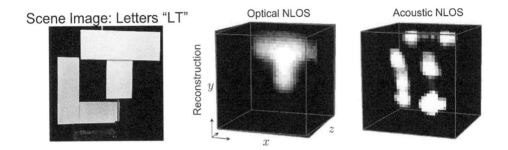

Figure 10.25 Comparing acoustic and visual NLOS imaging. The acoustic method *(right)* reproduces the L in the hidden scene *(left)*, whereas the visual method does not *(center)*. Reprinted from (Lindell et al., 2019).

of nonconfocal measurements, there is a significantly larger amount of data in comparison, which also leads to better image resolution. This requires separate processing due to the additional potential specular reflections returning outside of the confocal receiver positions. The non-confocal measurements are adjusted to emulate confocal measurements. This computational adjustment employs methods such as normal moveout (NMO) and dip moveout (DMO) corrections inspired by seismic imaging. The image is then reconstructed through deconvolution: the results (compared with O'Toole's optical confocal reconstruction) are shown in Figure 10.25.

(Scheiner et al., 2020) provides a Doppler radar based method for NLOS using radio waves. The setup uses a colocated emitter and receiver array (analogous to laser and camera for the optical case) to identify moving targets in the hidden scene, taking advantage of the increased specularity of many real-world surfaces.

The end result is a set of feature points in space (distance, radial velocity, angle, and amplitude) that are then fed into a neural network to detect and track features. The approach successfully identifies bicycles and pedestrians outside of line-of-sight. We refer the reader to (Scheiner et al., 2020) for details of how one can recover distance, radial velocity, and angular position of the hidden object by analyzing the received signal.

Fermat paths for ToF imaging. To extend the scope of transient NLOS imaging beyond Lambertian approximations and intensity constraints, we can approach the problem with a geometric intuition.

We do this by exclusively analyzing Fermat light paths (in the context of Fermat's principle), which are locally longest or locally shortest light paths. Additionally, Fermat light paths can be classified as reflected by the object of interest in either a specular fashion or at boundary locations (see Figure 10.27). Using a collocated emitter and detector, Fermat paths can be detected by locating points of discontinuity along the transient curve.

10.4 Non-Line-of-Sight Imaging

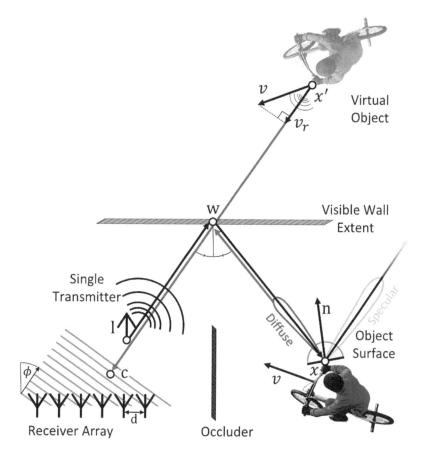

Figure 10.26 Doppler radar NLOS. Using radar, as with using sound, means that various real-world surfaces become more specular. As in the previous section on acoustic NLOS, radar reflections are captured by an array of receivers positioned at the same location as the transmitter, and the outgoing and incoming signals are mixed. We can recover information about distance, velocity, and angle from the received signal. Reprinted from (Scheiner et al., 2020).

Fermat paths have a number of useful properties that can be applied to NLOS object reconstruction. First, we note that these paths are invariant with respect to the BRDF of the surfaces involved. Assuming our Fermat path, reflected from a wall at v, specularly strikes the hidden object of interest at some point $x_{F,1}$ and the boundary at some point $x_{F,2}$. We can construct a sphere S around the the point v, with radius

$$r = \frac{\tau_F(v)}{2}$$

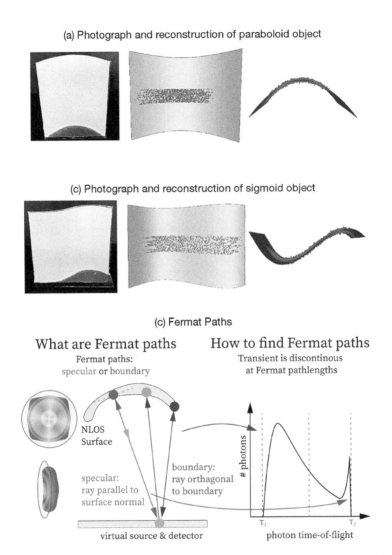

Figure 10.27 Fermat paths. (a) Experimental results for the reconstruction *(right)* of hidden topologies *(left)*. The objects on the left were 3D printed from ground truth meshes *(center)*, on top of which various reconstruction points (red) are overlaid. This is the reconstruction of a paraboloid object. (b) This is the reconstruction of the sigmoid object. (c) Transient light (as measured with a photon counter) exhibits discontinuities at Fermat pathlengths, which correspond to significant features on the hidden surface. Adapted from (Xin et al., 2019).

10.4 Non-Line-of-Sight Imaging

where $\tau_F(v)$ is the length of the Fermat path relative to v.

Furthermore, Fermat's flow constraint dictates that the direction of the light path between x_F and v is parallel to $\nabla_v \tau_F(v)$, where ∇_v is the spatial gradient operator. This operator can be measured by perturbing the incident light emitter slightly, and measuring the degree to which the length of the Fermat path changes. According to the Fermat flow constraint, if we know the length of the Fermat path and its gradient, we can reconstruct the point on the NLOS surface by intersecting the sphere S, with a line that is parallel to the gradient and passes through v. Using this geometric technique, we can generate a point cloud of the hidden surface. This point cloud can then be supplied to common surface reconstruction algorithms to create a uniform surface. This method can also be extended to surface reconstruction for an object that is in the line of sight, but is between a heavy diffuser (such as a sheet of paper).

The reconstruction (as shown in Figure 10.27a–b) is accurate to about 2 mm when compared to the ground truth. This geometric approach offers a novel perspective with which to tackle the NLOS problem and can potentially be complemented with intensity/BRDF data to create an even more optimal solution (Xin et al., 2019).

10.4.2 Intensity-Based Methods

Turning corners into cameras. The previous sections described a variety of approaches to NLOS that largely rely on the time-of-flight principle, making use of fast lasers and ToF cameras to obtain accurate temporal measurements of photon arrival times. However, we can instead use intensity and color of the observed scene for NLOS (Bouman et al., 2017). The approach leverages the fact that objects moving behind a corner (e.g., the edge of an occluding wall) will result in reflection patterns on the ground in front of the corner that depend on the angular position of the objects.

Whereas most NLOS approaches have traditionally used a vertical surface to reflect light off the hidden scene, this approach makes use of the ground (a horizontal surface). The camera is placed on one side of the occluding wall looking at the ground parallel to the wall, so that light on the same side of the wall as the camera comprises the visible scene and light on the opposing side comprises the hidden scene. The light from a hidden object cannot reach the camera directly but does hit the region on the ground not blocked by the wall.

The size of this region depends on the angular coordinates of the object within the hidden scene. Under the assumption that (1) the ground is Lambertian and (2) incoming light can be assumed to originate from a distant celestial sphere, the reflected light L'_o from point (r, θ) on the ground can be expressed

$$L'_o(r, \theta) = a(r, \theta) \int_{\alpha=0}^{2\pi} \int_{\delta=0}^{\pi/2} \gamma L'_i(\alpha, \delta) \, d\alpha \, d\delta,$$

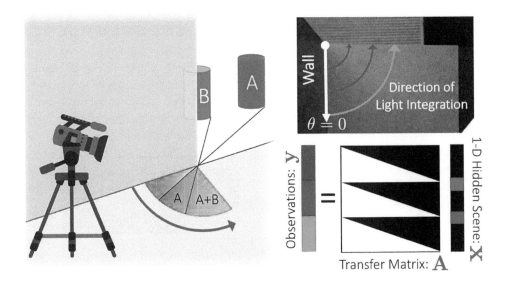

Figure 10.28 Hidden objects and shadows. In the presence of an occluding wall, objects hidden from the camera still influence the colors in the shadows cast by the wall *(left)*. Observations at a given angle from the wall *(upper right)* include light from only a portion of the background, resulting in a transfer matrix similar to that shown *(bottom right)*. Reprinted from (Bouman et al., 2017).

where $L'_i(\alpha, \delta)$ denotes incoming light at the ascension and declination (α, δ); $a(r, \theta)$ denotes albedo; and γ is the dot product of the incident ray and the surface normal.

Note that $\alpha \in (0, \pi)$ is the visible scene, $\alpha \in (\pi, 2\pi)$ is the hidden scene, and light from $\alpha \in (\pi + \theta, 2\pi)$ will be blocked by the occluding wall from reaching the ground. Therefore

$$L'_o(r, \theta) = a(r, \theta) \left[L_v + \int_{\phi=0}^{\theta} L_h(\phi) \, d\phi \right], \quad (10.3)$$

where L_v corresponds to the visible scene and is constant in (r, θ), and

$$L_h(\phi) = \int_{\delta=0}^{\pi/2} \gamma L'_i(\pi + \phi, \delta) \, d\delta.$$

Under the assumption that $a(r, \theta)$ is largely constant (or if one subtracts the background), (10.3) can be differentiated to establish a relationship between observed light and the angular change in lighting from a hidden object. We can then use spatial smoothness and a *maximum a posteriori (MAP)* optimization to obtain the angular projection from observed image intensities.

Using only commercial video cameras, (Bouman et al., 2017) demonstrated recovery of the angular motion of two people wearing red and blue shirts walking around within

10.4 Non-Line-of-Sight Imaging

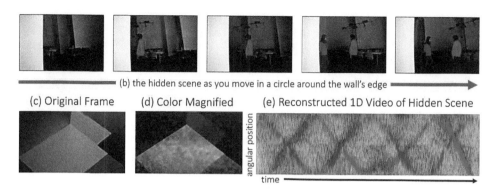

Figure 10.29 Motion from shadows. The (d) color-augmented version of (c) the shadow demonstrates the concept in Figure 10.28. This enables (e) a reconstruction of the motion of colored objects, or in this case, (b) people (Bouman et al., 2017).

the hidden scene. Note that while this method provides only 1D angular information, *two corners* (or wall edges), such as a doorway, effectively produce a stereo imaging system. The angles from each corner can then be used to triangulate a 2D position.

Polarization cues to supplement NLOS imaging. As with many other imaging techniques discussed in previous chapters, polarization can be instrumental in optimizing NLOS imaging. In passive NLOS imaging, if a camera is aimed at a wall patch that is reflecting the hidden object, the fundamental least-squares equation to be solved is

$$\widehat{\mathbf{I}} = \mathbf{T}^+ \mathbf{i}$$

where $\widehat{\mathbf{I}}$ is the array of estimated reflected scene intensities, $\mathbf{T}^+$ is the pseudoinverse of the light transport matrix, and $\mathbf{i}$ is the array of incoming intensities recorded by the camera.

Polarization cues can be used to "condition" the light transport matrix, in order to get a better estimate of the scene intensities. Polarizers have a property called the *effective angle*, where if a light ray strikes even a crossed polarizer obliquely, light is leaked (Figure 10.30).

When the camera is placed so that the incoming light has been reflected near Brewster's angle, the light becomes almost linearly polarized. This causes a light leakage pattern that can be incorporated into the light transport matrix, in order to further condition it. Consequently, this method becomes a viable supplement to alternate methods of NLOS imaging that have previously been discussed (Tanaka et al., 2020).

Computational periscopy. Inspiration can also be taken from the workings of a periscope and applied to the NLOS problem by treating the reflecting wall as a mirror in a periscope system. To achieve this end, an ordinary digital camera and passive imaging is sufficient.

The primary equation that encapsulates this situation is

$$\mathbf{y} = \mathbf{T}(\theta_{\text{occ}})\mathbf{f} + \mathbf{b},$$

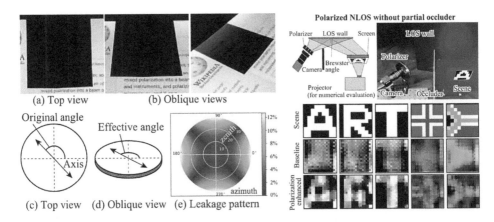

Figure 10.30 Polarized NLOS. (c)–(d) The effective polarization axis of a polarizer changes based on viewing angle, as demonstrated by (a)–(b) the polarizer placed on top of a monitor. This occurs even when (e) two polarizers are placed at 90° angles. Light from a projector *(top right row)* is captured by a camera placed at the Brewster angle with respect to the screen *(top)*. Placing a polarizer in front of the camera leads to better results *(bottom row)* than without it *(center row)* (Tanaka et al., 2020).

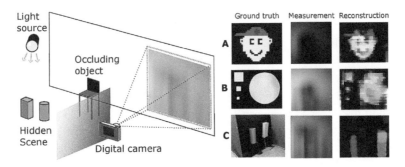

Figure 10.31 Periscopy NLOS. (a) The classic NLOS setup, replete with occluder, hidden object, light source, and sensor. (b) The results of the reconstruction algorithm *(right)* on various scenes *(left)*, with the raw camera image in the center. Reprinted from (Murray-Bruce et al., 2019).

where **y** is the vectorized array of camera pixels, **T** (θ_{occ}) is the light transport matrix given a parametrization θ_{occ} of the occluded object, **f** is the vectorized array of scene pixels, and **b** is the array of background contributions. Assuming that there is an opaque occluder of a known shape, but arbitrary position, in front of the hidden scene, the created penumbra can provide us with unique columns of **T** and thus a well-conditioned recovery of **f**.

10.5 Applications

An important aspect of this procedure is the computational field of view, which is a set of points on the hidden scene where the shadows cast by the occluder provide enough variability and do not completely obscure the field of view of the camera. $\mathbf{f}$ and θ_{occ} can be solved by using computational inversion with a two-step algorithm. First, solve for θ_{occ} by using an estimated (poorly conditioned) light transport matrix, $\mathbf{T}(\theta_{occ})$, built by assuming an unknown θ_{occ}. This nonlinear optimization can be expressed

$$\widehat{\theta}_{occ} = \arg\max_{\theta_{occ}} \left\| \mathbf{T}(\theta_{occ}) \left(\mathbf{T}(\theta_{occ})^\top \mathbf{T}(\theta_{occ}) \right)^{-1} \mathbf{T}(\theta_{occ})^\top \mathbf{y} \right\|_2^2$$

Second, use the estimated occluder parameterization to compute the well-conditioned light transport matrix $\mathbf{T}\left(\widehat{\theta}_{occ}\right)$. With this information the hidden scene's RGB content can finally be computed

$$\mathbf{f} = \left(\mathbf{T}\left(\widehat{\theta}_{occ}\right)^\top \mathbf{T}\left(\widehat{\theta}_{occ}\right) \right)^{-1} \mathbf{T}\left(\widehat{\theta}_{occ}\right)^\top \mathbf{y}.$$

Note that in the case of excessive unknown background $\mathbf{b}$, its contributions to the captured image $\mathbf{y}$ need to be canceled out. This can be accomplished by realizing that light originating from outside the computational field of view has minimal variation across space, and thus the neighboring background values are approximately equal.

After postprocessing as needed, a successful reconstruction image of the object will have been completed (Saunders et al., 2019). To tackle the task of tracking an object placed around a corner, without using the time-of-flight technology, (Klein et al., 2016) proposed the use of 2D intensity images. This was accomplished by devising an optimization framework based on an unsophisticated imaging model using a laser pointer as the light source. The next section looks at some more challenging imaging problems that can be solved using light transport.

10.5 Applications

10.5.1 Applications in ToF Imaging

Multi-path interference correction for time-of-flight imaging. Time-of-flight systems, as discussed in 5.4.2 are increasingly being incorporated into everyday products. However, the noise in real-world environments creates many sources of errors for a ToF camera that need to be corrected for. One such source of error is multipath interference.

This occurs when multiple light rays converge onto the same pixel in the camera sensor, as shown in Figure 10.32. This causes errors in ToF imaging systems that assume single optical reflection. This occurs because the estimated depth calculated through continuous ToF imaging hinges on the computed phase delay, in the relationship

$$z = \frac{c\phi}{4\pi f_m}$$

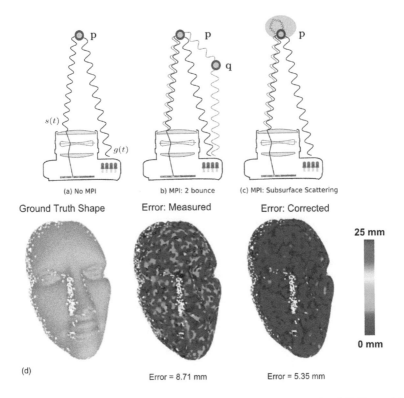

Figure 10.32 Multipath interference in ToF imaging. (a) An example of ToF in which a single light ray is emitted and captured after striking the scene surface at point p. (b) Here, a different light ray strikes the scene at point p after being reflected from q and also reaches the same sensor as the first light ray. This introduces interference in the ToF sensor computations. (c) Partial subsurface scattering of a light ray results in multiple light rays reaching the ToF sensor. (d) The first shape is the measured ground truth. The second shape is the error for the generated depth map using classical ToF imaging. Finally, we have the error of the depth map constructed using light transport optimization to lower noise. The corrected error is markedly lower than the original error. Reprinted from (Naik et al., 2015).

where z is the depth, ϕ is the phase delay, c is the speed of light, and f_m is the modulation frequency of the camera. This phase delay is calculated by cross correlating the emitted and detected signal (Naik et al., 2015). Light transport techniques discussed in the previous sections can serve to alleviate the problem of inaccurate phase delay computations, by separating the detected signal into its direct and global components.

10.5 Applications

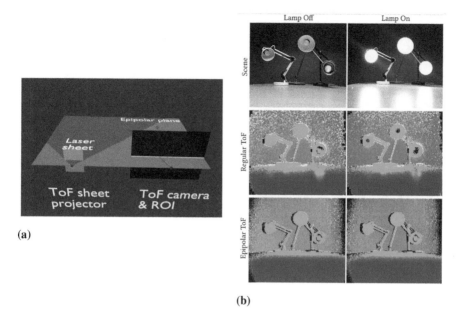

Figure 10.33 Epipolar ToF imaging. (a) In epipolar imaging, one row is imaged at a time, using a laser sheet. (b) Epipolar ToF imaging improves the depth measurements for even bright light bulbs. The errors caused by the surface reflectance of the light is suppressed in epipolar imaging. Reprinted from (Achar et al., 2017).

Using this separation, we can generate a closed-form solution for a more accurate phase angle. In this manner, light transport complements ToF imaging by offering more control over its environmental influences.

Epipolar time-of-flight for MPI correction. As discussed in the previous sections, continuous wave time-of-flight imaging methods provide accurate depth reconstructions but are often constrained by various assumptions. Incorporating epipolar geometry can help alleviate these disadvantages. Classic CW-ToF is very energy intensive, especially in outdoor settings, with bright ambient light. Comparatively, epipolar imaging converges the emitted light onto a single sheet, which increases its range threshold. Epipolar imaging also circumvents the problem of excessive global illumination interference, by blocking most global illumination prior to image capture. Epipolar ToF framework is also more effective at handling motion in a scene compared to regular CW-ToF, which suffers from corruption of depth reconstruction because its multiple input frames vary in time.

In the context of ToF, epipolar geometry is used to acquire strips of the image scene, sequentially. At a time, a sheet of laser light is projected onto the scene and is eventually captured by a camera with only the corresponding epipolar row of pixels active (see

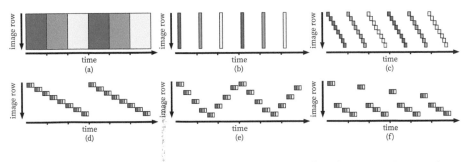

Figure 10.34 Epipolar scene sampling. (a) Capturing epipolar planes over time, in the manner of a rolling shutter camera. This reduces the effect of time varying motion blur, global illumination, and the ambient light on the image. (b) We can trade vertical resolution for higher temporal sampling by capturing every other epipolar plane. (c) Further optimization can be done for specific situations, by selectively increasing temporal resolution in different parts of the image. Reprinted from (Achar et al., 2017).

Figure 10.33a). Note that the row of camera pixels and the laser sheet are focused on the same scene location. Because CW-ToF requires two images to recover depth, there is some freedom to choose the order in which the epipolar planes are sampled.

The optimal method for image acquisition is to capture the set of modulations for a single row together, so that blur artifacts due to external motion are minimized (see Figure 10.34a). We can further optimize this by capturing the planes in a sawtooth manner to decrease misalignment of rows across an image, at the expense of vertical resolution (see Figure 10.34b). In certain applications such as ToF on a car, it may be better to have higher temporal resolution lower in the image, because lower portions of the field of view are moving faster, such as the road (see Figure 10.34c).

There is marked improvement in epipolar ToF with regard to global illumination and interreflections, as shown in Figure 10.33b. This method of ToF bridges the gap between improving the various situational performance deficiencies of CW-ToF and requiring less data collection than the point-by-point depth recovery of LiDAR.

10.5.2 Skin Imaging

Modeling the light transport in human skin has always been a problem of relevance for the imaging and graphics community. Human skin is composed of multiple layers with different optical behaviors, making it challenging to see through the skin or even render its appearance. A first step toward understanding light transport in human skin is to familiarize ourselves with skin anatomy. The three skin layers—epidermis, dermis, and subcutis—important from a graphics and imaging perspective. The optical behavior of the skin is governed by the unique optical behavior of each of these layers.

10.5 Applications

The epidermis, which is the outermost layer of the skin, is a transparent medium. It is followed by dermis, which is a semi-opaque or turbid medium. The subcutis, which is beneath the dermal layer, consists of fat cells. Noninvasive imaging of the dermal and subdermal layers of human skin is an important challenge for the medical community. Light undergoes a variety of optical phenomena such as interreflections, scattering, and selective absorption while passing through these layers.

To understand the optical behavior of skin, the community focuses largely on two pigments: melanin, the pigment responsible for skin color, and hemoglobin, the pigment that binds with oxygen in blood. Melanin is found in the epidermis and is responsible for the skin tone of a person. Higher concentrations of melanin result in darker skin tones. Hemoglobin is found in the dermal tissues. These two pigments are of importance because the absorption of light in the skin is largely dominated by melanin and hemoglobin compared with all other pigments combined. Deoxyhemoglobin, oxyhemoglobin, and melanin all have different wavelength or frequency of light at which they absorb maximally; all of them preferentially absorb light at shorter wavelengths. Melanin also has a markedly high refractive index compared with all other cellular level elements. As a result, when light hits the skin it is absorbed by melanin in the epidermis, with minimal scattering; within the dermis, the light is scattered by collagen fibers and eventually absorbed by hemoglobin. As a result, the volume of melanin in a person's skin drastically changes the optical behavior observed and the level of absorption. This results in a dependence of skin color on the performance of skin imaging systems, thus creating a melanin-dependent bias that should be accounted for. The spectral reflectance of different skin colors heavily depends on the absorption characteristics of melanin and hemoglobin (as well as carotene, which we do not discuss here). By extracting information about the melanin and hemoglobin contents of skin, it is possible to create photos in which a person has been reduced in age by several decades, among other things.

Using quantitative scattering and absorption parameters, we can develop models for light transport within the skin. There have been many different proposed models, ranging from the Lambert-Beer law, in which we are assuming almost no scattering and an exponential attenuation due to absorption; a modified Lambert-Beer model assuming light transport within a highly scattering medium; the Monte Carlo simulation for skin with a complex multilayered structure, particularly useful for modeling light transport in tissues; and the Kubelka-Munk theory for modeling light transport in tissues. Looking at the bidirectional reflectance distribution function (BRDF) and bidirectional surface scattering reflectance distribution function (BSSRDF) can also be useful in modeling the appearance of skin under different illuminations and in different viewing conditions.

Researchers have been looking not only at understanding light transport within skin but also at developing subsurface imaging systems for human skin. Using high-frequency coded illumination patterns, as mentioned previously in this chapter, it is possible to separate

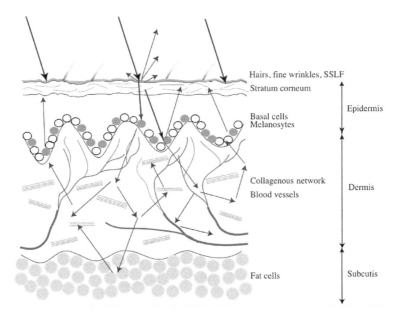

Figure 10.35 Optical behavior of skin. The epidermis, dermis, and subcutis skin layers all have their own unique optical behaviors based on their specific structures. Adapted from (Igarashi et al., 2007).

direct and global components of light; combining this with a multispectral sparsity-based approach enables us to look past the skin and see the veins and tissues beneath, as depicted in Figure 10.35. The key idea here is that spectral decomposition can be used to enhance certain components, for example, veins. The next step is to separate the global and direct components of the enhanced skin parts. For example, the scattering (global) component can yield an image of the tissues beneath the skin. An illustration of this approach is shown in Figure 10.36.

Some other skin imaging techniques include using a layered heterogeneous reflectance model to render a realistic human hand (Donner et al., 2008). This method pays special attention to interscattering between different skin layers. By taking into account the absorption presented by the melanin in the epidermis and hemoglobin in the dermis, it is possible not only to create a realistic rendering of human skin but also to render what that skin would look like if, for example, a person clenches a fist, resulting in temporary loss of blood flow to parts of the hand and a resultant lack of hemoglobin in those areas.

The complexity of human skin and handling multiple optical phenomena like scattering, absorption, and interreflections make skin imaging a challenging problem. However, light transport in human skin is governed by the same principles we studied previously in this chapter. We have seen that there are many promising models, as well as promising results—

10.5 Applications

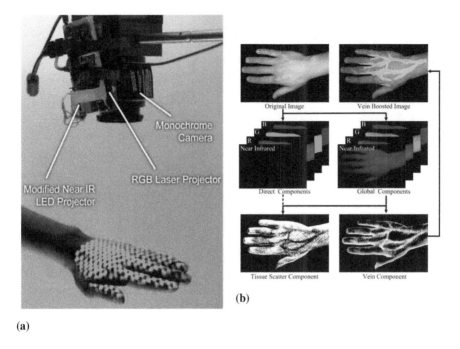

Figure 10.36 Separation of skin components. In (a), an RGB projector displays a green pattern on a hand, which is captured by a monochrome camera. In (b), we see the isolation of the veins and separation of global and direct components using the RGB and infrared spectra. Adapted from (Kadambi et al., 2013).

whether it is in rendering photorealistic images of skin or in actively performing subsurface imaging using the methods described here. Skin imaging is currently a field of great importance, owing to the increasing focus on virtual reality and telehealth.

10.5.3 Imaging through Scattering Media

The focus of this chapter is to introduce the light transport formulation. The formulation can also be used for imaging through scattering media. Imaging through scattering media finds applications in a variety of critical tasks such as biomedical imaging, underwater exploration, and improved transportation systems in challenging weather. Scattering of light from particles contributes to the global light paths emerging from the scenes. Early work in this area focused on using spatial light modulators (SLMs) for wavefront shaping for descattering (Katz et al., 2011). The techniques we have studied throughout this chapter have also been shown effective for this approach.

(Katz et al., 2012) introduced wavefront shaping and established its ability to facilitate wide-field imaging across turbid layers with incoherent illumination, and imaging of oc-

(a) (b) (c) (d) (e)

Figure 10.37 Confocal imaging and descattering. (a) Scene with objects in a 3D fish tank. (b) Original image of fish tank filled with diluted milk. (c) Partially descattered using confocal imaging. (d) Additional optimization removes more global scattering. (e) The recovered 3D structure is visualized for a different view. Reprinted from (Fuchs et al., 2008).

cluded objects using light scattered from diffuse walls. Interestingly, this approach did not require raster scanning, coherent sources, offline reconstruction, or interferometric detection. (Katz et al., 2014a) showed that a single high-resolution image of the scattered light, taken with a standard camera, encodes sufficient data due to the memory effect for speckle correlations. This information was shown to be adequate enough to image around visually obscure layers and around corners with diffraction-constrained resolution. (Chaigne et al., 2014), paved the path toward deep-tissue imaging and light delivery utilizing endogenous optical contrast by combining a transmission-matrix approach with the advantages of photo-acoustic imaging. Their approach allowed the noninvasive measurement of an optical transmission matrix over a large volume, inside complex samples, using a standard photoacoustic imaging setup. Optical nonlinearities were exploited to form a diffraction-limited focus inside or through a complex sample, even when the feedback signal is not localized by (Katz et al., 2014b). This system enabled imaging through strongly scattering turbid and visually opaque layers.

The idea of global and direct separation in combination with confocal imaging has also been applied to this problem, as shown in Figure 10.37, and challenges that we saw in the skin imaging section. (Hebden et al., 1991), (Satat et al., 2016) have demonstrated the use of transient imaging and SPAD sensors for seeing through scattering media. (Satat et al., 2018a) showed a technique for imaging through highly scattering media with a SPAD camera in optical reflection mode. One of the advantages of this method was the recovery of multiple targets at different scattering levels. Due to its superior noise performance in low lighting conditions, SPAD could be a good choice for imaging through volumetric scattering. These approaches are promising and have achieved success in small-scale experimental settings, but are still miles away from developing systems that can be deployed on a self-driving car to perfectly see through fog.

10.5 Applications

Chapter Appendix: Notations

Notation	Description
I	Scene
S	Source
P	Sensor (e.g., camera)
$\mathbf{T}$	Light transport matrix
$\mathbf{p}$	Irradiance measurements/camera pixels
$\mathbf{l}$	Independent source/illumination pixels
$\mathbf{I}$	Captured image
$\mathbf{I}_i$	Image containing ith-order interreflections
$\omega_\mathbf{x}^\mathbf{y}$	Rays originating from $\mathbf{x}$ and directed to $\mathbf{y}$
$\mathbf{L}_{in}(\omega_\mathbf{x}^\mathbf{y})$	Radiance as a function of all incident light rays $\omega_\mathbf{x}^\mathbf{y}$
$\mathbf{L}_{out}(\omega_\mathbf{x}^\mathbf{y})$	Radiance as a function of all outgoing light rays $\omega_\mathbf{x}^\mathbf{y}$
$\mathbf{L}_{in}^i, \mathbf{L}_{out}^i$	Component due to ith interreflection ($i = 1$ constitutes direct, and $i > 1$ constitutes indirect component)
$\mathbf{E}$	Identity matrix
$\mathbf{A}$	Matrix characterizing proportion of irradiance
$\mathbf{C}^1$	Cancellation matrix
$\mathbf{T}^1$	Components of $\mathbf{T}$ due to 1-bounce reflections
$\mathbf{L}_d$	Direct component
$\mathbf{L}_g$	Global component
$\mathbf{L}[c, i]$	Radiance of a patch i measured by a camera c
P	Set of patches in the scene
$\mathbf{A}[i, j]$	Reflectance distribution over the patch $[i, j]$
$\mathbf{L}_{gd}$	Direct component of radiation from scene patches
$\mathbf{L}_{gg}$	Global component of radiation from scene patches
$\mathbf{L}^+$	Image of the scene lit with high frequency illumination
$\mathbf{L}^-$	Image of the scene lit with a complementary illumination
b	Brightness of the deactivated source as a fraction of the activated element
$\mathbf{b}_{mat}$	Matrix that stores the value of b for each pixel
Π	Probing matrix
$\mathbb{1}$	Vector of all ones
$\odot$	Elementwise product
$\{\mathbf{p}_k\}$	Set of rank 1 matrix of illumination patterns

$\{\mathbf{m}_k\}$	Set of rank 1 matrix of masks for optical probing
$\mathbf{T}^D$	Transport matrix of direct image
$\mathbf{T}^{EI}$	Transport matrix of epipolar indirect image
$\mathbf{T}^{NE}$	Transport matrix of non-epipolar indirect
$\mathbf{T}^\tau$	Light transport matrix with path length τ
$\mathbf{M}$	Binary matrix
$\omega(\tau)$	Binary function of path length τ
d_r	Distance between reference arm and the beamsplitter
d_s	Distance between the target mirror and the beamsplitter
ε	Measurement noise
$I(x, y)$	Light intensity at pixel location (x, y)
$I(x, y, t)$	Light intensity at pixel location (x, y) and time t
STIR $(x_i, y_i, \theta_i, \Phi_i, t)$	Space time impulse response (STIR) for sensor pixel (x_i, y_i), direction of ray (θ_i, Φ_i), and time t
$\mathbf{O}^i$	Set of ith onsets (i.e., collection of time instants)
$\mathbf{D}$	Matrix of pairwise Euclidean distances between all patches
$\mathbf{d}$	Vectorized version of $\mathbf{D}$
c	Speed of light
$\mathbf{d}_H$	Distances to the hidden patches
$\mathbf{O}_h$	Arrival times corresponding to hidden patches
$L(w)$	Radiance at wall patch w
$L_e(l)$	Emitted radiance hitting a wall patch l
ρ	Diffuse albedo/orientation
V	Volume of the hidden scene
$g(x)$	Geometry term correcting for hidden patch location and albedo/orientation
$\tau(x)$	Total travel time
$\mathbf{v}$	Discrete hidden patch locations
$\mathbf{i}$	Transient image
$\mathbf{C}$	Correlation matrix corresponding to camera and image modulation
$\mathbf{h}$	Measurement from the ToF sensor
$\Gamma(\mathbf{v})$	Regularization term
ϕ_L, ϕ_C	Phase difference
ψ	Angle of reflection
$C(u, v, \psi)$	Phasor representing transport between the wall and the camera
$M(u, v)$	Total transport phasor
FWHM$^\angle$	Reconstruction resolution

10.5 Applications

$\gamma_\angle$	Diffusivity of sensor/wall
f_m	Modulation frequency of the camera
$\tau(x', y', t)$	Final 3D volume of measurements
$\mathbf{R}_t$	Transformation on the temporal domain
$\mathbf{R}_z$	Transformation on the spatial domain
$\mathbf{H}$	3D shift-invariant convolution operation
$\tau_F(v)$	Length of the Fermat path relative to v
∇_v	Spatial gradient operator
$a(r, \theta)$	Albedo at point (r, θ)
θ_{occ}	Occluded object

Exercises

For the computations in this exercise, you may find these vector derivatives helpful.

$$\nabla_x \left[x^T A \right] = A$$
$$\nabla_x \left[x^T c \right] = c$$
$$\nabla_x \left[x^T x \right] = 2x$$
$$\nabla_x \left[x^T A x \right] = 2Ax$$

1. Review of core concepts

   a) Consider the pinhole camera. When the camera aperture decreases in size, what happens to the image? (Mention one effect in the context of ray optics and one effect in the context of wave optics).

   b) Describe the plenoptic equation. You do not need to write the full equation, but offer a qualitative interpretation.

   c) Provide the name of an algorithm to rigidly register 3D point clouds.

   d) Write a regularized optimization program that deblurs an image, such that the deblurred result is piecewise smooth. In other words, the program should aim to smooth out noise, while retaining sharp edges. (Hint, look at question 3 for an example of formatting a regularized optimization program.)

2. Deblurring without regularization

   You are designing a deblurring algorithm for an imaging system. Let $x \in \mathbb{R}^N$ be a vectorized ground truth image and $y \in \mathbb{R}^M$ be a vectorized measured image. The process can be modeled as an LSI system

   $$x \longrightarrow H \longrightarrow y,$$

   which in matrix-vector form is $y = Hx$. Our goal is to recover an estimate of the ground truth image, $\hat{x}$. Because H is not necessarily square, we appeal to the notion of a pseudoinverse. In this two-part question, we first pose this as an optimization program to minimize the least-squares error, for which we will find a closed-form solution to $\hat{x}$. If done correctly, this closed-form solution should converge to the pseudoinverse.

   a) Formulate a least-squares optimization program for finding $\hat{x}$. It is not necessary to solve the program in this part.

   b) Solve the optimization program by finding the closed-form solution to $\hat{x}$. Show your work for the closed-form derivation.

Exercises

3. Tikhonov regularization

   Many inverse problems in imaging are ill posed and cannot be solved with a simple pseudoinverse. Hence, one appeals to the notion of *regularization*, in which additional information is used to prevent overfitting. In computer vision and imaging, this additional information manifests as a regularized constraint in the optimization program.

   One regularization approach is known as *Tikhonov regularization*. Alternate names for this formulation are *ridge regression* (statistics community) and *shrinkage* (machine learning community). Consider the deblurring example from question 2. As compared with the least-squares solution, the idea is to solve an optimization program of the form:

   $$\text{(P1): } \widehat{\mathbf{x}} = \arg\min_{\mathbf{x}} \|\mathbf{y} - \mathbf{H}\mathbf{x}\|_2^2 + \|\Gamma\mathbf{x}\|_2^2, \qquad (10.1)$$

   where Γ is a matrix that controls the type of regularization. For example, if Γ is the identity matrix, then we are directly shrinking the values of $\mathbf{x}$. However, if Γ is an operator that computes the gradients of $\mathbf{x}$, then we are looking for a smooth solution.

   a) Find the closed form solution to $\widehat{\mathbf{x}}$ for program (P1). Write only the solution for (P1).

   b) Consider the ℓ_1 norm, where $\|\mathbf{a}\|_1 = \sum_i |a_i|$. We can replace the ℓ_2 norm in the regularization term with an ℓ_1 norm, yielding:

   $$\text{(P2): } \widehat{\mathbf{x}} = \arg\min_{\mathbf{x}} \|\mathbf{y} - \mathbf{H}\mathbf{x}\|_2^2 + \|\Gamma\mathbf{x}\|_1 \qquad (10.2)$$

   Suppose Γ is assumed to be $\mathbf{I}$. Compare and contrast the estimate $\widehat{\mathbf{x}}$ from program (P1), (P2), and the unregularized solution (i.e., ordinary least squares). Hint: Distinguish the nuances in the solution from programs P1 and P2.

4. Separating light transport

   This question draws from the SIGGRAPH paper "Fast separation of direct and global components of a scene using high frequency illumination" (Nayar et al., 2006).

   a) Draw an experimental configuration to separate components of light transport using Nayar's method. Prepare a list of parts, and write pseudocode to compute the separation. (Hint: The basic idea of Nayar et al.'s work is to interpret global light transport as a low spatial frequency phenomenon, and direct light transport as a high spatial frequency phenomenon.)

   b) Imagine that you are working at a self-driving car company and your engineering team is encountering a problem. Shiny objects in the scene (like oil slicks or shop windows) act as secondary reflectors for car headlamps. Your computer vision algorithms are misidentifying these confounding reflections as cars. Meanwhile, the company sends you to SIGGRAPH, as they do every year, where you see the Nayar et al. paper on global/direct separation. On the flight back, you are considering tech transfer to your company to solve the problem noted here.

Assuming that the prototype in Nayar's paper can be streamlined (e.g., real-time performance, size, weight, etc.), should you engage in tech transfer to solve this problem? Provide justification to why or why not.

Glossary

Aberration	It is a property of lenses that makes light spread out over some region of space instead of being focused to a point. An example is the chromatic aberration, also called dispersion, which causes light rays of different wavelengths to focus at variable distances form the lens. Using a 3rd order approximation of the sine function, up to 5 different types of aberration can be identified.
Active Illumination	Active illumination is an engineered source of light in a scene. This type of illumination is either used to artificially brighten a dark room, or to encode spatial, temporal, or frequency information into the image measurement process. It differs from passive illumination, which is the ambient illumination already present in a scene.
Aliasing	It represents an image distortion caused by a low sampling rate acquisition. According to Shannon's sampling theory, the acquisition rate should be twice the highest frequency in the image. Aliasing can be prevented by low pass filtering the image before sampling.
Arnoldi Iteration	Arnoldi iteration is an iterative eigenvalue algorithm. Arnoldi finds an approximation to the eigenvalues and eigenvectors of general matrices by constructing an orthonormal basis of the Krylov subspace, which makes it particularly useful when dealing with large sparse matrices. The Arnoldi method gives a partial result after a small number of iterations, in contrast to direct methods which must complete all iterations to give useful results.
Augmented Reality	Augmented reality (AR) is an interactive experience of a real-world environment where the objects that reside in the real world are enhanced by computer-generated perceptual information, sometimes across multiple sensory modalities. The AR system fulfills three basic features: a combination of real and virtual worlds, real-time interaction, and accurate 3D registration of both virtual and real objects.
Bidirectional Reflectance Distribution Function	Bidirectional Reflectance Distribution Function (BRDF) is a general reflectance model used to describe the proportion of light reflecting in a certain direction, given the direction of the incoming light. It is a function of four angles (two for incident light, two for reflected light) and can take on values from 0 to 1. The BRDF is typically dependent on the illumination conditions of a scene, as well as the material properties of the object of interest.
Color Filter Array	A color filter array is placed in front of a sensor to obtain color/spectral information from a scene. Each filter has a specific (and known) spectral sensitivity, selectively allowing only certain wavelengths to pass freely. The most well known such filter is the ubiquitous Bayer filter.
Colorimetry	Colorimetry is the mathematical and psychological study of color. Retinal color discusses the manner in which humans perceive color, specifically RGB vision due to the L-, M-, and S-cones. Perceptual color is related to human adaptations to illumination and context cues, e.g., color constancy.

Compressive Imaging Compressive imaging is an image processing technique to efficiently capture an image with a reduced number of samples. This allows recovering a higher resolution image of the scene, by solving an ill-conditioned linear system. The method achieves good results assuming that the initial image is sparse, i.e., it can be expressed using a low number of samples in a different base.

Cross-Polarization Cross-polarization is defined as the angle of polarization orthogonal to the polarization state of interest. In cross-polarization imaging, a polarizer is placed in front of the camera aperture at an angle orthogonal to the glare polarization, enabling the removal of glare from photographs.

Dappled Photography Dappled photography is a new imaging method that uses an attenuation mask to capture the light field in the scene. For example, by placing a high frequency sinusoidal mask between the sensor and the optical elements of a camera, a wider region of the light field in the scene can be captured with one single shot of the camera.

Deblurring It is the process of eliminating the artifacts caused by blurring in an image. Blurring can have various causes. For example, motion blur is likely to appear when capturing fast objects in motion. This can be solved by a technique called *fluttered shutter*, which opens/closes the camera shutter very quickly in a predefined pattern.

Demosaicking Typically the sensors in modern cameras are coded such that each pixel is only sensitive to one of the RGB colors, thus creating a mosaic-like color pattern. The full resolution image is computed with an algorithm called demosaicking, which recovers the image for each color, via interpolation at the missing pixels.

Dichromatic Reflection Model The dichromatic reflection model predicts that scene reflectance can be modeled as a sum of two terms: a specular and diffuse term. It also suggests that specular reflection is related to the incident spectral illumination, while the spectrum of diffuse reflection is related to the medium itself.

Diffraction Diffraction is a light bending phenomenon occurring when a light wave encounters an obstacle or a slit. For example, when a light beam encounters a slit of dimensions comparable to the light wavelength, the light bends around the slit edges, creating a pattern given by a circular disk with rings around it.

Diffuse Interreflection Diffuse interreflection is a process whereby light reflected from an object strikes other objects in the surrounding area, illuminating them. Diffuse interreflection specifically describes light reflected from objects which are not shiny or specular to reach areas not directly in view of a light source. Based on the coloration of the surface, the reflected light incident on the surrounding objects is also colored.

Direct Illumination It represents the illumination caused directly by a light source, and it is one of two components of the scene illumination. The direct light component enhances the material properties of a given point, and is in contrast to the global component, revealing the optical properties of the scene.

Dual Photography Dual photography is a photographic technique that uses Helmholtz reciprocity to capture the light field of all light paths from a structured illumination source to a camera. Image processing software is then generally used to reconstruct the scene as it would have been seen from the viewpoint of the projector.

Epipolar Plane In terms of stereo vision, the plane formed by the three-dimensional object (which is being imaged) and the optical centers of the two cameras (that are imaging the object) is referred to as the epipolar plane. The plane intersects each of the two camera's image planes such that at the intersection, epipolar lines are formed. It is important to know that all the epipolar lines and epipolar planes intersect the epipole (the object) irrespective of where it is located with respect to the two cameras.

Focal Stack	A focal stack denotes a set of images captured with the camera focused at different depths. This gives a more comprehensive description of the light field, and allows computing extended depth of field photographs, which allow focusing on multiple points in the image.
Fourier Ptychography	Fourier ptychography is a method that increases the range of light field angles that can be captured with a microscope by recording images illuminated from a range of different angles. This leads to an increased image resolution and is much faster and simpler than moving the specimen for each capture.
Fluorescence Lifetime Imaging	Fluorescence lifetime imaging is an imaging technique capturing the differences in the exponential decay rate of fluorescent chemicals in a sample. Knowledge of the sample's fluorescence lifetime allows applications such as DNA sequencing, tumor detection, and high-resolution microscopy.
Fresnel Coefficients	Fresnel coefficients describe the proportion of light reflected and transmitted at the interface between two different media. These coefficients are defined separately for two polarization states: one orthogonal to the plane of incidence, and the other parallel. The total amount of light reflected at the interface can be determined by decomposing the light into its constituent polarization states and calculating the reflectance for each polarization state separately.
Global Illumination	It represents the illumination caused by points in the scene different from the light source, and it represents one of two components of the scene illumination. The global illumination reveals the optical properties of the scene, indicating how a certain point is illuminated by other points in the scene.
Gradient Descent	It is defined as a first-order iterative optimization algorithm for finding a local minimum of a differentiable function. Steps are taken which are proportional to the negative of the gradient of the function at the current point in order to find a local minimum of a function using gradient descent approach.
High Dynamic Range Imaging	This represents a method used to reproduce luminosity ranges much wider than possible with standard imaging techniques. It involves algorithms that combine images captured with different exposure values which contain details of the brighter or darker portions of the image.
Ideal Point Source Light	It represents a source of light that is infinitesimal in size and radiates light outward uniformly in all directions. The light rays therefore form a continuum and they are mapped one-to-one to an imaginary sphere centered in the point source. This ensures that quantities such as the radiant flux are transmitted equally across the sphere's surface.
Irradiance	The irradiance represents the radiant flux incident to an area on the imaginary sphere centered in an ideal point source. The irradiance is directly proportional to the radiant flux, and inversely proportional to the square of sphere.
Lambert's Law	Lambert's cosine law describes the attenuation of light reflected off of a diffuse object. It states that the reflected light is (approximately) isotropic, and the intensity reflecting off of the surface is proportional to the product of the incoming light intensity and the cosine of the angle between the surface normal and the incident light.
Lambertian Surface	The Lambertian surface is a surface which diffuses the light uniformly when illuminated. It represents an ideal *matte* surface, which means that the brightness is perceived the same irrespective of the observer's position.
Light Field	Light field is a mathematical function of one or more variables whose range is a set of multidimensional vectors that describes the amount of light flowing in every direction through every point in space. The magnitude of each ray is given by the radiance and the space of all possible light rays is given by the five-dimensional plenoptic function.
Light Ray	A line describing the trace that a photon might leave behind, which is considered infinitesimal in width and has an infinitesimal point of emergence. They are attenuated when passing through objects, and the overall attenuation is the same if the direction of the light ray is reversed, known as the reversibility property of light rays.

Light Stage	A light stage is a mechanical component used to illuminate an object from many directions. It has important applications in graphics, as it can fully capture the reflectance field of an object. It can be used to digitally relight an image and obtain the shape of an object with high accuracy.
Lock-in Sensor Imaging	A lock-in sensor is one of the most widely used sensor mechanisms in time-resolved imaging which measure phase differences between emitted and received signals. This makes it particularly useful for time-of-flight, which measures the round trip time of an artificially generated light signal.
Monochromatic	Light (electromagnetic radiation) can be said to be monochromatic when the optical spectrum contains only a single optical frequency. The associated electric field strength at a certain point in space generally exhibits a purely sinusoidal oscillation, having a constant instantaneous frequency and a zero bandwidth. Light sources can also be called monochromatic if they emit monochromatic light.
Non-line-of-sight imaging	A set of imaging techniques, usually active, aimed at recovering objects beyond the direct line-of-sight.
Parallax	Parallax which could be described as visual alternation is a displacement or difference in the apparent position of an object viewed along two different lines of sight and is measured by the angle or semi-angle of inclination between those two lines. Due to foreshortening, nearby objects show a larger parallax than farther objects when observed from different positions, so parallax can be used to determine distances of those objects from the viewer.
Parallax Barriers	Parallax barriers denote a technology used in traditional 3D displays based on series of occluding bars. This allows the spectator to see only one perspective of the object from each viewing angle. This technology creates a trade-off between angular and spatial resolution, and was replaced in newer 3D displays.
Pinhole Camera	It is a camera without lenses containing only a tiny hole in one of its walls. The light gets projected upside down on the opposite wall. It generally suffers from low light throughput, since a good image contrast requires a small hole.
Plenoptic Function	The plenoptic function is a high dimensional function which represents a detailed mathematical model of the light field. In a general setting it has 7 variables for position, angle, wavelength, and time. However, it is common to use a 5 variable simplification that considers light to be monochromatic and time-invariant.
Multilayer Display	A multilayer display is a 3D display technique comprising of multiple levels of LCD screens stacked in parallel and separated by predefined distances. Unlike parallax barriers, these displays have a better light throughput, and a higher spatial resolution.
Radiance	The radiance represents the ray strength, measuring the combined angular and spatial power densities. Radiance can be used to indicate how much of the power emitted by the light source that is reflected, transmitted, or absorbed by a surface will be captured by a camera facing that surface from a specified angle of view.
Radiant Flux	The energy emitted, reflected, transmitted, or received, per unit time, and is measured in watts, or joule per second. It characterizes an ideal point source, and is transmitted equally across the surface of an imaginary sphere centered in the point source.
Radiant Intensity	The radiant intensity measures the angular power density, and is the radiant flux emitted per unit solid angle. Therefore, a light beam has a higher radiant intensity if the power of the emitting source is focused on a narrower solid angle.
Ray Model	It represents a model of light that consists of single photon traces that don't interact with each other. This is in contrast to the wave model of light, which typically emerges in closed environments such as a pinhole camera.

Reflectance Map	A reflectance map is a contour map in gradient space used to express the reflectance of a surface under certain illumination. The reflectance is expressed with respect to the partial spatial derivatives of the object. The shape of the object can be determined by mapping the reflectance from multiple different maps to the respective point in gradient space.
Spatially Coded Imaging	Spatially coded imaging is a flexible alternative to the conventional imaging setup where spatial imaging parameters such as the aperture, sensor, and the illumination can be engineered to enhance the quality of the imaging system.
Spectral Unmixing	Spectral unmixing is an inverse problem that aims to express each pixel as a sum of certain materials. The idea is that each pixel contains a small number of materials, and each material has a distinct spectrum. If more than one material is in a pixel, the spectrum of each material present will linearly superimpose when measured by the camera. The goal of spectral unmixing is to recover these constituent spectras and their respective intensity in each pixel.
Spectrometry	Spectrometry is defined as the field analyzing the spectra of point sources. Most photographs taken capture the light intensity distributed in space. Analyzing the light in the frequency domain can say a lot about the material properties.
Specular Surface	A specular surface reflects the incoming light in a unique direction relative to the surface, and is common with objects having a glossy or polished texture. This is in contrast to Lambertian surfaces, which reflect light uniformly in all directions.
Stokes Vector	The Stokes parametrization of electromagnetic radiation is a compact notation used to denote its polarization state. It is a vector with four entries: (S_0, S_1, S_2, S_3). The first entry of Stokes vector describes the intensity of the light, while the last three entries are used to indicate the level of linear and circular polarization. A linear combination of linear and circular polarization can be expressed as an elliptical polarization.
Strobe Photography	It represents an imaging technique that uses light and sound to trigger the flash burst with precise timing. This allows capturing fast phenomena, and can lead to stunning photographs such as the famous *bullet through apple*.
Subsurface Scattering	Subsurface scattering is a mechanism of light transport in which light that penetrates the surface of a translucent object is scattered by interacting with the material (reflected a number of times at irregular angles inside the material) and exits the surface at a different point.
Superresolution Imaging	Superresolution imaging is an approach to boost the resolution of a camera by transcending the diffraction limit, or exceed the native resolution of the imaging sensors used.
Temporal Coherence	Temporal coherence is the measure of the average correlation between the value of a wave and itself delayed by some time, at any pair of times. Temporal coherence tells us how monochromatic a source is. It basically characterizes how well a wave can interfere with itself at different time instances.
Thin Lens	The thin lens is an ideal model of a lens with a thickness that is negligible compared to the radii of curvature of the lens surfaces. It is characterized by three parameters: the focal length, aperture diameter, and lens speed, and bends the light according to a simplified equation known as the thin lens equation.
Time-of-Flight Camera	A time-of-flight (ToF) camera is a range imaging camera system that employs time-of-flight techniques to resolve distance between the camera and the subject for each point of the image, by measuring the round trip time of an artificial light signal provided, generally by a laser.
Tone Mapping	Tone mapping is a technique to create a mapping between two sets of colors, in order to produce aesthetically pleasing images, enhance details, or generate a higher contrast photograph.

References

Abdi, Hervé, and Lynne J. Williams. 2010. Principal component analysis. *Wiley Interdisciplinary Reviews: Computational Statistics* 2 (4): 433–459.

Achar, Supreeth, Joseph R. Bartels, William L'red' Whittaker, Kiriakos N. Kutulakos, and Srinivasa G. Narasimhan. 2017. Epipolar time-of-flight imaging. *ACM Transactions on Graphics (ToG)* 36 (4): 1–8.

Ackermann, Jens, and Michael Goesele. 2015. A survey of photometric stereo techniques. *Foundations and Trends® in Computer Graphics and Vision* 9 (3-4): 149–254.

Adams, Ansel. 1980. *The camera, the Ansel Adams photography series*. Little, Brown and Company, 1981.

Adelson, Edward H., and James R. Bergen, et al. 1991. Computational models of visual processing, eds. Michael Landy and J. Anthony Movshon, Vol. 2.

Adelson, Edward H., and John Y. A. Wang. 1992. Single lens stereo with a plenoptic camera. *IEEE Transactions on Pattern Analysis and Machine Intelligence* 14 (2): 99–106.

Adib, Fadel, and Dina Katabi. 2013. See through walls with WiFi! In *Proceedings of the ACM SIGCOMM*, 75–86.

Agard, David A. 1984. Optical sectioning microscopy: Cellular architecture in three dimensions. *Annual Review of Biophysics and Bioengineering* 13 (1): 191–219.

Agrawal, Amit, and Ramesh Raskar. 2007. Resolving objects at higher resolution from a single motion-blurred image. In *2007 IEEE Conference on Computer Vision and Pattern Recognition*, 1–8.

Agrawal, Amit, Ramesh Raskar, and Rama Chellappa. 2006. What is the range of surface reconstructions from a gradient field? In *European Conference on Computer Vision (ECCV)*, 578–591.

Agrawal, Amit, Ramesh Raskar, Shree K. Nayar, and Yuanzhen Li. 2005. Removing photography artifacts using gradient projection and flash-exposure sampling. In *ACM SIGGRAPH 2005 Papers*, 828–835.

Akers, David, Frank Losasso, Jeff Klingner, Maneesh Agrawala, John Rick, and Pat Hanrahan. 2003. Conveying shape and features with image-based relighting. In *IEEE Visualization, 2003. VIS 2003*, 349–354.

Alvarez-Cortes, Sara, Timo Kunkel, and Belen Masia. 2016. Practical low-cost recovery of spectral power distributions. In *Computer Graphics Forum*, Vol. 35, 166–178.

Andreou, Andreas G., and Zaven Kevork Kalayjian. 2002. Polarization imaging: principles and integrated polarimeters. *IEEE Sensors Journal* 2 (6): 566–576.

Anrys, Frederik, and Philip Dutré. 2004. Image-Based Lighting Design. In *Proceedings of the 4th IASTED International Conference on Visualization, Imaging, and Image Processing*.

Antipa, Nicholas, Sylvia Necula, Ren Ng, and Laura Waller. 2016. Single-shot diffuser-encoded light field imaging. In *Proceedings of IEEE Intl. Conf. on Computational Photography (IEEE ICCP)*, 1–11.

Arellano, Victor, Diego Gutierrez, and Adrian Jarabo. 2017. Fast back-projection for non-line of sight reconstruction. *Optics Express* 25 (10): 11574–11583.

Asghari, Mohammad H., and Bahram Jalali. 2015. Edge detection in digital images using dispersive phase stretch transform. *International Journal of Biomedical Imaging* 2015.

Au, Whitlow W. L., and Kelly J. Benoit-Bird. 2003. Automatic gain control in the echolocation system of dolphins. *Nature* 423 (6942): 861–863.

Ba, Yunhao, Alex Gilbert, Franklin Wang, Jinfa Yang, Rui Chen, Yiqin Wang, Lei Yan, Boxin Shi, and Achuta Kadambi. 2020. Deep shape from polarization. In *European Conference on Computer Vision (ECCV)*, 554–571.

Bach, Francis, Rodolphe Jenatton, Julien Mairal, and Guillaume Obozinski. 2011. Optimization with sparsity-inducing penalties. *Foundations and Trends® in Machine Learning* 4 (1): 1–106. doi:10.1561/2200000015.

Bae, Soonmin, and Frédo Durand. 2007. Defocus magnification. In *Computer Graphics Forum*, Vol. 26, 571–579.

Baek, Seung-hwan, Incheol Kim, Diego Gutierrez, and Min H. Kim. 2017. Compact single-shot hyperspectral imaging using a prism. *ACM Transactions on Graphics (ToG)* 36 (6): 1–12.

Banks, Martin S., David M. Hoffman, Joohwan Kim, and Gordon Wetzstein. 2016. 3D Displays. *Annual Review of Vision Science* 2: 397–435.

Baraniuk, Richard G. 2007. Compressive sensing [lecture notes]. *IEEE Signal Processing Magazine* 24 (4): 118–121.

Baronti, S., A. Casini, F. Lotti, and S. Porcinai. 1998. Multispectral imaging system for the mapping of pigments in works of art by use of principal-component analysis. *Applied Optics* 37 (8): 1299–1309.

Barsky, Svetlana, and Maria Petrou. 2003. The 4-source photometric stereo technique for three-dimensional surfaces in the presence of highlights and shadows. *IEEE Transactions on Pattern Analysis and Machine Intelligence* 25 (10): 1239–1252.

Basri, Ronen, David Jacobs, and Ira Kemelmacher. 2007. Photometric stereo with general, unknown lighting. *International Journal of Computer Vision* 72 (3): 239–257.

Beck, Amir, and Marc Teboulle. 2009. A fast iterative shrinkage-thresholding algorithm for linear inverse problems. *SIAM Journal on Imaging Sciences* 2 (1): 183–202.

Beckmann, Matthias, Felix Krahmer, and Ayush Bhandari. 2020. HDR tomography via modulo Radon transform. In *IEEE International Conference on Image Processing (ICIP)*, 3025–3029. IEEE. doi:10.1109/icip40778.2020.9190878.

Betremieux, Yan, Timothy A. Cook, Daniel M. Cotton, and Supriya Chakrabarti. 1993. SPINR: Two-dimensional spectral imaging through tomographic reconstruction. *Optical Engineering* 32 (12): 3133–3139.

Bhandari, A., Micha Feigin, Shahram Izadi, Christoph Rhemann, Mirko Schmidt, and Ramesh Raskar. 2014a. Resolving multipath interference in Kinect: An inverse problem approach. In *Proc. of IEEE SENSORS*, 614–617.

Bhandari, A., A. Kadambi, R. Whyte, C. Barsi, M. Feigin, A. Dorrington, and R. Raskar. 2014b. Resolving multipath interference in time-of-flight imaging via modulation frequency diversity and sparse regularization. *Optics Letters* 39 (7).

Bhandari, Ayush. 2018. Sampling time-resolved phenomena. PhD diss, Massachusetts Institute of Technology.

Bhandari, Ayush, and Felix Krahmer. 2020. HDR imaging from quantization noise. In *IEEE International Conference on Image Processing (ICIP)*, 101–105. IEEE. doi:10.1109/icip40778.2020.9190872.

Bhandari, Ayush, and Ramesh Raskar. 2016. Signal Processing for Time-of-Flight Imaging Sensors: An introduction to inverse problems in computational 3-D imaging. *IEEE Signal Processing Magazine* 33 (5): 45–58.

Bhandari, Ayush, Christopher Barsi, and Ramesh Raskar. 2015. Blind and Reference-free Fluorescence Lifetime Estimation via Consumer Time-of-Flight Sensors. *Optica* 2 (11): 965–973.

Bhandari, Ayush, Matthias Beckmann, and Felix Krahmer. 2020a. The Modulo Radon Transform and its inversion. In *European Signal Processing Conference (EUSIPCO)*, 770–774.

Bhandari, Ayush, Miguel Heredia Conde, and Otmar Loffeld. 2020b. One-bit time-resolved imaging. *IEEE Transactions on Pattern Analysis and Machine Intelligence* 42 (7): 1630–1641. doi:10.1109/tpami.2020.2986950.

Bhandari, Ayush, Felix Krahmer, and Thomas Poskitt. 2021. Unlimited sampling from theory to practice: Fourier-Prony recovery and prototype ADC. *IEEE Transactions on Signal Processing* 70: 1131–1141. doi:10.1109/TSP.2021.3113497.

Bhandari, Ayush, Felix Krahmer, and Ramesh Raskar. 2017. On unlimited sampling. In *International conference on sampling theory and applications (SampTA)*. doi:10.1109/sampta.2017.8024471.

Bhandari, Ayush, Felix Krahmer, and Ramesh Raskar. 2020. On unlimited sampling and reconstruction. *IEEE Transactions on Signal Processing*. doi:10.1109/TSP.2020.3041955.

Bhandari, Ayush, Christopher Barsi, Refael Whyte, Achuta Kadambi, Anshuman J. Das, Adrian Dorrington, and Ramesh Raskar. 2014. Coded time-of-flight imaging for calibration free fluorescence lifetime estimation. In *Imaging Systems and App.*, 2–5.

Björck, Ake. 1996. *Numerical Methods for Least Squares Problems*.

Blackwell, H. Richard. 1946. Contrast thresholds of the human eye. *Journal of the Optical Society of America* 36 (11): 624.

Blu, T., and M. Unser. 1999. Quantitative Fourier analysis of approximation techniques. I. Interpolators and projectors. *IEEE Transactions on Signal Processing* 47 (10): 2783–2795. doi:10.1109/78.790659.

Bolles, Robert C., H. Harlyn Baker, and David H. Marimont. 1987. Epipolar-plane image analysis: An approach to determining structure from motion. *International Journal of Computer Vision* 1 (1): 7–55.

Bouman, Katherine L., Vickie Ye, Adam B. Yedidia, Frédo Durand, Gregory W. Wornell, Antonio Torralba, and William T. Freeman. 2017. Turning corners into cameras: Principles and methods. In *Proceedings of the IEEE International Conference on Computer Vision*, 2270–2278.

Boyd, Stephen P., and Lieven Vandenberghe. 2004. *Convex Optimization*.

Bromberg, Yaron, Ori Katz, and Yaron Silberberg. 2009. Ghost imaging with a single detector. *Physical Review A* 79 (5): 053840.

Brown, Steven W., Joseph P. Rice, Jorge E. Neira, B. C. Johnson, and J. D. Jackson. 2006. Spectrally tunable sources for advanced radiometric applications. *Journal of Research of the National Institute of Standards and Technology* 111 (5): 401.

Buehler, Chris, Michael Bosse, Leonard McMillan, Steven Gortler, and Michael Cohen. 2001. Unstructured lumigraph rendering. In *Proceedings of the 28th Annual Conference on Computer Graphics and Interactive Techniques*, 425–432.

Buller, Gerald S., and Andrew M. Wallace. 2007. Ranging and three-dimensional imaging using time-correlated single-photon counting and point-by-point acquisition. *IEEE Journal of Selected Topics in Quantum Electronics* 13 (4): 1006–1015.

Buttafava, Mauro, Jessica Zeman, Alberto Tosi, Kevin Eliceiri, and Andreas Velten. 2015. Non-line-of-sight imaging using a time-gated single photon avalanche diode. *Optics Express* 23 (16): 20997–21011.

Buttgen, B., and P. Seitz. 2008. Robust optical time-of-flight range imaging based on smart pixel structures. *IEEE Transactions on Circuits and Systems I: Regular Papers* 55 (6): 1512–1525.

Cao, Xun, Xin Tong, Qionghai Dai, and Stephen Lin. 2011. High resolution multispectral video capture with a hybrid camera system. In *Conference on Computer Vision and Pattern Recognition (CVPR)*, 297–304.

Caorsi, S., G. L. Gragnani, and M. Pastorino. 1994. An electromagnetic imaging approach using a multi-illumination technique. *IEEE Transactions on Biomedical Engineering* 41 (4): 406–409.

Carpenter, William Benjamin. 1856. *The Microscope: And Its Revelations*.

Chael, Andrew A., Michael D. Johnson, Ramesh Narayan, Sheperd S. Doeleman, John F. C. Wardle, and Katherine L. Bouman. 2016. High-resolution linear polarimetric imaging for the event horizon telescope. *The Astrophysical Journal* 829 (1): 11.

Chai, Jin-xiang, Xin Tong, Shing-chow Chan, and Heung-yeung Shum. 2000. Plenoptic sampling. In *Proceedings of the 27th Annual Conference on Computer Graphics and Interactive Techniques*, 307–318.

Chaigne, Thomas, Ori Katz, A. Claude Boccara, Mathias Fink, Emmanuel Bossy, and Sylvain Gigan. 2014. Controlling light in scattering media non-invasively using the photoacoustic transmission matrix. *Nature Photonics* 8 (1): 58–64.

Chakrabarti, Ayan. 2016. Learning sensor multiplexing design through back-propagation. In *Advances in Neural Information Processing Systems*, 3081–3089.

Chakrabarti, Ayan, and Todd Zickler. 2011. Statistics of real-world hyperspectral images. In *Conference on Computer Vision and Pattern Recognition (CVPR)*, 193–200.

Chang, Julie, Vincent Sitzmann, Xiong Dun, Wolfgang Heidrich, and Gordon Wetzstein. 2018. Hybrid optical-electronic convolutional neural networks with optimized diffractive optics for image classification. *Scientific Reports* 8 (1): 1–10.

Chatterjee, P., and P. Milanfar. 2010. Is denoising dead? *IEEE Transactions on Image Processing* 19 (4): 895–911.

Chen, Jiawen, Sylvain Paris, and Frédo Durand. 2007. Real-time edge-aware image processing with the bilateral grid. *ACM Transactions on Graphics (ToG)* 26 (3): 103.

Chen, Ping, Yan Han, and Jinxiao Pan. 2015. High-dynamic-range CT reconstruction based on varying tube-voltage imaging. *PLOS ONE* 10 (11): 0141789. doi:10.1371/journal.pone.0141789.

Chen, Xiaomin, and S. Kiaei. 2002. Monocycle shapes for ultra wideband system. In *IEEE International Symposium on Circuits and Systems*.

Chi, Cui, Hyunjin Yoo, and Moshe Ben-ezra. 2010. Multi-spectral imaging by optimized wide band illumination. *International Journal of Computer Vision* 86 (2-3): 140.

Choi, Inchang, Daniel S. Jeon, Giljoo Nam, Diego Gutierrez, and Min H. Kim. 2017. High-quality hyperspectral reconstruction using a spectral prior. *ACM Transactions on Graphics (ToG)* 36 (6): 1–13.

Choy, Christopher B, Danfei Xu, JunYoung Gwak, Kevin Chen, and Silvio Savarese. 2016. 3D-R2N2: A unified approach for single and multi-view 3D object reconstruction. In *European Conference on Computer Vision (ECCV)*, 628–644.

Coakley, J. A. 2003. Reflectance and albedo, surface. *Encyclopedia of the Atmosphere*.

Cohen, Adam Lloyd. 1982. Anti-pinhole imaging. *Optica Acta: International Journal of Optics* 29 (1): 63–67.

Comaniciu, Dorin, Visvanathan Ramesh, and Peter Meer. 2001. The variable bandwidth mean shift and data-driven scale selection. In *IEEE International Conference on Computer Vision. ICCV*, Vol. 1, 438–445.

Combettes, Patrick L., and Jean-Christophe Pesquet. 2011. Proximal splitting methods in signal processing. In *Springer Optimization and Its Applications*, 185–212. Springer.

Cooke, Dennis, and John Cant. 2010. Model-based seismic inversion: Comparing deterministic and probabilistic approaches. *Cseg Recorder* 35 (4): 29–39.

Cossairt, Oliver, Nathan Matsuda, and Mohit Gupta. 2014. Digital refocusing with incoherent holography. In *IEEE International Conference on Computational Photography (ICCP)*, 1–9.

Cossairt, Oliver S., Daniel Miau, and Shree K. Nayar. 2011. Gigapixel computational imaging. In *Proceedings of IEEE International Conference on Computational Photography (IEEE ICCP)*, 1–8.

Curless, Brian, David Salesin, and Michael Cohen. 2004. Interactive digital photomontage. *ACM Transactions on Graphics (siggraph)* 23 (3): 294302.

Damask, Jay N. 2004. *Polarization Optics in Telecommunications*, Vol. 101.

Daubechies, I., M. Defrise, and C. De Mol. 2004. An iterative thresholding algorithm for linear inverse problems with a sparsity constraint. *Communications on Pure and Applied Mathematics* 57 (11): 1413–1457.

Debevec, Paul. 2002. The Light Stage: Photorealistically Integrating Real Actors into Virtual Environments. *Svenska Föreningen För Grafisk Databehandling*.

Debevec, Paul. 2008. Rendering synthetic objects into real scenes: Bridging traditional and image-based graphics with global illumination and high dynamic range photography. In *ACM SIGGRAPH 2008 Classes*, 1–10.

Debevec, Paul E., and Jitendra Malik. 1997. Recovering high dynamic range radiance maps from photographs. In *Proceedings of the 24th Annual Conference on Computer Graphics and Interactive Techniques–SIGGRAPH '97*. ACM Press. doi:10.1145/258734.258884.

Debevec, Paul E., Camillo J. Taylor, and Jitendra Malik. 1996. Modeling and rendering architecture from photographs: A hybrid geometry-and image-based approach. In *Proceedings of the 23rd Annual Conference on Computer Graphics and Interactive Techniques*, 11–20.

Debevec, Paul, Tim Hawkins, Chris Tchou, Haarm-pieter Duiker, Westley Sarokin, and Mark Sagar. 2000. Acquiring the reflectance field of a human face. In *Proceedings of the 27th Annual Conference on Computer Graphics and Interactive Techniques*, 145–156.

Demro, James C., Richard Hartshorne, Loren M. Woody, Peter A. Levine, and John R. Tower. 1995. Design of a multispectral, wedge filter, remote-sensing instrument incorporating a multiport, thinned, CCD area array. In *Imaging Spectrometry*, Vol. 2480, 280–286.

Dicarlo, Jeffrey M., Feng Xiao, and Brian A. Wandell. 2001. Illuminating illumination. In *Color and Imaging Conference*, Vol. 2001, 27–34.

Donner, Craig, Tim Weyrich, Eugene d'Eon, Ravi Ramamoorthi, and Szymon Rusinkiewicz. 2008. A layered, heterogeneous reflectance model for acquiring and rendering human skin. *ACM Transactions on Graphics (ToG)* 27 (5): 1–12.

Duarte, Marco F., Mark A. Davenport, Dharmpal Takhar, Jason N. Laska, Ting Sun, Kevin F. Kelly, and Richard G. Baraniuk. 2008. Single-pixel imaging via compressive sampling. *IEEE Signal Processing Magazine* 25 (2): 83–91.

Durand, Frédo, and Julie Dorsey. 2002. Fast bilateral filtering for the display of high-dynamic-range images. In *Proceedings of the 29th Annual Conference on Computer Graphics and Interactive Techniques*, 257–266.

Eigen, David, and Rob Fergus. 2015. Predicting depth, surface normals and semantic labels with a common multi-scale convolutional architecture. In *Conference on Computer Vision and Pattern Recognition (CVPR)*, 2650–2658.

Elad, Michael. 2010. *Sparse and Redundant Representations: From Theory to Applications in Signal and Image Processing. Mathematics and Statistics*.

Elad, Michael, and Arie Feuer. 1997. Restoration of a single superresolution image from several blurred, noisy, and undersampled measured images. *IEEE Transactions on Image Processing* 6 (12): 1646–1658.

Esteban, Carlos Hernandez, George Vogiatzis, and Roberto Cipolla. 2008. Multiview photometric stereo. *IEEE Transactions on Pattern Analysis and Machine Intelligence* 30 (3): 548–554.

Fang, Shuai, Xiushan Xia, Xing Huo, and Changwen Chen. 2014. Image dehazing using polarization effects of objects and airlight. *Optics Express* 22 (16): 19523–19537.

Farid, Hany, and Eero P. Simoncelli. 1998. Range estimation by optical differentiation. *Journal of the Optical Society of America A* 15 (7): 1777.

Farup, Ivar, Jan Henrik Wold, Thorstein Seim, and Torkjel Søndrol. 2007. Generating light with a specified spectral power distribution. *Applied Optics* 46 (13): 2411–2422.

Fattal, Raanan, Maneesh Agrawala, and Szymon Rusinkiewicz. 2007. Multiscale shape and detail enhancement from multi-light image collections. *ACM Transactions on Graphics (ToG)* 26 (3): 51.

Fattal, Raanan, Dani Lischinski, and Michael Werman. 2002. Gradient domain high dynamic range compression. In *Proceedings of the 29th Annual Conference on Computer Graphics and Interactive Techniques*, 249–256.

Fenimore, Edward E., and Thomas M. Cannon. 1978. Coded aperture imaging with uniformly redundant arrays. *Applied Optics* 17 (3): 337–347.

Feris, Rogerio, Matthew Turk, Ramesh Raskar, Karhan Tan, and Gosuke Ohashi. 2004. Exploiting depth discontinuities for vision-based fingerspelling recognition. In *2004 Conference on Computer Vision and Pattern Recognition Workshop*, 155–155.

Ferri, F., D. Magatti, L. A. Lugiato, and A. Gatti. 2010. Differential ghost imaging. *Physical Review Letters* 104 (25): 253603.

Figueiredo, M. A. T., J. M. Bioucas-Dias, and R. D. Nowak. 2007. Majorization–minimization algorithms for wavelet-based image restoration. *IEEE Transactions on Image Processing* 16 (12): 2980–2991.

Foi, Alessandro, Mejdi Trimeche, Vladimir Katkovnik, and Karen Egiazarian. 2008. Practical Poissonian-Gaussian noise modeling and fitting for single-image raw-data. *IEEE Transactions on Image Processing* 17 (10): 1737–1754.

Foix, S., G. Alenya, and C. Torras. 2011. Lock-in time-of-flight (ToF) cameras: A survey. *IEEE Sensors Journal* 11 (9): 1917–1926.

Forsyth, David A., and Jean Ponce. 2012. *Computer vision: A modern approach*, 2nd edn. Pearson.

Foster, D., K. Amano, S. Nascimento, and M. Foster. 2006. Frequency of metamerism in natural scenes. *Journal of the Optical Society of America A* 23 (10): 2359–2372.

Freedman, Daniel, Yoni Smolin, Eyal Krupka, Ido Leichter, and Mirko Schmidt. 2014. SRA: Fast removal of general multipath for ToF sensors. In *Proc. of ECCV 2014*, 234–249.

Frieder, Gideon, Dan Gordon, and R. Reynolds. 1985. Back-to-front display of voxel based objects. *IEEE Computer Graphics and Applications*.

Fuchs, Christian, Michael Heinz, Marc Levoy, Hans-Peter Seidel, and Hendrik P. A. Lensch. 2008. Combining confocal imaging and descattering. In *Computer Graphics Forum*, Vol. 27, 1245–1253.

Gan, Lu, Thong T. Do, and Trac D. Tran. 2008. Fast compressive imaging using scrambled block Hadamard ensemble. In *European Signal Processing Conference (EUSIPCO)*, 1–5.

Gatti, A., Morten Bache, D. Magatti, E. Brambilla, F. Ferri, and L. A. Lugiato. 2006. Coherent imaging with pseudo-thermal incoherent light. *Journal of Modern Optics* 53 (5–6): 739–760.

Gatti, Alessandra, Enrico Brambilla, Morten Bache, and Luigi A. Lugiato. 2004. Ghost imaging with thermal light: Comparing entanglement and classicalcorrelation. *Physical Review Letters* 93 (9): 093602.

Georgeiv, Todor, and Chintan Intwala. 2006. *Light Field Camera Design for Integral View Photography*.

Gershun, Arun. 1939. The Light Field. *Journal of Mathematical Physics* 18: 51–151.

Ghodsi, Ali. 2006. Dimensionality reduction a short tutorial. *Department of Statistics and Actuarial Science, University of Waterloo, Ontario, Canada* 37 (38): 2006.

Ghosh, Abhijeet, Tongbo Chen, Pieter Peers, Cyrus A. Wilson, and Paul Debevec. 2010. Circularly polarized spherical illumination reflectometry. In *ACM SIGGRAPH Asia 2010 papers*, 1–12.

Ghosh, Abhijeet, Graham Fyffe, Borom Tunwattanapong, Jay Busch, Xueming Yu, and Paul Debevec. 2011. Multiview face capture using polarized spherical gradient illumination. In *Proceedings of the 2011 SIGGRAPH Asia Conference*, 1–10.

Girshick, Ross. 2015. Fast R-CNN. In *Proceedings of IEEE International Conference on Computer Vision*, 1440–1448.

Gkioulekas, Ioannis. 2018. Lecture 11: Focal Stacks and Lightfields in Computational Photography (Course). http://graphics.cs.cmu.edu/courses/15-463/2018_fall/lectures/lecture11.pdf.

Gkioulekas, Ioannis, Anat Levin, Frédo Durand, and Todd Zickler. 2015. Micron-scale light transport decomposition using interferometry. *ACM Transactions on Graphics (ToG)* 34 (4): 1–14.

Goldman, Dan B., Brian Curless, Aaron Hertzmann, and Steven M. Seitz. 2009. Shape and Spatially-varying BRDFs from Photometric Stereo. *IEEE Transactions on Pattern Analysis and Machine Intelligence* 32 (6): 1060–1071.

Goodfellow, Ian J., Jean Pouget-Abadie, Mehdi Mirza, Bing Xu, David Warde-Farley, Sherjil Ozair, Aaron Courville, and Yoshua Bengio. 2014. Generative adversarial nets. In *Proceedings of the 27th International Conference on Neural Information Processing Systems - Volume 2. Nips' 14*, 2672–2680. Cambridge, MA.

Goodman, Joseph W. 2005. *Introduction to Fourier Optics*, 3rd edn. Ben Roberts.

Gorthi, Sai Siva, Diane Schaak, and Ethan Schonbrun. 2013. Fluorescence imaging of flowing cells using a temporally coded excitation. *Optics Express* 21 (4): 5164–5170.

Gortler, Steven J. 2012. *Foundations of 3D Computer Graphics*. MIT Press.

Gortler, Steven J., Radek Grzeszczuk, Richard Szeliski, and Michael F. Cohen. 1996. The lumigraph. In *Proceedings of the 23rd Annual Conference on Computer Graphics and Interactive Techniques*, 43–54.

Gottesman, Stephen R., and E. E. Fenimore. 1989. New family of binary arrays for coded aperture imaging. *Applied Optics* 28 (20): 4344–4352.

Goy, Alexandre, Kwabena Arthur, Shuai Li, and George Barbastathis. 2018. Low photon count phase retrieval using deep learning. *Physical Review Letters* 121 (24): 243902.

Gruev, Viktor, and Ralph Etienne-Cummings. 2002. Implementation of steerable spatiotemporal image filters on the focal plane. *IEEE Transactions on Circuits and Systems II: Analog and Digital Signal Processing* 49 (4): 233–244.

Gu, Jinwei, Toshihiro Kobayashi, Mohit Gupta, and Shree K. Nayar. 2011. Multiplexed illumination for scene recovery in the presence of global illumination. In *Proceedings of IEEE International Conference on Computer Vision*, 691–698.

Guarnera, Giuseppe Claudio, Pieter Peers, Paul Debevec, and Abhijeet Ghosh. 2012. Estimating Surface Normals from Spherical Stokes Reflectance Fields. In *European Conference on Computer Vision (ECCV)*, 340–349.

Gupta, Mohit, Srinivasa G Narasimhan, and Yoav Y Schechner. 2008. On controlling light transport in poor visibility environments. In *Proceedings of the IEEE Conference on Computer Vision and Pattern Recognition*, 1–8.

Gupta, Mohit, Yuandong Tian, Srinivasa G. Narasimhan, and Li Zhang. 2009. (de) focusing on global light transport for active scene recovery. In *Proceedings of the IEEE Conference on Computer Vision and Pattern Recognition*, 2969–2976.

Gupta, Mohit, Shree K. Nayar, Matthias B. Hullin, and Jaime Martin. 2015. Phasor imaging: A generalization of correlation-based time-of-flight imaging. *ACM Transactions on Graphics* 34 (5): 1–18.

Gupta, Mohit, Andreas Velten, Shree K. Nayar, and Eric Breitbach. 2018. What are optimal coding functions for time-of-flight imaging? *ACM Transactions on Graphics* 37 (2): 1–18.

Gupta, Otkrist, Thomas Willwacher, Andreas Velten, Ashok Veeraraghavan, and Ramesh Raskar. 2012. Reconstruction of hidden 3D shapes using diffuse reflections. *Optics Express* 20 (17): 19096–19108.

Haeberli, Paul. 1992. Synthetic lighting for photography. *http://www.graficaobscura.com/synth/index.html*.

Hammernik, Kerstin, Teresa Klatzer, Erich Kobler, Michael P. Recht, Daniel K. Sodickson, Thomas Pock, and Florian Knoll. 2018. Learning a variational network for reconstruction of accelerated MRI data. *Magnetic Resonance in Medicine* 79 (6): 3055–3071.

Han, Yan, and Bahram Jalali. 2003. Photonic time-stretched analog-to-digital converter: Fundamental concepts and practical considerations. *Journal of Lightwave Technology* 21 (12): 3085.

Hartley, Richard, and Andrew Zisserman. 2004. *Multiple View Geometry in Computer Vision*. Cambridge University Press.

Harvey, Andrew Robert, John E. Beale, Alain H. Greenaway, Tracy J. Hanlon, and John W. Williams. 2000. Technology options for imaging spectrometry. In *Imaging Spectrometry VI*, Vol. 4132, 13–24.

Hasinoff, Samuel W., and Kiriakos N. Kutulakos. 2007. A layer-based restoration framework for variable-aperture photography. In *2007 IEEE 11th International Conference on Computer Vision*, 1–8.

Hau, Lene Vestergaard. 2011. Slowing single photons. *Nature Photonics* 5 (4): 197.

Hawkins, Tim, Jonathan Cohen, and Paul Debevec. 2001. A photometric approach to digitizing cultural artifacts. In *Proceedings of the Conference on Virtual Reality, Archeology, and Cultural Heritage*, 333–342.

Hayakawa, Hideki. 1994. Photometric stereo under a light source with arbitrary motion. *Journal of the Optical Society of America A* 11 (11): 3079–3089.

Hebden, Jeremy C., Robert A. Kruger, and K. S. Wong. 1991. Time resolved imaging through a highly scattering medium. *Applied Optics* 30 (7): 788–794.

Hecht, Eugene. 1998. Hecht optics. *Addison Wesley* 997: 213–214.

Hecht, Eugene. 2012. *Optics*, 5th edn. Pearson.

Hecht, Selig, Simon Shlaer, and Maurice Henri Pirenne. 1942. Energy, quanta, and vision. *Journal of General Physiology* 25 (6): 819–840.

Hee, Michael R., Joseph A. Izatt, Joseph M. Jacobson, James G. Fujimoto, and Eric A. Swanson. 1993. Femtosecond transillumination optical coherence tomography. *Optics Letters* 18 (12): 950–952.

Heide, Felix, Wolfgang Heidrich, and Gordon Wetzstein. 2015. Fast and flexible convolutional sparse coding. In *Proceedings of the IEEE Conference on Computer Vision and Pattern Recognition*, 5135–5143.

Heide, Felix, Matthias B. Hullin, James Gregson, and Wolfgang Heidrich. 2013. Low-budget transient imaging using photonic mixer devices. *ACM Transactions on Graphics (ToG)* 32 (4): 1–10.

Heide, Felix, Lei Xiao, Wolfgang Heidrich, and Matthias B. Hullin. 2014a. Diffuse mirrors: 3D reconstruction from diffuse indirect illumination using inexpensive time-of-flight sensors. In *Proceedings of the IEEE Conference on Computer Vision and Pattern Recognition CVPR*, 3222–3229.

Heide, Felix, Markus Steinberger, Yun-ta Tsai, Mushfiqur Rouf, Dawid Pająk, Dikpal Reddy, Orazio Gallo, Jing Liu, Wolfgang Heidrich, and Karen Egiazarian, et al. 2014b. FlexISP: A flexible camera image processing framework. *ACM Transactions on Graphics (ToG)* 33 (6): 1–13.

Heide, Felix, Lei Xiao, Andreas Kolb, Matthias B. Hullin, and Wolfgang Heidrich. 2014c. Imaging in scattering media using correlation image sensors and sparse convolutional coding. *Optics Express* 22 (21): 26338–26350.

Heide, Felix, Wolfgang Heidrich, Matthias Hullin, and Gordon Wetzstein. 2015. Doppler time-of-flight imaging. *ACM Transactions on Graphics (ToG)* 34 (4): 1–11.

Heide, Felix, Matthew O'Toole, Kai Zang, David B. Lindell, Steven Diamond, and Gordon Wetzstein. 2019. Non-line-of-sight imaging with partial occluders and surface normals. *ACM Transactions on Graphics (ToG)* 38 (3): 1–10.

Herman, Gabor, and Jayaram Udupa. 1983. Display of 3-D Digital Images: Computational Foundations and Medical Applications. *IEEE Computer Graphics and Applications*.

Hernandez-Marin, Sergio, Andrew M. Wallace, and Gavin J. Gibson. 2007. Bayesian analysis of lidar signals with multiple returns. *IEEE Transactions on Pattern Analysis and Machine Intelligence* 29 (12): 2170–2180.

Hertzmann, Aaron, and Steven M. Seitz. 2005. Example-based photometric stereo: Shape reconstruction with general, varying BRDFs. *IEEE Transactions on Pattern Analysis and Machine Intelligence* 27 (8): 1254–1264.

Hicks, R. Andrew, and Ruzena Bajcsy. 2001. Reflective surfaces as computational sensors. *Image and Vision Computing* 19 (11): 773–777.

Hitomi, Yasunobu, Jinwei Gu, Mohit Gupta, Tomoo Mitsunaga, and Shree K. Nayar. 2011. Video from a single coded exposure photograph using a learned over-complete dictionary. In *2011 International Conference on Computer Vision*, 287–294.

Hiura, Shinsaku, and Takashi Matsuyama. 1998. Depth measurement by the multi-focus camera. In *Conference on Computer Vision and Pattern Recognition (CVPR)*, 953–959.

Hoppe, Hugues, and Kentaro Toyama. 2003. Continuous flash, Technical Report MSR-TR-2003-63, Microsoft Corporation.

Horn, Berthold K. P. 1975. Obtaining shape from shading information. *The Psychology of Computer Vision*.

Horn, Berthold K. P., and Michael J. Brooks. 1989. *Shape from Shading*.

Horn, Berthold K. P., and Robert W. Sjoberg. 1979. Calculating the reflectance map. *Applied Optics* 18 (11): 1770–1779.

Hsu, Wei-liang, Graham Myhre, Kaushik Balakrishnan, Neal Brock, Mohammed Ibn-Elhaj, and Stanley Pau. 2014. Full-Stokes imaging polarimeter using an array of elliptical polarizer. *Optics Express* 22 (3): 3063–3074.

Huang, David, Eric A. Swanson, Charles P. Lin, Joel S. Schuman, William G. Stinson, Warren Chang, Michael R. Hee, Thomas Flotte, Kenton Gregory, and Carmen A. Puliafito, et al. 1991. Optical coherence tomography. *Science* 254 (5035): 1178–1181.

Huang, Fu-chung, David P. Luebke, and Gordon Wetzstein. 2015. The light field stereoscope. In *SIGGRAPH Emerging Technologies*, 24–1.

Huang, Xiao, Jian Bai, Kaiwei Wang, Qun Liu, Yujie Luo, Kailun Yang, and Xianjing Zhang. 2017. Target enhanced 3D reconstruction based on polarization-coded structured light. *Optics Express* 25 (2): 1173–1184.

Hughes, John, Andries van Dam, Morgan McGuire, David Sklar, James Foley, Steven Feiner, and Kurt Akeley. 2013. *Computer graphics: Principles and practice*, 3rd edn. Addison-Wesley.

Igarashi, Takanori, Ko Nishino, and Shree K. Nayar. 2007. *The appearance of human skin: A survey*.

Ikeuchi, Katsushi. 1981. Determining surface orientations of specular surfaces by using the photometric stereo method. *IEEE Transactions on Pattern Analysis and Machine Intelligence*.

Inagaki, Yasutaka, Yuto Kobayashi, Keita Takahashi, Toshiaki Fujii, and Hajime Nagahara. 2018. Learning to capture light fields through a coded aperture camera. In *European Conference on Computer Vision (ECCV)*, 418–434.

Irani, Michal, and Shmuel Peleg. 1990. Super resolution from image sequences. In *[1990] Proceedings. 10th International Conference on Pattern Recognition*, Vol. 2, 115–120.

Irani, Michal, and Shmuel Peleg. 1991. Improving resolution by image registration. *CVGIP: Graphical Models and Image Processing* 53 (3): 231–239.

Jaaskelainen, T., J. Parkkinen, and S. Toyooka. 1990. Vector-subspace model for color representation. *Journal of the Optical Society of America A* 7 (4): 725–730.

James, John. 2007. *Spectrograph Design Fundamentals*.

Jarabo, Adrian, Julio Marco, Adolfo Muñoz, Raul Buisan, Wojciech Jarosz, and Diego Gutierrez. 2014. A framework for transient rendering. *ACM Transactions on Graphics (ToG)* 33 (6): 1–10.

Jarabo, Adrian, Belen Masia, Julio Marco, and Diego Gutierrez. 2017. Recent advances in transient imaging: A computer graphics and vision perspective. *Visual Informatics* 1 (1): 65–79.

Jayasuriya, Suren, Adithya Pediredla, Sriram Sivaramakrishnan, Alyosha Molnar, and Ashok Veeraraghavan. 2015. Depth fields: Extending light field techniques to time-of-flight imaging. In *2015 International Conference on 3D Vision*, 1–9.

Jiang, Yunshan, Sebastian Karpf, and Bahram Jalali. 2020. Time-stretch LiDAR as a spectrally scanned time-of-flight ranging camera. *Nature Photonics* 14 (1): 14–18.

Jones, Andrew, Ian Mcdowall, Hideshi Yamada, Mark Bolas, and Paul Debevec. 2007. Rendering for an interactive 360 light field display. In *ACM SIGGRAPH 2007 papers*, 40.

Joshi, Manjunath V., Subhasis Chaudhuri, and Rajkiran Panuganti. 2004. Super-resolution imaging: Use of zoom as a cue. *Image and Vision Computing* 22 (14): 1185–1196.

Joshi, Neel, Bennett Wilburn, Vaibhav Vaish, Marc Levoy, and Mark Alan Horowitz. 2005. *Automatic Color Calibration for Large Camera Arrays*.

Kadambi, A., R. Whyte, A. Bhandari, L. Streeter, C. Barsi, A. Dorrington, and R. Raskar. 2013. Coded time of flight cameras: Sparse deconvolution to address multipath interference and recover time profiles. *ACM Transactions on Graphics (TOG)* 32 (6): 167.

Kadambi, Achuta, and Ramesh Raskar. 2017. Rethinking machine vision time of flight with GHz heterodyning. *IEEE Access* 5: 26211–26223.

Kadambi, Achuta, Hayato Ikoma, Xing Lin, Gordon Wetzstein, and Ramesh Raskar. 2013. Subsurface enhancement through sparse representations of multispectral direct/global decomposition. In *Computational Optical Sensing and Imaging*, 1–4.

Kadambi, Achuta, Vage Taamazyan, Boxin Shi, and Ramesh Raskar. 2015. Polarized 3D: High-quality depth sensing with polarization cues. In *Proceedings of the IEEE International Conference on Computer Vision*, 3370–3378.

Kadambi, Achuta, Hang Zhao, Boxin Shi, and Ramesh Raskar. 2016. Occluded imaging with time-of-flight sensors. *ACM Transactions on Graphics (ToG)* 35 (2): 1–12.

Katz, Ori, Yaron Bromberg, and Yaron Silberberg. 2009. Compressive ghost imaging. *Applied Physics Letters* 95 (13): 131110.

Katz, Ori, Eran Small, and Yaron Silberberg. 2012. Looking around corners and through thin turbid layers in real time with scattered incoherent light. *Nature Photonics* 6 (8): 549–553.

Katz, Ori, Eran Small, Yaron Bromberg, and Yaron Silberberg. 2011. Focusing and compression of ultrashort pulses through scattering media. *Nature Photonics* 5 (6): 372–377.

Katz, Ori, Pierre Heidmann, Mathias Fink, and Sylvain Gigan. 2014a. Non-invasive single-shot imaging through scattering layers and around corners via speckle correlations. *Nature Photonics* 8 (10): 784–790.

Katz, Ori, Eran Small, Yefeng Guan, and Yaron Silberberg. 2014b. Noninvasive nonlinear focusing and imaging through strongly scattering turbid layers. *Optica* 1 (3): 170–174.

Kawakami, Rei, Yasuyuki Matsushita, John Wright, Moshe Ben-ezra, Yu-wing Tai, and Katsushi Ikeuchi. 2011. High-resolution hyperspectral imaging via matrix factorization. In *Conference on Computer Vision and Pattern Recognition (CVPR)*, 2329–2336.

Keller, Alexander, and Wolfgang Heidrich. 2001. Interleaved sampling. In *Eurographics Workshop on Rendering Techniques*, 269–276.

Kellman, Michael R., Emrah Bostan, Nicole A. Repina, and Laura Waller. 2019. Physics-based learned design: Optimized coded-illumination for quantitative phase imaging. *IEEE Transactions on Computational Imaging* 5 (3): 344–353.

Keren, Danny, Shmuel Peleg, and Rafi Brada. 1988. Image sequence enhancement using sub-pixel displacements. In *Conference on Computer Vision and Pattern Recognition (CVPR)*, 742–743.

Keshava, Nirmal, and John F. Mustard. 2002. Spectral unmixing. *IEEE Signal Processing Magazine* 19 (1): 44–57.

Kiku, Daisuke, Yusuke Monno, Masayuki Tanaka, and Masatoshi Okutomi. 2013. Residual interpolation for color image demosaicking. In *2013 IEEE International Conference on Image Processing*, 2304–2308.

Kim, S. P., Nirmal K. Bose, and Hector M. Valenzuela. 1990. Recursive reconstruction of high resolution image from noisy undersampled multiframes. *IEEE Transactions on Acoustics, Speech, and Signal Processing* 38 (6): 1013–1027.

Kim, Seung P., and W. Y. Su. 1993. Recursive high-resolution reconstruction of blurred multiframe images. *IEEE Transactions on Image Processing* 2 (4): 534–539.

Kingma, Diederik P., and Max Welling. 2019. An introduction to variational autoencoders. *arXiv preprint arXiv:1906.02691*.

Kirmani, Ahmed, Arrigo Benedetti, and Philip A. Chou. 2013. SPUMIC: Simultaneous phase unwrapping and multipath interference cancellation in time-of-flight cameras using spectral methods. In *IEEE International Conference on Multimedia and Expo (ICME)*.

Kirmani, Ahmed, Tyler Hutchison, James Davis, and Ramesh Raskar. 2009. Looking around the corner using transient imaging. In *2009 IEEE 12th International Conference on Computer Vision*, 159–166.

Klein, Jonathan, Christoph Peters, Jaime Martín, Martin Laurenzis, and Matthias B. Hullin. 2016. Tracking objects outside the line of sight using 2d intensity images. *Scientific Reports* 6 (1): 1–9.

Kolb, Andreas, Erhardt Barth, Reinhard Koch, and Rasmus Larsen. 2010. Time-of-flight cameras in computer graphics. In *Computer Graphics Forum*, Vol. 29, 141–159.

Komatsu, Takashi, Toru Igarashi, Kiyoharu Aizawa, and Takahiro Saito. 1993. Very high resolution imaging scheme with multiple different-aperture cameras. *Signal Processing: Image Communication* 5 (5-6): 511–526.

Koshikawa, Kazutada. 1979. A polarimetric approach to shape understanding of glossy objects. In *Proceedings of the 6th International Joint Conference on Artificial Intelligence-Volume 1*, 493–495.

Kotwal, Alankar, Anat Levin, and Ioannis Gkioulekas. 2020. Interferometric transmission probing with coded mutual intensity. *ACM Transactions on Graphics (ToG)* 39 (4): 74–1.

Krishnan, Dilip, and Rob Fergus. 2009. Dark flash photography. *ACM Transactions on Graphics* 28 (3): 96.

Kulce, Onur, Deniz Mengu, Yair Rivenson, and Aydogan Ozcan. 2021. All-optical information-processing capacity of diffractive surfaces. *Light: Science & Applications* 10 (1): 1–17.

Kundur, D., and D. Hatzinakos. 1996. Blind image deconvolution. *IEEE Signal Processing Magazine* 13 (3): 43–64.

Laina, Iro, Christian Rupprecht, Vasileios Belagiannis, Federico Tombari, and Nassir Navab. 2016. Deeper depth prediction with fully convolutional residual networks. In *2016 Fourth International Conference on 3D Vision (3DV)*, 239–248.

Lecun, Yann, Léon Bottou, Yoshua Bengio, and Patrick Haffner. 1998. Gradient-based learning applied to document recognition. *Proceedings of the IEEE* 86 (11): 2278–2324.

Lee, Alan WeiMin, and Qing Hu. 2005. Real-time, continuous-wave terahertz imaging by use of a microbolometer focal-plane array. *Optics Letters* 30 (19): 2563–2565.

Levin, Anat, Rob Fergus, Frédo Durand, and William T. Freeman. 2007. Image and depth from a conventional camera with a coded aperture. *ACM Transactions on Graphics (ToG)* 26 (3): 70.

Levoy, Marc. 2006. Light fields and computational imaging. *Computer* 39 (8): 46–55.

Levoy, Marc, and Pat Hanrahan. 1996. Light field rendering. In *Proceedings of the 23rd Annual Conference on Computer Graphics and Interactive Techniques*, 31–42.

Levoy, Marc, Billy Chen, Vaibhav Vaish, Mark Horowitz, Ian McDowall, and Mark Bolas. 2004. Synthetic aperture confocal imaging. *ACM Transactions on Graphics* 23 (3).

Levoy, Marc, Ren Ng, Andrew Adams, Matthew Footer, and Mark Horowitz. 2006. Light field microscopy. In *ACM SIGGRAPH 2006 Papers*, 924–934.

Lewis, J. P. 1995. *Fast Normalized Cross-Correlation*.

Li, Fengqiang, Huaijin Chen, Adithya Pediredla, Chiakai Yeh, Kuan He, Ashok Veeraraghavan, and Oliver Cossairt. 2017a. CS-ToF: High-resolution compressive time-of-flight imaging. *Optics Express* 25 (25): 31096–31110.

Li, Fengqiang, Joshua Yablon, Andreas Velten, Mohit Gupta, and Oliver Cossairt. 2017b. High-depth-resolution range imaging with multiple-wavelength superheterodyne interferometry using 1550-nm lasers. *Applied Optics* 56 (31): 51–56.

Li, Wenchong, and Chunhua Ma. 1991. Imaging spectroscope with an optical recombination system. In *Three-Dimensional Bioimaging Systems and Lasers in the Neurosciences*, Vol. 1428, 242–248.

Li, Xiaodong, Feng Ling, Yun Du, and Yihang Zhang. 2013. Spatially adaptive superresolution land cover mapping with multispectral and panchromatic images. *IEEE Transactions on Geoscience and Remote Sensing* 52 (5): 2810–2823.

Liang, Chia-kai, Gene Liu, and Homer H. Chen. 2007. Light field acquisition using programmable aperture camera. In *2007 IEEE International Conference on Image Processing*, Vol. 5, 233.

Liang, Chia-kai, Tai-hsu Lin, Bing-yi Wong, Chi Liu, and Homer H. Chen. 2008. Programmable aperture photography: Multiplexed light field acquisition. In *Proceedings of ACM SIGGRAPH*, 1–10.

Likas, Aristidis, Nikos Vlassis, and Jakob J Verbeek. 2003. The global k-means clustering algorithm. *Pattern Recognition* 36 (2): 451–461.

Lin, Jingyu, Yebin Liu, Jinli Suo, and Qionghai Dai. 2017. Frequency-Domain Transient Imaging. *IEEE Transactions on Pattern Analysis and Machine Intelligence* 39 (5): 937–950.

Lin, Xing, Yair Rivenson, Nezih T. Yardimci, Muhammed Veli, Yi Luo, Mona Jarrahi, and Aydogan Ozcan. 2018. All-optical machine learning using diffractive deep neural networks. *Science* 361 (6406): 1004–1008.

Lin, Zhouchen, and Heung-yeung Shum. 2004. A geometric analysis of light field rendering. *International Journal of Computer Vision* 58 (2): 121–138.

Lindell, David B., Gordon Wetzstein, and Vladlen Koltun. 2019. Acoustic non-line-of-sight imaging. In *Proceedings of the IEEE Conference on Computer Vision and Pattern Recognition CVPR*, 6780–6789.

Lippmann, Gabriel. 1908. *Epreuves Reversibles Donnant La Sensation Du Relief*.

Lischinski, Dani, Zeev Farbman, Matt Uyttendaele, and Richard Szeliski. 2006. Interactive local adjustment of tonal values. *ACM Transactions on Graphics (ToG)* 25 (3): 646–653.

Liu, Dengyu, Jinwei Gu, Yasunobu Hitomi, Mohit Gupta, Tomoo Mitsunaga, and Shree K. Nayar. 2013. Efficient space-time sampling with pixel-wise coded exposure for high-speed imaging. *IEEE Transactions on Pattern Analysis and Machine Intelligence* 36 (2): 248–260.

Liu, Fayao, Chunhua Shen, and Guosheng Lin. 2015. Deep convolutional neural fields for depth estimation from a single image. In *Conference on Computer Vision and Pattern Recognition (CVPR)*, 5162–5170.

Lohit, Suhas, Kuldeep Kulkarni, Ronan Kerviche, Pavan Turaga, and Amit Ashok. 2018. Convolutional neural networks for noniterative reconstruction of compressively sensed images. *IEEE Transactions on Computational Imaging* 4 (3): 326–340.

Lombardi, Stephen, Tomas Simon, Jason Saragih, Gabriel Schwartz, Andreas Lehrmann, and Yaser Sheikh. 2019. Neural volumes: Learning dynamic renderable volumes from images. *Arxiv Preprint Arxiv:1906.07751*.

Lu, Renfu. 2004. Multispectral imaging for predicting firmness and soluble solids content of apple fruit. *Postharvest Biology and Technology* 31 (2): 147–157.

Lumsdaine, Andrew, and Todor Georgiev. 2009. The focused plenoptic camera. In *IEEE International Conference on Computational Photography (ICCP)*, 1–8.

Lyu, Meng, Wei Wang, Hao Wang, Haichao Wang, Guowei Li, Ni Chen, and Guohai Situ. 2017. Deep-learning-based ghost imaging. *Scientific Reports* 7 (1): 1–6.

Ma, Wan-Chun, Tim Hawkins, Pieter Peers, Charles-Felix Chabert, Malte Weiss, and Paul E. Debevec. 2007. Rapid Acquisition of Specular and Diffuse Normal Maps from Polarized Spherical Gradient Illumination. *Rendering Techniques* 2007 (9): 10.

MacKinnon, Nicholas, Ulrich Stange, Pierre Lane, Calum Macaulay, and Mathieu Quatrevalet. 2005. Spectrally programmable light engine for in vitro or in vivo molecular imaging and spectroscopy. *Applied Optics* 44 (11): 2033–2040.

Macleod, Angus. 2005. Polarization in optical coatings. In *Novel Optical Systems Design and Optimization VIII*, Vol. 5875, 587504.

Maeda, Tomohiro, Achuta Kadambi, Yoav Y. Schechner, and Ramesh Raskar. 2018. Dynamic heterodyne interferometry. In *Proceedings of IEEE International Conference on Computational Photography (IEEE ICCP)*, 1–11.

Maeda, Tomohiro, Guy Satat, Tristan Swedish, Lagnojita Sinha, and Ramesh Raskar. 2019. Recent advances in imaging around corners. *Arxiv Preprint Arxiv:1910.05613*.

Mahjoubfar, Ata, Dmitry V. Churkin, Stéphane Barland, Neil Broderick, Sergei K. Turitsyn, and Bahram Jalali. 2017. Time stretch and its applications. *Nature Photonics* 11 (6): 341.

Mallat, Stéphane. 2009. *A wavelet tour of signal processing: The sparse way*, 3rd edn. Academic Press.

Malzbender, Tom, Dan Gelb, and Hans Wolters. 2001. Polynomial texture maps. In *Proceedings of the 28th Annual Conference on Computer Graphics and Interactive Techniques*, 519–528.

Marcia, Roummel F., Zachary T. Harmany, and Rebecca M. Willett. 2009. Compressive coded aperture imaging. In *Computational Imaging VII*, Vol. 7246, 72460.

References

Marco, Julio, Quercus Hernandez, Adolfo Munoz, Yue Dong, Adrian Jarabo, Min H. Kim, Xin Tong, and Diego Gutierrez. 2017. DeepToF: Off-the-shelf real-time correction of multipath interference in time-of-flight imaging. *ACM Transactions on Graphics (ToG)* 36 (6): 1–12.

Marschner, Steve, and Peter Shirley. 2018. *Fundamentals of computer graphics*, 4th edn. A K Peters/CRC Press.

Martel, Julien N. P., Lorenz Mueller, Stephen J Carey, Piotr Dudek, and Gordon Wetzstein. 2020. Neural Sensors: Learning Pixel Exposures for HDR Imaging and Video Compressive Sensing with Programmable Sensors. *IEEE Transactions on Pattern Analysis and Machine Intelligence*.

Marwah, Kshitij, Gordon Wetzstein, Yosuke Bando, and Ramesh Raskar. 2013. Compressive light field photography using overcomplete dictionaries and optimized projections. *ACM Transactions on Graphics (ToG)* 32 (4): 1–12.

Masia, Belen, Lara Presa, Adrian Corrales, and Diego Gutierrez. 2012. Perceptually optimized coded apertures for defocus deblurring. In *Computer Graphics Forum*, Vol. 31, 1867–1879.

Masia, Belen, Gordon Wetzstein, Piotr Didyk, and Diego Gutierrez. 2013. A survey on computational displays: Pushing the boundaries of optics, computation, and perception. *Computers & Graphics* 37 (8): 1012–1038.

Masselus, Vincent, Philip Dutré, and Frederik Anrys. 2002. The free-form light stage. In *ACM SIGGRAPH 2002 Conference Abstracts and Applications*, 262–262.

Masselus, Vincent, Pieter Peers, Philip Dutré, and Yves D. Willems. 2003. Relighting with 4d incident light fields. *ACM Transactions on Graphics (ToG)* 22 (3): 613–620.

Matusik, Wojciech. 2003. A data-driven reflectance model. PhD Thesis, Massachusetts Institute of Technology.

McGuire, Morgan, Wojciech Matusik, Hanspeter Pfister, Billy Chen, John F. Hughes, and Shree K. Nayar. 2007. Optical splitting trees for high-precision monocular imaging. *IEEE Computer Graphics and Applications* 27 (2): 32–42.

Mertz, L., and N. O. Young. 1961. *Fresnel Transformations of Images*. London: Chapman and Hall.

Mescheder, Lars, Michael Oechsle, Michael Niemeyer, Sebastian Nowozin, and Andreas Geiger. 2019. Occupancy networks: Learning 3D reconstruction in function space. In *Proceedings of the IEEE Conference on Computer Vision and Pattern Recognition CVPR*, 4460–4470.

Meyer, Carl. 2000. *Matrix Analysis and Applied Linear Algebra*.

Meyers, Ron, Keith S. Deacon, and Yanhua Shih. 2008. Ghost-imaging experiment by measuring reflected photons. *Physical Review A* 77 (4): 041801.

Mildenhall, Ben, Pratul P. Srinivasan, Matthew Tancik, Jonathan T. Barron, Ravi Ramamoorthi, and Ren Ng. 2020. Nerf: Representing scenes as neural radiance fields for view synthesis. *Arxiv Preprint Arxiv:2003.08934*.

Mohan, Ankit, Ramesh Raskar, and Jack Tumblin. 2008. Agile spectrum imaging: Programmable wavelength modulation for cameras and projectors. In *Computer Graphics Forum*, Vol. 27, 709–717.

Mohan, Ankit, Jack Tumblin, Bobby Bodenheimer, Reynold Bailey, and Cindy Grimm. 2005. Table-top computed lighting for practical digital photography. In *ACM SIGGRAPH 2005 Sketches*, 76.

Mohan, Ankit, Douglas Lanman, Shinsaku Hiura, and Ramesh Raskar. 2009. Image destabilization: Programmable defocus using lens and sensor motion. In *Proceedings of IEEE International Conference on Computational Photography (IEEE ICCP)*, 1–8.

Monno, Yusuke, Daisuke Kiku, Masayuki Tanaka, and Masatoshi Okutomi. 2017. Adaptive residual interpolation for color and multispectral image demosaicking. *Sensors* 17 (12): 2787.

Murray-Bruce, John, Charles Saunders, and Vivek K. Goyal. 2019. Occlusion-based computational periscopy with consumer cameras. In *Wavelets and Sparsity XVIII*, Vol. 11138, 111380.

Nagahara, Hajime, Sujit Kuthirummal, Changyin Zhou, and Shree K. Nayar. 2008. Flexible depth of field photography. In *European Conference on Computer Vision (ECCV)*, 60–73.

Naik, Nikhil, Shuang Zhao, Andreas Velten, Ramesh Raskar, and Kavita Bala. 2011. Single view reflectance capture using multiplexed scattering and time-of-flight imaging. In *Proceedings of the 2011 SIGGRAPH Asia Conference*, 1–10.

Naik, Nikhil, Achuta Kadambi, Christoph Rhemann, Shahram Izadi, Ramesh Raskar, and Sing Bing Kang. 2015. A light transport model for mitigating multipath interference in time-of-flight sensors. In *Proceedings of the IEEE Conference on Computer Vision and Pattern Recognition CVPR*, 73–81.

Nam, Giljoo, Joo Ho Lee, Hongzhi Wu, Diego Gutierrez, and Min H. Kim. 2016. Simultaneous acquisition of microscale reflectance and normals. *ACM Transactions on Graphics* 35 (6): 185–1.

NASA and Ball Aerospace. 2008. Kepler Focal Plane Array. https://www.nasa.gov/mission_pages/kepler/multimedia/images/kepler-focal-plane-assembly.html. Last Updated: November 10, 2008.

Nayar, S. K., and T. Mitsunaga. 2000. High dynamic range imaging: Spatially varying pixel exposures. In *Conference on Computer Vision and Pattern Recognition (CVPR)*.

Nayar, Shree K., Vlad Branzoi, and Terry E. Boult. 2006. Programmable imaging: Towards a flexible camera. *International Journal of Computer Vision* 70 (1): 7–22.

Nayar, Shree K., Xi-sheng Fang, and Terrance Boult. 1997. Separation of reflection components using color and polarization. *International Journal of Computer Vision* 21 (3): 163–186.

Nayar, Shree K., Katsushi Ikeuchi, and Takeo Kanade. 1991. Shape from interreflections. *International Journal of Computer Vision* 6 (3): 173–195.

Nayar, Shree K., Masahiro Watanabe, and Minori Noguchi. 1996. Real-time focus range sensor. *IEEE Transactions on Pattern Analysis and Machine Intelligence* 18 (12): 1186–1198.

Nayar, Shree K., Gurunandan Krishnan, Michael D. Grossberg, and Ramesh Raskar. 2006. Fast separation of direct and global components of a scene using high frequency illumination. In *ACM SIGGRAPH*.

Ng, Ren. 2006. *Digital light field photography*. Stanford University.

Ng, Ren, Marc Levoy, Mathieu Brédif, Gene Duval, Mark Horowitz, Pat Hanrahan, and Others. 2005. Light field photography with a hand-held plenoptic camera. *Computer Science Technical Report* 2 (11): 1–11.

Nguyen, Quang Minh, Peter M. Atkinson, and Hugh G. Lewis. 2011. Super-resolution mapping using Hopfield neural network with panchromatic imagery. *International Journal of Remote Sensing* 32 (21): 6149–6176.

Nimeroff, Jeffry S., Eero Simoncelli, and Julie Dorsey. 1995. Efficient re-rendering of naturally illuminated environments. In *Photorealistic Rendering Techniques*, 373–388.

Nomura, Yoshikuni, Li Zhang, and Shree K. Nayar. 2007. Scene collages and flexible camera arrays. In *Proceedings of the 18th Eurographics Conference on Rendering Techniques*, 127–138.

Okamoto, Takayuki, and Ichirou Yamaguchi. 1991. Simultaneous acquisition of spectral image information. *Optics Letters* 16 (16): 1277–1279.

Okawara, Tadashi, Michitaka Yoshida, Hajime Nagahara, and Yasushi Yagi. 2020. Action recognition from a single coded image. In *Proceedings of IEEE International Conference on Computational Photography (IEEE ICCP)*, 1–11.

O'Toole, Matthew, and Kiriakos N. Kutulakos. 2010. Optical computing for fast light transport analysis. *ACM Transactions on Graphics* 29 (6): 164.

O'Toole, Matthew, David B. Lindell, and Gordon Wetzstein. 2018. Confocal non-line-of-sight imaging based on the light-cone transform. *Nature* 555 (7696): 338–341.

O'Toole, Matthew, John Mather, and Kiriakos N. Kutulakos. 2014. 3D shape and indirect appearance by structured light transport. In *Proceedings of the IEEE Conference on Computer Vision and Pattern Recognition CVPR*, 3246–3253.

O'Toole, Matthew, Ramesh Raskar, and Kiriakos N. Kutulakos. 2012. Primal-dual coding to probe light transport. *ACM Transactions on Graphics* 31 (4): 39–1.

O'Toole, Matthew, Felix Heide, Lei Xiao, Matthias B. Hullin, Wolfgang Heidrich, and Kiriakos N. Kutulakos. 2014. Temporal frequency probing for 5D transient analysis of global light transport. *ACM Transactions on Graphics* 33 (4): 1–11.

O'Toole, Matthew, Felix Heide, David B. Lindell, Kai Zang, Steven Diamond, and Gordon Wetzstein. 2017. Reconstructing transient images from single-photon sensors. In *Proceedings of the IEEE Conference on Computer Vision and Pattern Recognition CVPR*, 1539–1547.

Pandharkar, Rohit, Andreas Velten, Andrew Bardagjy, Everett Lawson, Moungi Bawendi, and Ramesh Raskar. 2011. Estimating motion and size of moving non-line-of-sight objects in cluttered environments. In *Conference on Computer Vision and Pattern Recognition CVPR*, 265–272.

Parikh, Neal, and Stephen Boyd. 2014. Proximal algorithms. *Foundations and Trends® in Optimization* 1 (3): 127–239. doi:10.1561/2400000003.

Park, Jeong Joon, Peter Florence, Julian Straub, Richard Newcombe, and Steven Lovegrove. 2019. DeepSDF: Learning continuous signed distance functions for shape representation. In *Proceedings of the IEEE Conference on Computer Vision and Pattern Recognition CVPR*, 165–174.

Park, Jong-il, Moon-hyun Lee, Michael D. Grossberg, and Shree K. Nayar. 2007. Multispectral Imaging Using Multiplexed Illumination. In *2007 IEEE 11th International Conference on Computer Vision*, 1–8.

Park, Sung Cheol, Min Kyu Park, and Moon Gi Kang. 2003. Super-resolution image reconstruction: A technical overview. *IEEE Signal Processing Magazine* 20 (3): 21–36.

Pediredla, Adithya Kumar, Mauro Buttafava, Alberto Tosi, Oliver Cossairt, and Ashok Veeraraghavan. 2017. Reconstructing rooms using photon echoes: A plane based model and reconstruction algorithm for looking around the corner. In *Proceedings of IEEE International Conference on Computational Photography (IEEE ICCP)*, 1–12.

Peers, Pieter, Dhruv K. Mahajan, Bruce Lamond, Abhijeet Ghosh, Wojciech Matusik, Ravi Ramamoorthi, and Paul Debevec. 2009. Compressive light transport sensing. *ACM Transactions on Graphics (ToG)* 28 (1): 1–18.

Peng, Yifan, Xiong Dun, Qilin Sun, and Wolfgang Heidrich. 2017. Mix-and-match holography. *ACM Transactions on Graphics* 36 (6): 191–1.

Peng, Yifan, Suyeon Choi, Nitish Padmanaban, Jonghyun Kim, and Gordon Wetzstein. 2020. Neural holography. In *ACM SIGGRAPH 2020 Emerging Technologies*, 1–2.

Petschnigg, Georg, Richard Szeliski, Maneesh Agrawala, Michael Cohen, Hugues Hoppe, and Kentaro Toyama. 2004. Digital photography with flash and no-flash image pairs. *ACM Transactions on Graphics (ToG)* 23 (3): 664–672.

Phong, Bui Tuong. 1975. Illumination for computer generated pictures. *Communications of the ACM* 18 (6): 311–317.

Pitsianis, N. P., D. J. Brady, A. Portnoy, X. Sun, T. Suleski, M. A. Fiddy, M. R. Feldman, and R. D. Tekolste. 2006. Compressive imaging sensors. In *Intelligent Integrated Microsystems*, Vol. 6232, 62320.

Popescu, Dan P., Costel Flueraru, Youxin Mao, Shoude Chang, John Disano, Sherif Sherif, Michael G. Sowa, et al.. 2011. Optical coherence tomography: Fundamental principles, instrumental designs and biomedical applications. *Biophysical Reviews* 3 (3): 155–169.

Posdamer, Jeffrey L., and M. D. Altschuler. 1982. Surface measurement by space-encoded projected beam systems. *Computer Graphics and Image Processing* 18 (1): 1–17.

Powell, Samuel B., Roman Garnett, Justin Marshall, Charbel Rizk, and Viktor Gruev. 2018. Bioinspired polarization vision enables underwater geolocalization. *Science Advances* 4 (4): 6841.

Prevedel, Robert, Young-gyu Yoon, Maximilian Hoffmann, Nikita Pak, Gordon Wetzstein, Saul Kato, Tina Schrödel, Ramesh Raskar, Manuel Zimmer, Edward S. Boyden, and Others. 2014. Simultaneous whole-animal 3D imaging of neuronal activity using light-field microscopy. *Nature Methods* 11 (7): 727–730.

Rahman, Md Sadman Sakib, Jingxi Li, Deniz Mengu, Yair Rivenson, and Aydogan Ozcan. 2021. Ensemble learning of diffractive optical networks. *Light: Science & Applications* 10 (1): 1–13.

Rajan, Deepu, Subhasis Chaudhuri, and Manjunath V. Joshi. 2003. Multi-objective super resolution: Concepts and examples. *IEEE Signal Processing Magazine* 20 (3): 49–61.

Ramanath, Rajeev, Wesley E. Snyder, Youngjun Yoo, and Mark S. Drew. 2005. Color image processing pipeline. *IEEE Signal Processing Magazine* 22 (1): 34–43.

Raskar, Ramesh, and James Davis. 2008. 5d time-light transport matrix: What can we reason about scene properties?, Technical report, Massachusetts Institute of Technology.

Raskar, Ramesh, and Jack Tumblin. 2011. *Computational Photography: Mastering New Techniques for Lenses, Lighting, and Sensors*.

Raskar, Ramesh, Amit Agrawal, and Jack Tumblin. 2006. Coded exposure photography: Motion deblurring using fluttered shutter. In *ACM SIGGRAPH 2006 Papers*, 795–804.

Raskar, Ramesh, Kar-han Tan, Rogerio Feris, Jingyi Yu, and Matthew Turk. 2004. Non-photorealistic camera: Depth edge detection and stylized rendering using multi-flash imaging. *ACM Transactions on Graphics (ToG)* 23 (3): 679–688.

Raskar, Ramesh, Amit Agrawal, Cyrus A. Wilson, and Ashok Veeraraghavan. 2008. Glare aware photography: 4D ray sampling for reducing glare effects of camera lenses. In *ACM SIGGRAPH 2008 papers*, 1–10.

Rhee, Jehyuk, and Youngjoong Joo. 2003. Wide dynamic range CMOS image sensor with pixel level ADC. *Electronics Letters* 39 (4): 360. doi:10.1049/el:20030246.

Rivenson, Yair, Yibo Zhang, Harun Günaydın, Da Teng, and Aydogan Ozcan. 2018. Phase recovery and holographic image reconstruction using deep learning in neural networks. *Light: Science & Applications* 7 (2): 17141–17141.

Romberg, Justin. 2008. Imaging via compressive sampling. *IEEE Signal Processing Magazine* 25 (2): 14–20.

Ronchi, Vasco, and Edward Rosen. 1991. *Optics: The Science of Vision*.

Rusinkiewicz, Szymon, Olaf Hall-Holt, and Marc Levoy. 2002. Real-time 3D model acquisition. *ACM Transactions on Graphics (ToG)* 21 (3): 438–446.

Sahin, Furkan E., and Rajiv Laroia. 2017. Light L16 computational camera. In *Applied Industrial Optics: Spectroscopy, Imaging and Metrology*, 5–20.

Saleh, Bahaa E. A., and Malvin Carl Teich. 1991. *Fundamentals of Photonics*. John Wiley & Sons.

Sarafraz, Amin, Shahriar Negahdaripour, and Yoav Y. Schechner. 2009. Enhancing images in scattering media utilizing stereovision and polarization. In *2009 Workshop on Applications of Computer Vision (WACV)*, 1–8.

Saragadam, Vishwanath, and Aswin C. Sankaranarayanan. 2019. KRISM—krylov subspace-based optical computing of hyperspectral images. *ACM Transactions on Graphics (ToG)* 38 (5): 1–14.

Saragadam, Vishwanath, Jian Wang, Xin Li, and Aswin C. Sankaranarayanan. 2017. Compressive spectral anomaly detection. In *Proceedings of IEEE International Conference on Computational Photography (IEEE ICCP)*, 1–9.

Sasagawa, Kiyotaka, Takahiro Yamaguchi, Makito Haruta, Yoshinori Sunaga, Hironari Takehara, Hiroaki Takehara, Toshihiko Noda, Takashi Tokuda, and Jun Ohta. 2016. An implantable CMOS image sensor with self-reset pixels for functional brain imaging. *IEEE Transactions on Electron Devices* 63 (1): 215–222. doi:10.1109/ted.2015.2454435.

Satat, Guy, Matthew Tancik, and Ramesh Raskar. 2018a. Imaging through volumetric scattering with a single photon sensitive camera. In *Mathematics in Imaging*, 5–2.

Satat, Guy, Matthew Tancik, and Ramesh Raskar. 2018b. Towards photography through realistic fog. In *Proceedings of IEEE International Conference on Computational Photography (IEEE ICCP)*, 1–10.

Satat, Guy, Barmak Heshmat, Dan Raviv, and Ramesh Raskar. 2016. All photons imaging through volumetric scattering. *Scientific Reports* 6 (1): 1–8.

Satat, Guy, Matthew Tancik, Otkrist Gupta, Barmak Heshmat, and Ramesh Raskar. 2017. Object classification through scattering media with deep learning on time resolved measurement. *Optics Express* 25 (15): 17466–17479.

Satkin, Scott, Jason Lin, and Martial Hebert. 2012. Data-driven scene understanding from 3D models. In *European Conference on Computer Vision (ECCV)*.

Saunders, Charles, John Murray-Bruce, and Vivek K. Goyal. 2019. Computational periscopy with an ordinary digital camera. *Nature* 565 (7740): 472–475.

Schechner, Yoav Y., and Nir Karpel. 2005. Recovery of underwater visibility and structure by polarization analysis. *IEEE Journal of Oceanic Engineering* 30 (3): 570–587.

Schechner, Yoav Y., and Shree K. Nayar. 2002. Generalized mosaicing: Wide field of view multispectral imaging. *IEEE Transactions on Pattern Analysis and Machine Intelligence* 24 (10): 1334–1348.

Schechner, Yoav Y., Srinivasa G. Narasimhan, and Shree K. Nayar. 2003. Polarization-based vision through haze. *Applied Optics* 42 (3): 511–525.

Schechner, Yoav Y., Shree K. Nayar, and Peter N. Belhumeur. 2007. Multiplexing for optimal lighting. *IEEE Transactions on Pattern Analysis and Machine Intelligence* 29 (8): 1339–1354.

Schechner, Yoav Y., Joseph Shamir, and Nahum Kiryati. 1999. Polarization-based decorrelation of transparent layers: The inclination angle of an invisible surface. In *Proceedings of the Seventh IEEE International Conference on Computer Vision*, Vol. 2, 814–819.

Scheiner, Nicolas, Florian Kraus, Fangyin Wei, Buu Phan, Fahim Mannan, Nils Appenrodt, Werner Ritter, Jurgen Dickmann, Klaus Dietmayer, Bernhard Sick, and Others. 2020. Seeing around street corners: Non-line-of-sight detection and tracking in-the-wild using doppler radar. In *Proceedings of the IEEE/CVF Conference on Computer Vision and Pattern Recognition*, 2068–2077.

Schnars, Ulf, Claas Falldorf, John Watson, and Werner Jüptner. 2015. Digital holography. In *Digital Holography and Wavefront Sensing*, 39–68.

Schotland, John C. 1997. Continuous-wave diffusion imaging. *Journal Optical Society of America A* 14 (1): 275–279.

Seetzen, Helge, Wolfgang Heidrich, Wolfgang Stuerzlinger, Greg Ward, Lorne Whitehead, Matthew Trentacoste, Abhijeet Ghosh, and Andrejs Vorozcovs. 2004. High dynamic range display systems. In *ACM SIGGRAPH 2004 Papers*, 760–768.

Seitz, Steven M., Yasuyuki Matsushita, and Kiriakos N. Kutulakos. 2005. A theory of inverse light transport. In *Tenth IEEE International Conference on Computer Vision (ICCV'05) Volume 1, Volume 2*, 1440–1447.

Sen, Pradeep, Billy Chen, Gaurav Garg, Stephen R. Marschner, Mark Horowitz, Marc Levoy, and Hendrik P. A. Lensch. 2005. Dual photography. In *ACM SIGGRAPH 2005 Papers*, 745–755.

Serrano, Ana, Felix Heide, Diego Gutierrez, Gordon Wetzstein, and Belen Masia. 2016. Convolutional sparse coding for high dynamic range imaging. In *Computer Graphics Forum*, Vol. 35, 153–163.

Serrano, Ana, Elena Garces, Belen Masia, and Diego Gutierrez. 2017. Convolutional Sparse Coding for Capturing High-Speed Video Content. In *Computer Graphics Forum*, Vol. 36, 380–389.

Shapiro, Jeffrey H. 2008. Computational ghost imaging. *Physical Review A* 78 (6): 061802.

Shechtman, Eli, Yaron Caspi, and Michal Irani. 2005. Space-time super-resolution. *IEEE Transactions on Pattern Analysis and Machine Intelligence* 27 (4): 531–545.

Shedligeri, Prasan A, Sreyas Mohan, and Kaushik Mitra. 2017. Data driven coded aperture design for depth recovery. In *2017 IEEE International Conference on Image Processing (ICIP)*, 56–60.

Shi, Boxin. 2019. Data-driven photometric 3D modeling. In *SIGGRAPH Asia 2019 Courses*, 1–139.

Shi, Boxin, Ping Tan, Yasuyuki Matsushita, and Katsushi Ikeuchi. 2013. Bi-polynomial modeling of low-frequency reflectances. *IEEE Transactions on Pattern Analysis and Machine Intelligence* 36 (6): 1078–1091.

Shi, Boxin, Zhe Wu, Zhipeng Mo, Dinglong Duan, Sai-kit Yeung, and Ping Tan. 2016. A benchmark dataset and evaluation for non-lambertian and uncalibrated photometric stereo. In *Proceedings of the IEEE Conference on Computer Vision and Pattern Recognition CVPR*, 3707–3716.

Shogenji, Rui, Yoshiro Kitamura, Kenji Yamada, Shigehiro Miyatake, and Jun Tanida. 2004. Multispectral imaging using compact compound optics. *Optics Express* 12 (8): 1643–1655.

Shrestha, Shikhar, Felix Heide, Wolfgang Heidrich, and Gordon Wetzstein. 2016. Computational imaging with multi-camera time-of-flight systems. *ACM Transactions on Graphics (ToG)* 35 (4): 1–11.

Simonyan, Karen, and Andrew Zisserman. 2014. Very deep convolutional networks for large-scale image recognition. *arXiv preprint arXiv:1409.1556*.

Sinha, Ayan, Justin Lee, Shuai Li, and George Barbastathis. 2017. Lensless computational imaging through deep learning. *Optica* 4 (9): 1117–1125.

Smith, T., and J. Guild. 1931. The C.I.E. colorimetric standards and their use. *Transactions of the Optical Society* 33 (3): 73–134.

Som, Subhojit, and Philip Schniter. 2012. Compressive imaging using approximate message passing and a Markov-tree prior. *IEEE Transactions on Signal Processing* 60 (7): 3439–3448.

Srinivasan, Pratul P., Ren Ng, and Ravi Ramamoorthi. 2017. Light field blind motion deblurring. In *Proceedings of the IEEE Conference on Computer Vision and Pattern Recognition CVPR*, 3958–3966.

Srinivasan, Pratul P., Rahul Garg, Neal Wadhwa, Ren Ng, and Jonathan T. Barron. 2018. Aperture supervision for monocular depth estimation. In *Proceedings of the IEEE Conference on Computer Vision and Pattern Recognition CVPR*, 6393–6401.

Srinivasan, Pratul P., Ben Mildenhall, Matthew Tancik, Jonathan T. Barron, Richard Tucker, and Noah Snavely. 2020. Lighthouse: Predicting lighting volumes for spatially-coherent illumination. In *Proceedings of the IEEE/CVF Conference on Computer Vision and Pattern Recognition*, 8080–8089.

Strang, Gilbert. 2016. *Introduction to Linear Algebra*, 5th edn. Wellesley-Cambridge Press.

Sturmfels, Pascal, Saige Rutherford, Mike Angstadt, Mark Peterson, Chandra Sripada, and Jenna Wiens. 2018. A domain guided CNN architecture for predicting age from structural brain images. *arXiv preprint arXiv:1808.04362*.

Su, Shuochen, Felix Heide, Robin Swanson, Jonathan Klein, Clara Callenberg, Matthias Hullin, and Wolfgang Heidrich. 2016. Material classification using raw time-of-flight measurements. In *Proceedings of the IEEE Conference on Computer Vision and Pattern Recognition CVPR*, 3503–3511.

Sun, He, Adrian V. Dalca, and Katherine L. Bouman. 2020. Learning a probabilistic strategy for computational imaging sensor selection. In *Proceeding of IEEE International Conference on Computational Photography (IEEE ICCP)*, 1–12.

Sutton, Oliver. 2012. Introduction to k nearest neighbour classification and condensed nearest neighbour data reduction. *University Lectures,* University of Leicester.

Sutton, Richard S., and Andrew G. Barto. 2018. *Reinforcement learning: An introduction*.

Swinehart, Donald F. 1962. The Beer-lambert law. *Journal of Chemical Education* 39 (7): 333.

Szeliski, Richard. 2011. *Computer Vision*, 1st edn. Springer London.

Tadano, Ryuichi, Adithya Kumar Pediredla, Kaushik Mitra, and Ashok Veeraraghavan. 2016. Spatial phase-sweep: Increasing temporal resolution of transient imaging using a light source array. In *2016 IEEE International Conference on Image Processing (ICIP)*, 1564–1568.

Taguchi, Yuichi. 2014. Rainbow flash camera: Depth edge extraction using complementary colors. *International Journal of Computer Vision* 110 (2): 156–171.

Takhar, Dharmpal, Jason N. Laska, Michael B. Wakin, Marco F. Duarte, Dror Baron, Shriram Sarvotham, Kevin F. Kelly, and Richard G. Baraniuk. 2006. A new compressive imaging camera architecture using optical-domain compression. In *Computational Imaging IV*, Vol. 6065, 606509.

Talvala, Eino-Ville, Andrew Adams, Mark Horowitz, and Marc Levoy. 2007. Veiling glare in high dynamic range imaging. *ACM Transactions on Graphics (ToG)* 26 (3): 37.

Tanaka, Kenichiro, Yasuhiro Mukaigawa, and Achuta Kadambi. 2020. Polarized non-line-of-sight imaging. In *Proceedings of the IEEE/CVF Conference on Computer Vision and Pattern Recognition*, 2136–2145.

Tanaka, Kenichiro, Yasuhiro Mukaigawa, Takuya Funatomi, Hiroyuki Kubo, Yasuyuki Matsushita, and Yasushi Yagi. 2017. Material classification using frequency-and depth-dependent time-of-flight distortion. In *Proceedings of the IEEE Conference on Computer Vision and Pattern Recognition CVPR*, 79–88.

Tao, Michael W., Pratul P. Srinivasan, Jitendra Malik, Szymon Rusinkiewicz, and Ravi Ramamoorthi. 2015. Depth from shading, defocus, and correspondence using light-field angular coherence. In *Proceedings of the IEEE Conference on Computer Vision and Pattern Recognition CVPR*, 1940–1948.

Taylor, Dayton. 1996. Virtual camera movement: The way of the future. *American Cinematographer* 77 (9): 93–100.

Themelis, George, Jung Sun Yoo, and Vasilis Ntziachristos. 2008. Multispectral imaging using multiple-bandpass filters. *Optics Letters* 33 (9): 1023–1025.

Tian, Lei, and Laura Waller. 2015. Quantitative differential phase contrast imaging in an LED array microscope. *Optics Express* 23 (9): 11394–11403.

Tian, Lei, Xiao Li, Kannan Ramchandran, and Laura Waller. 2014. Multiplexed coded illumination for Fourier ptychography with an LED array microscope. *Biomedical Optics Express* 5 (7): 2376–2389.

Tian, Lei, Ziji Liu, Li-hao Yeh, Michael Chen, Jingshan Zhong, and Laura Waller. 2015. Computational illumination for high-speed in vitro Fourier ptychographic microscopy. *Optica* 2 (10): 904–911.

Tomasi, Carlo, and Roberto Manduchi. 1998. Bilateral filtering for gray and color images. In *Proceedings of IEEE International Conference on Computer Vision*, 839–846.

Treibitz, Tali, and Yoav Y. Schechner. 2008. Active polarization descattering. *IEEE Transactions on Pattern Analysis and Machine Intelligence* 31 (3): 385–399.

Tsai, Chia-yin, Kiriakos N. Kutulakos, Srinivasa G. Narasimhan, and Aswin C. Sankaranarayanan. 2017. The geometry of first-returning photons for non-line-of-sight imaging. In *Proceedings of the IEEE Conference on Computer Vision and Pattern Recognition CVPR*, 7216–7224.

Tsai, R. Y., and T. S. Huang. 1984. Multiframe image restoration and register. *Advances in Computer Vision and Image Processing,* Jal Press Inc.

Unser, Michael. 1999. Splines: a perfect fit for signal and image processing. *IEEE Signal Processing Magazine* 16 (6): 22–38. doi:10.1109/79.799930.

Unser, Michael. 2020. A note on BIBO stability. *IEEE Transactions on Signal Processing* 68: 5904–5913. doi:10.1109/tsp.2020.3025029.

Vaish, Vaibhav, Bennett Wilburn, Neel Joshi, and Marc Levoy. 2004. Using plane + parallax for calibrating dense camera arrays. In *Conference on Computer Vision and Pattern Recognition (CVPR)*, Vol. 1.

Vandewalle, Patrick, Sabine Süsstrunk, and Martin Vetterli. 2006. A frequency domain approach to registration of aliased images with application to super-resolution. *Eurasip Journal on Advances in Signal Processing* 2006 (1): 071459.

Vedel, Mathieu, Sebastien Breugnot, and Nick Lechocinski. 2011. Full Stokes polarization imaging camera. In *Polarization Science and Remote Sensing V*, Vol. 8160, 81600.

Veeraraghavan, Ashok, Ramesh Raskar, Amit Agrawal, Ankit Mohan, and Jack Tumblin. 2007. Dappled photography: Mask enhanced cameras for heterodyned light fields and coded aperture refocusing. 26: 69.

Velten, Andreas, Thomas Willwacher, Otkrist Gupta, Ashok Veeraraghavan, Moungi G. Bawendi, and Ramesh Raskar. 2012a. Recovering three-dimensional shape around a corner using ultrafast time-of-flight imaging. *Nature Communications* 3 (1): 1–8.

Velten, Andreas, Di Wu, Adrian Jarabo, Belen Masia, Christopher Barsi, Everett Lawson, Chinmaya Joshi, Diego Gutierrez, Moungi G Bawendi, and Ramesh Raskar. 2012b. Relativistic ultrafast rendering using time-of-flight imaging. In *ACM SIGGRAPH 2012 Talks*, 1–1.

Velten, Andreas, Di Wu, Adrian Jarabo, Belen Masia, Christopher Barsi, Chinmaya Joshi, Everett Lawson, Moungi Bawendi, Diego Gutierrez, and Ramesh Raskar. 2013. Femto-photography: capturing and visualizing the propagation of light. *ACM Transactions on Graphics (ToG)* 32 (4): 1–8.

Vetterli, Martin, Jelena Kovačević, and Vivek K Goyal. 2014. *Foundations of Signal Processing*. Cambridge University Press.

Vonesch, Cédric, and Michael Unser. 2008. A fast thresholded landweber algorithm for wavelet-regularized multidimensional deconvolution. *IEEE Transactions on Image Processing* 17 (4): 539–549.

Waechter, Michael, Nils Moehrle, and Michael Goesele. 2014. Let there be color! Large-scale texturing of 3D reconstructions. In *European Conference on Computer Vision (ECCV)*, 836–850.

Wakin, Michael B., Jason N. Laska, Marco F. Duarte, Dror Baron, Shriram Sarvotham, Dharmpal Takhar, Kevin F. Kelly, and Richard G. Baraniuk. 2006a. An architecture for compressive imaging. In *2006 International Conference on Image Processing*, 1273–1276.

Wakin, Michael, Jason N. Laska, Marco F. Duarte, Dror Baron, Shriram Sarvotham, Dharmpal Takhar, Kevin F. Kelly, and Richard G. Baraniuk. 2006b. Compressive imaging for video representation and coding. In *Picture Coding Symposium*, Vol. 1.

Wall, Christine F., Andrew R. Hanson, and Julie A. F. Taylor. 2001. Construction of a programmable light source for use as a display calibration artifact. In *Flat Panel Display Technology and Display Metrology II*, Vol. 4295, 259–266.

Waller, Laura, Lei Tian, and George Barbastathis. 2010. Transport of intensity phase-amplitude imaging with higher order intensity derivatives. *Optics Express* 18 (12): 12552–12561.

Walraven, Robert. 1977. Polarization imagery. In *Optical Polarimetry: Instrumentation and Applications*, Vol. 112, 164–167.

Wang, Suyu, Li Zhuo, and Xiaoguang Li. 2010. Spectral imagery super resolution by using of a high resolution panchromatic image. In *2010 3rd International Conference on Computer Science and Information Technology*, Vol. 4, 220–224.

Wang, Ting-chun, Alexei A. Efros, and Ravi Ramamoorthi. 2016. Depth estimation with occlusion modeling using light-field cameras. *IEEE Transactions on Pattern Analysis and Machine Intelligence* 38 (11): 2170–2181.

Wang, Zhou, Alan C. Bovik, Hamid R. Sheikh, and Eero P. Simoncelli. 2004. Image quality assessment: from error visibility to structural similarity. *IEEE Transactions on Image Processing* 13 (4): 600–612.

Wanner, Sven, and Bastian Goldluecke. 2012. Globally consistent depth labeling of 4D light fields. In *2012 IEEE Conference on Computer Vision and Pattern Recognition*, 41–48.

Watts, Claire M., David Shrekenhamer, John Montoya, Guy Lipworth, John Hunt, Timothy Sleasman, Sanjay Krishna, David R. Smith, and Willie J. Padilla. 2014. Terahertz compressive imaging with metamaterial spatial light modulators. *Nature Photonics* 8 (8): 605.

Wetzstein, Gordon, Ivo Ihrke, and Wolfgang Heidrich. 2013. On plenoptic multiplexing and reconstruction. *International Journal of Computer Vision* 101 (2): 384–400.

Wetzstein, Gordon, Douglas Lanman, Wolfgang Heidrich, and Ramesh Raskar. 2011. Layered 3D: Tomographic image synthesis for attenuation-based light field and high dynamic range displays. In *ACM SIGGRAPH 2011 Papers*, 1–12.

Wetzstein, Gordon, Douglas Lanman, Matthew Hirsch, Wolfgang Heidrich, and Ramesh Raskar. 2012. Compressive light field displays. *IEEE Computer Graphics and Applications* 32 (5): 6–11.

Wilburn, Bennett, Neel Joshi, Vaibhav Vaish, Eino-Ville Talvala, Emilio Antunez, Adam Barth, Andrew Adams, Mark Horowitz, and Marc Levoy. 2005. High performance imaging using large camera arrays. In *ACM SIGGRAPH 2005 Papers*, 765–776.

Willard, Jared, Xiaowei Jia, Shaoming Xu, Michael Steinbach, and Vipin Kumar. 2020. Integrating physics-based modeling with machine learning: A survey. *arXiv preprint arXiv:2003.04919*.

Willett, Rebecca M., Michael E. Gehm, and David J. Brady. 2007. Multiscale reconstruction for computational spectral imaging. In *Computational Imaging V*, Vol. 6498, 64980.

Wood, Daniel N., Daniel I. Azuma, Ken Aldinger, Brian Curless, Tom Duchamp, David H. Salesin, and Werner Stuetzle. 2000. Surface light fields for 3D photography. In *Proceedings of the 27th Annual Conference on Computer Graphics and Interactive Techniques*, 287–296.

Woodham, Robert J. 1980. Photometric method for determining surface orientation from multiple images. *Optical Engineering* 19 (1): 191139.

Wu, Di, Andreas Velten, Matthew O'Toole, Belen Masia, Amit Agrawal, Qionghai Dai, and Ramesh Raskar. 2013. Decomposing global light transport using time of flight imaging. *International Journal of Computer Vision* 107 (2): 123–138.

Wu, Gaochang, Belen Masia, Adrian Jarabo, Yuchen Zhang, Liangyong Wang, Qionghai Dai, Tianyou Chai, and Yebin Liu. 2017. Light field image processing: An overview. *IEEE Journal of Selected Topics in Signal Processing* 11 (7): 926–954.

Wu, Rihui, Adrian Jarabo, Jinli Suo, Feng Dai, Yongdong Zhang, Qionghai Dai, and Diego Gutierrez. 2018. Adaptive polarization-difference transient imaging for depth estimation in scattering media. *Optics Letters* 43 (6): 1299–1302.

Wu, Yicheng, Vivek Boominathan, Huaijin Chen, Aswin Sankaranarayanan, and Ashok Veeraraghavan. 2019. PhaseCam3D—learning phase masks for passive single view depth estimation. In *Proceedings of IEEE International Conference on Computational Photography (IEEE ICCP)*, 1–12.

Xiao, Lei, Felix Heide, Matthew O'Toole, Andreas Kolb, Matthias B. Hullin, Kyros Kutulakos, and Wolfgang Heidrich. 2015. Defocus deblurring and superresolution for time-of-flight depth cameras. In *Proceedings of the IEEE Conference on Computer Vision and Pattern Recognition*, 2376–2384.

Xin, Shumian, Sotiris Nousias, Kiriakos N. Kutulakos, Aswin C. Sankaranarayanan, Srinivasa G. Narasimhan, and Ioannis Gkioulekas. 2019. A theory of fermat paths for non-line-of-sight shape reconstruction. In *Proceedings of the IEEE Conference on Computer Vision and Pattern Recognition CVPR*, 6800–6809.

Xiong, Ying, Ayan Chakrabarti, Ronen Basri, Steven J. Gortler, David W. Jacobs, and Todd Zickler. 2014. From shading to local shape. *IEEE Transactions on Pattern Analysis and Machine Intelligence* 37 (1): 67–79.

Yang, Jason C., Matthew Everett, Chris Buehler, and Leonard McMillan. 2002. A real-time distributed light field camera. *Rendering Techniques* 2002: 77–86.

Yariv, Amnon, and Pochi Yeh. 2006. *Photonics: Optical Electronics in Modern Communications (The Oxford Series in Electrical and Computer Engineering)*.

Young, Matt. 1989. The pinhole camera: Imaging without lenses or mirrors. *The Physics Teacher* 27 (9): 648–655.

Zhang, Cha, and Tsuhan Chen. 2004. A self-reconfigurable camera array. In *ACM SIGGRAPH 2004 Sketches*, 151.

Zhang, Ruo, Ping-sing Tsai, James Edwin Cryer, and Mubarak Shah. 1999. Shape-from-shading: A survey. *IEEE Transactions on Pattern Analysis and Machine Intelligence* 21 (8): 690–706.

Zhang, Ying, Huijie Zhao, and Na Li. 2013. Polarization calibration with large apertures in full field of view for a full Stokes imaging polarimeter based on liquid-crystal variable retarders. *Applied Optics* 52 (6): 1284–1292.

Zhao, Xiaojin, Amine Bermak, Farid Boussaid, and Vladimir G. Chigrinov. 2010. Liquid-crystal micropolarimeter array for full Stokes polarization imaging in visible spectrum. *Optics Express* 18 (17): 17776–17787.

Zheng, Qian, Boxin Shi, and Gang Pan. 2020. Summary study of data-driven photometric stereo methods. *Virtual Reality & Intelligent Hardware* 2 (3): 213–221.

Zhou, Changyin, Stephen Lin, and Shree K. Nayar. 2010. Coded aperture pairs for depth from defocus and defocus deblurring. *International Journal of Computer Vision* 93 (1): 53–72.

Zhu, Xiaoqing, Anne Aaron, and Bernd Girod. 2003. Distributed compression for large camera arrays. In *IEEE Workshop on Statistical Signal Processing, 2003*, 30–33.

Zickler, Todd E., Peter N. Belhumeur, and David J. Kriegman. 2002. Helmholtz stereopsis: Exploiting reciprocity for surface reconstruction. *International Journal of Computer Vision* 49 (2-3): 215–227.

Zomet, Assaf, and Shree K. Nayar. 2006. Lensless imaging with a controllable aperture. In *Conference on Computer Vision and Pattern Recognition (CVPR)*, Vol. 1, 339–346.

Index

absorption, 289
airlight, 270
albedo, 317, 378
   diffuse, 317
   specular, 317
angular dimension, 215
aperture
   coded aperture, 128
   diameter, 21
Arnoldi iteration, 370
azimuthal ambiguity, 263
   azimuthal model mismatch, 264

backprojection, 194
backpropagation, 108
beamsplitter, 301
Beer-Lambert, 289
bidirectional reflectance distribution function (BRDF), 319, 361, 405
   Lambert's law, 319
   Phong, 320
   specular, 320
bidirectional surface scattering reflectance distribution function (BSSRDF), 405
bilateral filter, 145, 341
Brewster angle, 255

calibration, 227
classification, 100
clustering, 100
color, 287
   colorimetry, 293
   constancy, 295, 331
   perceptual, 295
   retinal, 293
computer graphics, 338, 358
convolution, 70
cross-polarization, 259

dappled photography, 151
data
   testing, 100
   training, 100
   validation, 100
deconvolution, 68
   3D deconvolution, 233
degree of polarization, 278
demosaicing, 139
   multispectral, 309
demosaicking, 47
denoising, 68
depolarization, 271
depth from defocus, 132, 133
detector
   bucket detector, 159
   SPAD detector, 176
   streak-tube, 176
dichromatic reflection model, 276, 329
diffraction, 18, 297
   gratings, 161, 297
digital micromirror device, 157, 161
direct-global separation, 365
dispersion, 27, 297
display
   compressive, 239
   High Rank 3D, 240
   Layered 3D, 240
   liquid crystal, 238, 257

tensor, 242
dual photography, 360, 369, 380
   dual image, 360
   primal image, 360

epipolar plane, 215
exposure, 87
   coded exposure, 132
   coded exposure photography, 89

facet, 334
falloff function, 228, 270
filter
   color filter array, 138, 299
   Bayer filter, 141, 299
   interference filter, 160
   LCTF, 299
   multispectral filter array, 299
   polarizing, 255
fixed pattern scanning, 369
flicker fusion, 239
flux, 13
focal stacks, 232
foreshortening, 319
four bucket method, 181
Fourier pytchography, 154
Fourier transform, 72
Fresnel coefficients, 255

generalized minimal residual, 370
ghost imaging, 158
global-direct separation, 137, 153, 178, 358, 363, 368, 373
gradien
   tgradient space, 322
gradient
   gradient field, 147
gradient descent, 222

Hadamard criteria, 86
Helmholtz reciprocity principle, 319, 359, 360
Hermitian, 82

illumination
   active, 301
   adaptive multiplexed, 369
   coded, 151, 302
   dark flash photography, 303
   direct, 364–366
   flash/no-flash, 152
   global, 364–366
   multiplexed, 302, 336
   passive, 301
   structured light, 266
image plane, 233
image projection
   orthographic, 323
   perspective, 323
imaging
   compound, 300
   compressive, 156, 221
   continuous wave, 181
   depth, 177, 275
   diffuse, 185
   fluorescence lifetime, 177
   multiplexed, 225
   NLOS, 193, 293, 359, 378
   Coherence based, 379
   Intensity based, 379
   time-of-flight, 379
   skin, 358, 365
   spatially coded, 127
   spectral, 290
   thermal, 293
   time-of-flight, 169, 170, 224, 275, 358, 365
   time-resolved, 173
   transient, 380, 408
   Wi-Fi, 292
   X-ray, 293
intensity
   radiant, 14
interferometry, 358, 376
   Michelson, 376
interferometry:Michelson, 201
interpolation, 141
   adaptive, 310
   difference, 309
   iterative, 310
   residual, 309
   minimized-Laplacian, 310
interreflection
   equation, 333
irradiance, 14, 15, 215, 318, 359, 363

Index

ISO, 149

Kubelka-Munk theory, 405

Lambert-Beer law, 405
Lambertian surface, 138, 219, 277, 319, 363
learning
   deep learning, 102, 221
   machine learning, 99
   supervised, 100
   unsupervised, 100
lens, 21
   condenser, 233
   objective, 233
   ocular, 233
   thin lens, 20
   tube, 233
lenslet array, 233
light field, 151, 211, 339, 358, 361
   direct, 361, 363
   general, 363
   indirect, 361
light slab parametrization, 211
light stage, 338
light transpor
   tequation, 359, 360, 371
   tforward, 358, 360
   tinverse, 358, 360
   tmatrix, 358, 359, 369–371
   tsteady-state, 379
light transport, 178, 357
   inverse, 363
   matrix, 363
local shading adaptation, 341

metamer, 148, 295
micropolarimeter, 261
model inversion
   seismic, 98
modeling
   forward, 67, 331
   inverse, 67, 331
Monte Carlo, 405
Mueller Matrix, 261
multiplexing
   Fourier, 229

   illumination, 154, 302, 336
   light field, 236
   spatial, 138, 229
   time, 240

nearest neighbor, 97
neural network, 102
   convolutional, 110
   feedforward, 107
   recurrent, 107
noise, 98
   read, 98
   shot, 98
Nyquist frequency, 77

optical computing, 370
optical manhole, 268
optical probing, 363, 368, 370
   matrix, 370, 371
   path isolation, 371
   transport probing equation, 370, 371
orientation consistency, 334

parallax, 218
   parallax barriers, 238
perceptron, 107, 221
photometric stereo
   4-source color, 330
   color, 329
   example-based, 334
photometry, 318
pinhole, 128
   camera, 16
   pinspeck, 18
pixel, 80
Planck
   Planck's constant, 12, 289
   Planck's relation, 289
plane of incidence, 255
   p-polarized, 255
   s-polarized, 255
plenoptic function, 169, 287, 358
Poincaré sphere, 261
point spread function (PSF), 97, 128, 268
polarization, 253
   circular, 253

elliptical, 254
linear, 253
unpolarized, 255
power density
    angular, 14
    spatial, 14
primal-dual coding, 370
    dual domain, 370
    primal domain, 370
principal component analysis, 307
prism, 297
probing function, 176
projection slice theorem, 96, 215

quantization, 78

radiance, 318, 361
radiometry, 318
Radon transform, 96
ray tracing, 334, 358
reflectance map, 322
    Stokes reflectance field, 266
reflection, 289
    diffuse, 193, 276
    specular, 191, 219, 276
refraction, 19
refractive distortion, 264
regression, 101
regularization, 98, 105
relighting, 338
rotational symmetry, 319

scalar, 81
scanner
    push-broom, 160, 298
    whiskbroom camera, 160
scattering, 196, 267, 270, 289
    coefficient, 188, 268
    de-scattering, 358
    mechanism, 267
    media
        fog, 271
        haze, 270
        turbid, 198
        water, 268
    Rayleigh, 271, 289

subsurface, 178, 276
semireflector, 280
Shannon's sampling theorem, 75
shape
    photometric stereo, 323
    shape from color, 329
    shape from intensity, 322
    shape from interreflections, 331
    shape from polarization, 264
    shape from shading, 325
signal, 68
    Dirac impulse, 70
singular value decomposition, 326
Snell's Law, 20, 255
soft-thresholding, 95
solid angle, 14, 320
    steradian, 14
source
    extended, 328
    point, 319
Space Time Impulse Response (STIR), 380
sparsity, 156
spatial dimension, 215
spectrometer
    wedge imaging spectrometer, 160
spectrometry, 160
spectroscopy, 290
spectrum, 287
    hyperspectral imaging, 291
    multispectral imaging, 160, 291
    spectral unmixing, 304
stereo transport matrix, 375
    direct, 375
    epipolar, 375
    non-epipolar, 375
stereo vision, 270
Stokes vector, 260
super-resolution, 149
superposition, 359
superresolution, 68
support vector machine, 102
system
    impulse response, 69
    linear, 69
    linear time-invariant, 69

tomography, 95

Index

chromatography, 160
tone mapping, 145
transmission, 289

vector, 81
visibility function, 332

wavelength, 12, 287
waveplate, 255
    quarter-wave, 256
wiregrid polarizer, 256